The Cypress Inn Group. From left to right: Aris, Somorjai, Hegedus, Wei, Boudart, Bell, and Gates. Carmel, California, June 1985.

CATALYST DESIGN

CATALYST DESIGN
Progress and Perspectives

L. LOUIS HEGEDUS (Editor), *W. R. Grace & Co., Research Division*

RUTHERFORD ARIS, *University of Minnesota*

ALEXIS T. BELL, *University of California, Berkeley*

MICHEL BOUDART, *Stanford University*

N. Y. CHEN, *Mobil Research and Development Corp.*

BRUCE C. GATES, *University of Delaware*

WERNER O. HAAG, *Mobil Research and Development Corp.*

GABOR A. SOMORJAI, *University of California, Berkeley*

JAMES WEI, *Massachusetts Institute of Technology*

A Wiley-Interscience Publication
JOHN WILEY & SONS
New York · Chichester · Brisbane · Toronto · Singapore

Copyright © 1987 by John Wiley & Sons, Inc.

All rights reserved. Published simultaneously in Canada.

Reproduction or translation of any part of this work beyond that permitted by Section 107 or 108 of the 1976 United States Copyright Act without the permission of the copyright owner is unlawful. Requests for permission or further information should be addressed to the Permissions Department, John Wiley & Sons, Inc.

Library of Congress Cataloging in Publication Data:
Catalyst design.

"A Wiley-Interscience publication."
Includes index.
1. Catalysts. I. Hegedus, L. Louis, 1941–
II. Aris, Rutherford.
TP159.C3C38 1987   660.2'995   86-28133
ISBN 0-471-85138-8

Printed in the United States of America

10 9 8 7 6 5 4 3 2 1

PREFACE

The idea for this book originated at a symposium on catalyst design at the Annual Meeting of the American Institute of Chemical Engineers in New Orleans, November 1981, followed by another one in San Francisco, November 1984. The large interest in the subject motivated the participants of the San Francisco symposium to address the topic of catalyst design by preparing a self-contained book. For this purpose, we met at the Cypress Inn, Carmel, California in June 1985 (see the photograph facing the title page) to discuss how this could best be accomplished. One decision we made was to add a chapter on zeolites, which is now part of this book.

Catalysis is a complex field that has, to date, largely defied attempts toward a systematic approach to the design of new catalysts (at least in the sense that systematic design is practiced in some other disciplines). Nevertheless we felt that a growing body of knowledge exists which can be utilized as the basis for the design of new catalysts, supplemented with the customary enlightened empiricism. We felt the need to provide some firm basis for the hope that someday catalysts could be designed in a more effective manner. However, we were also careful to avoid overstating the current capabilities of the field. Hence the subtitle of the book.

The bulk of the catalytic literature deals with the description and elucidation of catalysts and catalytic reactions. A smaller but rapidly growing literature deals with the preparation of catalysts; several recent symposia were devoted to this subject of widespread interest. In comparison, the literature that is specifically devoted to catalyst design is of a relatively recent vintage, even though the desire for catalyst design has at least implicitly motivated quite a lot of research.

This book addresses various aspects of catalyst design, from different points of view, reflecting the backgrounds and interests of the authors of the individual chapters. Nevertheless, an attempt was made to integrate these chapters with a common thread: we set out to critically examine our knowledge base for its suitability for tne rational design of workable catalysts and to point out gaps in our existing knowledge which we felt are worthwhile filling.

The book follows a bottoms-up structure, beginning with a microscopic view of catalytic surfaces: G. A. Somorjai's chapter deals with the effects of surface morphology and surface composition on catalyst performance, how our current understanding can be used, and how it could be extended to make it more useful for catalyst design.

The chapter by B. C. Gates addresses the opportunities in catalyst design via anchoring organometallic complexes to solid surfaces, together with the problems that need to be solved before such catalysts have the stability required for commercial applications.

Metal catalysts are usually dispersed over large-surface-area supports; a relatively new and quite lively area of research deals with the interactions between these metals and supports. The chapter by A. T. Bell discusses how supported metal systems might lend themselves to a more systematic design effort.

Surface-catalyzed reactions consist of a number of sequential or parallel elementary steps; catalytic selectivity can often be manipulated by selectively influencing these molecular pathways. The chapter by M. Boudart, drawing a parallel with homogeneous reaction systems, addresses the opportunities for designing selective heterogeneous catalysts by interfering with the appropriate elementary steps.

Zeolite-based catalysts are of major industrial interest; their unique selectivity combines the chemical selectivity of their active sites with the shape selectivity of their highly organized but molecular-size pore structures. Our understanding of shape-selective catalysis is now considerable; the chapter by W. O. Haag and N. Y. Chen attempts to organize this field from the standpoint of catalyst design.

Chemical reaction engineering, a relatively young discipline that deals with the mathematical modeling of catalytic systems, has already demonstrated its utility in the design and optimization of industrial catalysts. The chapter by R. Aris discusses how chemical reactivity and transport phenomena (diffusion, flow) can be coupled into quantitative models and how these can be used for the optimal design of porous catalysts. The theory is now useful enough to promise a broader applicability than heretofore achieved.

The book concludes with a chapter by J. Wei on the design of a specific, large-scale catalyst used for the removal of metal impurities from distillation residues in oil refineries. These catalysts operate under extreme conditions of temperature, pressure, and feed impurities. The combination of our understanding in catalytic chemistry and reaction engineering appears useful for the purposeful manipulation of catalyst design variables to achieve longer catalyst life, improved selectivity, and improved activity.

Although the book covers a wide area, including many aspects of chemistry and chemical engineering, it nevertheless omits some important aspects of catalysis. Notable among these is the absence of quantum-chemical considerations of molecular interactions with surfaces. Just as in homogeneous systems, the molecular-level, quantum-chemical interpretation of heterogeneous catalysis offers great promise and is thus of considerable research interest, but despite rapid advances in that area, it has not yet demonstrated its applicability to the design of working catalysts. The ongoing revolution in supercomputers will soon have a large impact here.

There are some additional qualifiers that need to be mentioned. The book is not a comprehensive review of all aspects of catalyst design or catalyst selection. For example, the catalytic literature contains numerous time-tested semiempirical rules and observations for catalyst selection (volcano plots for various reaction systems, d-electron character selection rules, etc.). These are not systematically reviewed. Instead, we emphasize more recent developments in selected areas, as enumerated above.

We also would like to acknowledge previous attempts at classifying catalysts and catalytic phenomena from the standpoint of catalyst selection and design. The recent book of Trimm (*Design of Industrial Catalysts*, Elsevier, New York, 1980) comes readily to mind.

Although the book is restricted to the discussion of heterogeneous catalytic systems, in several areas we have employed homogeneous concepts. Nevertheless, the design of homogeneous catalysts per se is not covered. Another omitted area is enzymatic catalysis, which is largely homogeneous in nature.

The book is intended primarily for researchers and practitioners in the field of catalysis. We hope that the researcher will be motivated to push catalytic science closer to the point where the ab initio, rational design of catalysts becomes a reality; we also hope that the industrial practitioner will find ideas in this book to shorten the tedious empirical work that is required today for the development of commercial catalysts. Although we had no intention of writing a textbook, we nevertheless hope that advanced graduate students will find this volume helpful as supplementary reading to textbooks on applied heterogeneous catalysis.

After having completed this manuscript, we have asked the question once again: "Can heterogeneous catalysts be designed on a rational basis?" The answer remains: not yet, at least not to a satisfactory extent. Progress toward that objective remains one of the prime research goals of the field and we expect continued significant developments in the years to come. It is also clear, however, that the growing body of catalytic knowledge is eminently suitable for the design of at least some aspects of catalytic performance; this belief united us in our efforts to publish this text.

L. L. HEGEDUS

Columbia, Maryland
February 1987

ACKNOWLEDGMENTS

L. L. Hegedus would like to express thanks to W. R. Grace & Co., N. D. Spencer, and J. A. Barrow.

G. A. Somorjai would like to express thanks to the U.S. Department of Energy.

B. C. Gates would like to express thanks to the National Science Foundation and students and collaborators whose names appear in the references.

A. T. Bell would like to express thanks to the U.S. Department of Energy, R. F. Hicks, J. S. Rieck, M. L. Levin, and R. P. Underwood.

M. Boudart would like to express thanks to the National Science Foundation.

W. O. Haag and **N. Y. Chen** would like to express thanks to Mobil Research and Development Corporation, F. Dwyer, and M. Snyder.

R. Aris would like to express thanks to the Petroleum Research Fund, National Science Foundation, S.-Y. Lee, and K. Kyriacos.

J. Wei would like to express thanks to the National Science Foundation, Mobil Oil Co., C. Hung, R. Agrawal, R. A. Ware, I. A. Webster, B. J. Smith, and K. W. Limbach.

CONTENTS

1. Introduction (*L. L. Hegedus*) — 1
2. The Building of Catalysts: A Molecular Surface Science Approach (*G. A. Somorjai*) — 11
3. Organometallic Chemistry: Basis for the Design of Supported Catalysts (*B. C. Gates*) — 71
4. Supports and Metal–Support Interactions in Catalyst Design (*A. T. Bell*) — 103
5. Kinetics-Assisted Design of Catalytic Cycles (*M. Boudart*) — 141
6. Catalyst Design with Zeolites (*W. O. Haag and N. Y. Chen*) — 163
7. Mathematical Models in Catalyst Design (*R. Aris*) — 213
8. Toward the Design of Hydrodemetallation Catalysts (*J. Wei*) — 245

Index — 273

CATALYST DESIGN

INTRODUCTION

L. L. HEGEDUS, *W. R. Grace & Company, Research Division, Columbia, Maryland*

1. ECONOMICS

The economic importance of catalysis is quite significant but hard to quantify in a precise manner because of the diversity of the products and processes that employ catalysts. As an example, let us examine the use of catalysts in conjunction with automotive transportation.

A wide variety of catalytically produced components are being used in automobiles today (the "organic" content of an automobile is about 13% of its weight). These include various polymers for structural components, paints, synthetic fibers, and synthetic elastomers, in addition to various solvents, detergents, and other chemicals employed in the manufacturing process. The polymeric materials are produced from monomers that are the products of a chain of catalytic processes beginning with raw materials such as crude oil and natural gas.

Catalysts are also widely used for refining crude oil into automotive fuels; various refinery streams may pass through a number of large-scale catalytic processes.

Finally, each carbon atom of the gasoline (in the form of uncombusted hydrocarbons or carbon oxides) passes once again through a catalyst that is used for the control of automotive emissions.

The above example for the interwoven nature of catalytic technologies illustrates the complexity associated with quantifying their economic impact. Nevertheless, a variety of breakdowns can be made. It is of interest to contemplate the following facts.

It has been estimated that about 90% of the currently practiced chemical and refinery processes are catalytic in nature. The worldwide usage of catalysts in 1984 cost approximately $2.7 billion (Strategic Analysis, Inc., 1985). Fifty-three percent was consumed by the United States, 17% by western Europe, and the remaining 30% by the rest of the world. Three main areas of application can be identified: catalysts for the manufacture of chemicals (43% of the total), catalysts for petroleum refining (35%), and catalysts for emission control (22%).

The three largest uses of catalysts in the United States are for automobile emission control (about $445 million in 1984, including noble metals, or about $285 million, excluding noble metals), for fluid catalytic cracking ($255 million), and for a variety of polymerization catalysts ($235 million). Table 1 lists the U.S. catalyst markets in more detail. The numbers add up to a U.S. catalyst consumption in 1984 worth $1.33 billion. The U.S. Department of Commerce (1986) estimates that the dollar volume of product shipments in 1984 in the categories "Petroleum Refining" and "Petrochemicals" was $176 billion and $83 billion, respectively. We can divide the sum of these items (as they are largely catalytically produced) by the 1984 U.S. catalyst sales to arrive at $195 worth of products for each dollar of catalyst consumed. This illustrates the tremendous economic leverage of catalytic technologies.

The world of catalyst manufacture is a complex one; it includes (a) companies that primarily sell catalysts to others and (b) companies that are captive users of catalysts they manufacture. In the former category, Stinson (1986) lists W. R. Grace, Engelhard, and Harshaw-Filtrol as the three largest suppliers based on their estimated dollar sales, making appropriate reservations for accuracy. Stinson's article lists the major U.S. catalyst suppliers to the chemical and petroleum refining industries.

Catalysts are typical specialty chemicals sold mostly on the basis of performance and involving a high level of contact between manufacturers and users. Since margins are driven primarily by performance advantage and only secondarily by cost, the catalyst industry tends to be highly research-intensive, often spending over 10% of catalyst sales on research (Stone, 1985). Catalyst developers are diverse in nature; they include major end-users such as oil and chemical companies (which may have both captive and merchant catalyst businesses), process developers, and specialized catalyst companies. The industry is highly proprietary in nature; composition and method of manufacture of complex industrial catalysts are rarely published. Nevertheless, there are references that allow a glimpse into this

2. TECHNOLOGY

TABLE 1. U.S. Consumption of Catalysts in 1984, Including Captive Consumption

	Consumption ($ Million)
Chemical Processing[a]	
Polymerization	235
Organic syntheses	85
Oxidation, ammoxidation, and oxychlorination	80
Hydrogen, ammonia, and methanol syntheses	50
Hydrogenation	30
Dehydrogenation	10
Total	490
Petroleum Refining[a,b]	
Catalytic cracking	255
Hydrotreating	75
Hydrocracking	40
Catalytic reforming	20
Total	390
Emission Control[a]	
Automotive[c,d]	445
Other[d]	5
Total	450
Total Catalyst Consumption	1330

[a]Strategic Analysis, Inc. (1985).
[b]Excluding $155 million for HF and H_2SO_4 alkylation catalysts.
[c]Including an estimated $160 million for noble metals.
[d]Estimated.

aspect of catalysis (e.g., Thomas, 1970; Sittig, 1978; Stiles, 1983). An interesting listing of commercial catalysts for the petroleum refining industry has been published recently (Aalund, 1984).

2. TECHNOLOGY

Catalytic technologies continue to advance rapidly, driven by strong economic pressures to improve the selectivity, activity, lifetime, and cost of catalysts. In a typical process, the catalyst is surrounded by expensive process steps aimed at processing the feed to the catalytic reactor or at processing the effluent from it. Thus, more selective, more active, and more durable catalysts have a large impact not only on feedstock utilization but also on overall plant capital and operating economics. Correspondingly, catalyst research activities are divided between those aimed at new catalyst–process combinations and those aimed at new or improved catalysts for existing processes.

Let us consider an example to illustrate the dynamic nature of commercial catalyst research and development. In the late 1960s and early 1970s it became evident that the emissions of automobiles would have to be considerably reduced in order to meet mounting public demands for strict emission standards. It also became evident that mere modifications to the gasoline-powered internal combustion engine as it existed at that time (or, as a matter of fact, as it exists today) were incapable of meeting these demands. Therefore, catalytic aftertreatment emerged as the only promising technology. As no catalyst existed at that time which could meet the required activity and durability specifications, a new catalyst–process combination had to be developed. A major industry-wide research effort was mounted which, in the case of General Motors alone, involved the testing of over 1500 catalyst formulations, submitted by some 82 prospective catalyst manufacturers, involving over 5000 General Motors employees and 22 million test miles before the catalyst was commercialized in the fall of 1974 (Hegedus and Gumbleton, 1980).

Once commercialized, further development of the catalyst continued: Among other factors, the need for improved fuel economy placed new demands on catalyst performance. Hence, improved catalysts emerged for an existing process. This example is especially interesting because the cycle did not stop there: new, more stringent environmental regulations required the catalytic control of not only hydrocarbons and carbon monoxide but also of nitrogen oxides. A new catalyst–process combination emerged in 1978 to meet this new demand: three-way catalysts (which can simultaneously oxidize hydrocarbons and carbon monoxide while reducing nitrogen oxides) in a catalytic converter controlled by an electronic computer to keep the feed stoichiometry in the desired range (some of the related developments were reviewed by Hegedus and Gumbleton, 1980). The cycle still continues: present-day demands on the catalyst require continued improvements in conversion performance, reactor volume, noble metal usage, durability, cost, and some other parameters. Thus, the research effort goes on at a high level, aimed yet again at improved catalysts for an existing process.

Other examples of significant technological developments in catalysis during the past 20 or so years include (a) zeolite-based fluid cracking catalysts [which, according to Mills (1983) save over $1 billion per year in imported oil as a result of improved gasoline yields from crude]; (b) a host of imaginative catalytic processes based on ZSM-5, a new zeolite material; (c) a large-scale catalytic technology for the reduction of nitrogen oxide emissions from power plant stack gases currently practiced in Japan and being introduced in West Germany (these are huge monolith-type reactors with up to 1000 m^3 in volume, employing base metal catalysts that selectively react nitrogen oxides with ammonia into nitrogen and water); and (d) a variety of important developments in the partial oxidation of hydrocarbons (e.g., acrylonitrile via ammoxidation, or improved catalysts for the selective oxidation of ethylene). Bimetallic hydrocarbon-reforming catalysts with vastly improved performance have emerged; hydroprocessing catalysts have been developed which can process heavier crudes with higher metals contents; methanol synthesis catalysts have been developed which allow improved process economics; and new polymerization catalysts have emerged which produce high-density and linear low-density polyethylene with substantially reduced capital and operating costs. The list of important recent developments would not be complete without mentioning the substantial catalytic technology base that was generated during the past 10–15 years in the synthetic fuels area. Although many of these processes are not economical at present, they may prove to be important in the future. Among these are several significant new catalysts and processes for the gasification or liquefaction of coal and for the downstream processing of coal-derived gaseous or liquid feedstreams.

3. SCIENCE

Catalytic science is in a state of rapid progress (e.g., Somorjai, 1985). Let us review briefly some of the important scientific developments of the past 10–15 years.

The evolution of modern instrumental–analytical techniques made a significant impact on the understanding of catalytic behavior. Surface structural information can now be readily gained by low-energy electron diffraction (LEED); surface compositional information by Auger electron spectroscopy (AES); both spatial and compositional information by the scanning Auger microprobe (SAM); and bonding, oxidation state, and related surface-chemical information by x-ray photoelectron spectroscopy

(XPS). Infrared (IR) spectroscopy and Fourier-transform infrared (FTIR) spectroscopy yield qualitative and quantitative information of catalyst–reactant interactions under actual reaction conditions, while modern variants such as photoacoustic infrared spectroscopy (PAS), diffuse reflectance spectroscopy (DRS) and reflectance–absorbance infrared (RAIR) spectroscopy have greatly expanded the utility of this time-tested technique. Vibrational features to which infrared is not sensitive can often be studied by laser Raman spectroscopy (LRS). A somewhat complementary and also relatively new technique for probing molecule–surface interactions is represented by high-resolution electron-energy loss spectroscopy (HREELS). Thermal desorption spectroscopy (TDS), a technique which provides for the time-resolved desorption of surface-bound species, usually into a mass spectrometer, allows the study of the energetics of highly complex, multicomponent catalytic surfaces. Variants of TDS are temperature-programmed desorption (TPD), which employs a carrier gas, and temperature-programmed reduction (TPR), which uses a reducing carrier gas.

In addition to surface-sensitive information, the bulk or near-surface properties of catalysts are also of significant importance. Long-range order can be studied by x-ray diffraction (XRD); shorter-range order can be studied by extended x-ray absorption fine structure (EXAFS) analysis, by transmission electron microscopy (TEM), and by magic angle-spinning nuclear magnetic resonance spectrometry (MASNMR), another new technique that has found important use for the near-order structural analysis of zeolites. The Mössbauer effect has also been successfully employed in catalytic studies to yield information about oxidation states and local coordination numbers.

The rapid penetration of these techniques into industrial catalytic research has completely changed the nature of the way in which new catalysts are being developed today. In addition, modern methods of analysis have proved to be invaluable for ensuring reproducibility and quality control, which are key commercial considerations.

Another area of rapid development in catalytic science is the synthesis and utilization of a number of important zeolites. The technical revolution associated with zeolite-based fluid cracking catalysts has been well documented; subsequent and ongoing work on ZSM-5 and other high silica–alumina-ratio materials has the promise of similar industrial impact. The unique transport properties of zeolites allow researchers to tailor them to achieve high catalytic selectivity (Weisz, 1973). As one of the interesting recent developments, Thomas et al. (1982) discovered ZSM-11 intergrowths in ZSM-5 zeolite crystals which can only be detected by TEM and which significantly influence the product distribution of methanol conversion to gasoline. While 40 zeolite minerals have been discovered and about 150 synthetic zeolites have been reported to date (Whyte and Dalla Betta, 1982), the majority of commercial catalytic zeolite applications employs only three of these: faujasite (zeolite Y), mordenite, and ZSM-5 (Heinemann, 1981). Thus, the catalytic exploitation of zeolites has a huge untapped potential.

Our understanding of catalytic mechanisms has been considerably enhanced by advances in organometallic homogeneous catalysis (e.g., Maugh, 1983); organometallic analogies of heterogeneous catalysis have also provided much new understanding (e.g., Canning and Madix, 1984). Isotope techniques have continued to represent an important key to the determination of catalytic mechanisms. Metal alloys and bimetallic clusters (Sinfelt, 1979) represent another area of scientific progress with technological potential. In light of these new concepts, the old notions of electronic and geometric factors in catalytic action are undergoing new interpretations (Sachtler, 1983).

Last but not least, important advances are being made in theoretical chemistry with potentially huge impact on catalytic science. Computer-aided materials simulation techniques are evolving (e.g., Goddard, 1985), which may have the ultimate potential for the ab initio design of high-selectivity catalytic surfaces. Similarly important future advances in catalytic kinetics are indicated by the increasing availability of supercomputers (e.g., Dagani, 1985) for rate calculations from quantum-mechanical principles.

As a result of this rapidly growing scientific understanding, new catalytic technologies are on the horizon; among these are various C–H bond activation reactions in alkanes, chirally selective catalysts, catalysts for the partial oxidation of methane into more useful chemicals, fuel cells, high-performance emission control catalysts, and a host of new catalytic materials that will be tailored to achieve catalytic selectivity approaching that of biological catalytic systems.

4. CATALYST DESIGN

The development of new or improved catalysts is a tedious, time-consuming process involving a combination of good science, hard work, experience, and considerable serendipity. It would, however, be misleading to claim that catalysts are developed entirely in an empirical way; in fact, progress in catalytic science is reflected by the decreasing role of empiricism as a function of time. More recently developed catalysts are better understood than their predecessors were at their time of commercialization, and a greater degree of preselection was employed in their development (e.g., Trimm, 1980).

Beyond the obvious driving forces toward a rational basis for catalyst design (i.e., maximum possible catalyst performance and thus improved margins), there are other pragmatic considerations. One of these is the sometimes extraordinary amount of testing required before a workable catalyst is identified and before this catalyst is optimized. As an example, new automobile exhaust catalysts have to pass several stages of laboratory-scale screening, followed by engine dynamometer and vehicle screening experiments by the automobile manufacturers, all involving lengthy durability runs. Once a catalyst has survived these, it enters the certification fleet of the car manufacturer, adding considerable time and expense. The process does not end there; the catalyst manufacturer continues to screen his various manufacturing batches, the car manufacturer faces end-of-the-assembly-line test requirements, and the catalyst, once in the field, is monitored by various agencies for performance in the customers' hands. It is clear that a more rigorous catalyst design procedure, emanating from a deeper understanding of the chemistry, physics, and engineering involved, would be of considerable benefit. Similar examples are provided by various petroleum refining catalysts (e.g., hydroprocessing), where lengthy durability testing requirements tend to limit the number of catalyst configurations that can be considered.

What is needed, then, is a more useful level of detailed understanding of the factors that influence catalyst performance. The list of such items begins at the catalyst surface: We need to have a better understanding of the relationships between the structure and composition of surface and its reactivity. The complexity in practical situations often arises because industrial catalysts may contain quite a large list of modifiers, dopants, selectivity enhancers, structure stabilizers, dispersion stabilizers, and so on, which make the difference between a workable and a nonworkable catalyst. Much of the investigative research to date has been focused on primary catalyst surfaces; chemically modified catalytic surfaces involving several components are still poorly understood. Thus, for a more effective catalyst design, more fundamental studies are needed which involve multicomponent catalytic surfaces.

4. CATALYST DESIGN

Another area in need of deeper understanding is metal–support interactions. This is a lively area of current catalyst research with a number of important recent developments.

Our quantitative understanding of molecular-level transport in molecular-size cavities (zeolite pores) lags considerably behind the rich detail that has been observed in the catalytic behavior of these materials. Quantification of these phenomena, suitable for kinetic-engineering analysis, is fortunately in progress today.

The mechanism of surface–molecular interactions determines activity and selectivity via intricate control of the often numerous combination of thermodynamically feasible elementary steps. Analogy with homogeneous catalytic reaction systems (which are much better understood) appears to provide useful clues in this area, with considerable promise for the coming years.

Once the catalytic process is in a "reasonable" shape at the molecular level, a barrage of reaction engineering tools becomes available for the design of a workable catalyst particle and for the design of an optimal reactor to house these particles. The reaction engineering literature has now several success stories for catalyst design once the chemistry has been established. Unfortunately, the application of chemical reaction engineering techniques to catalyst design is still lagging behind the capabilities of this field. Among the reasons might be the need for familiarity with complex, nonlinear differential equations and with advanced computer techniques for their solution. Another impediment has been the need for very large computers to solve "meaningfully complex" problems. Improvement is on the horizon in all these areas as a result of the increasing availability of canned software and of supercomputers.

Lastly, the design of new catalysts is predicated on the design of new catalytic materials. Rapid advances in materials science may ultimately lead to entirely new types of catalytic materials. Given the finite number of elements in the periodic table, all having been duly explored for many reactions, the nature of these materials will have to be multicomponent in composition and complex in structure.

Today's catalyst developer is a sophisticated formulator, with ever-improving tools at his or her disposal to replace trial and error with rational design. The trend toward "designer catalysts" (Cusumano, 1985) is now well recognizable and will contribute to further significant advances in catalytic technologies which are so vital for the world's economy.

REFERENCES

L. R. Aalund, *Oil Gas J.*, 55 (October 8, 1984).
N. D. S. Canning and R. J. Madix, *J. Phys. Chem.* **88,** 2437 (1984).
J. A. Cusumano, *Science* **85,** 120 (November 1985).
R. Dagani, *Chem. Eng. News*, 7 (August 12, 1985).
W. A. Goddard III, *Eng. Sci.*, 2 (September 1985).
L. L. Hegedus and J. J. Gumbleton, *Chemtech* **10,** 630 (1980).
H. Heinemann, *Catal. Rev. Sci. Eng.* **23,** 315 (1981).
T. H. Maugh II, *Science* **220,** 1032 (1983).
G. A. Mills, "Catalysis New Profit Opportunities Briefing," organized by Chemical Week, New York, September 1983.
W. M. H. Sachtler, *Chemtech*, 434 (July 1983).
J. H. Sinfelt, *Rev. Mod. Phys.* **51,** 569 (1979).
M. Sittig, *Handbook of Catalyst Manufacture*, Noyes Data Corp., Park Ridge, New Jersey, 1978.
G. A. Somorjai, *Science* **227,** 902 (1985).
A. B. Stiles, *Catalyst Manufacture*, Marcel Dekker, New York, 1983.
S. C. Stinson, *Chem. Eng. News*, 27 (February 17, 1986).
M. Stone, *Eur. Chem. News*, Specialty Chemicals Supplement (May 1985).
Strategic Analysis, Inc., *Chemical Week*, SAS 3 (June 26, 1985).
C. L. M. Thomas, *Catalytic Processes and Proven Catalysts*, Academic Press, New York, 1970.
J. Thomas, S. Ramdas, and B. Millward, *New Scientist*, 435 (November 1982).
D. L. Trimm, *Design of Industrial Catalysts*, Elsevier, New York, 1980.
U.S. Department of Commerce, *1986 U.S. Industrial Outlook*, January 1986.
P. B. Weisz, *Chemtech*, 498 (August 1973).
T. E. Whyte and R. A. Dalla Betta, *Catal. Rev. Sci. Eng.* **24,** 567 (1982).

2

THE BUILDING OF CATALYSTS: A Molecular Surface Science Approach

G. A. SOMORJAI, *Department of Chemistry and Materials and Molecular Research Division, Lawrence Berkeley Laboratory, University of California, Berkeley, California*

1. INTRODUCTION

The dream of every catalytic scientist or engineer is to build a catalyst that carries out a desired chemical reaction selectively and at high rates to maximize the conversion from the reactants to the products. This is becoming a reality for several heterogeneous catalyst systems through the atomic scale scrutiny of their structure, composition, and bonding properties along with careful investigations of their synthesis (Somorjai, 1981). Transition metal particles (iron, platinum) transition metal compounds (molybdenum sulfide), and zeolites (aluminosilicates) have been prepared with the desired atomic structure and composition to carry out selective catalysis of reactions ranging from ammonia synthesis (Emmett, 1975; Dumesic, Topsoe, and Boudart, 1975) and isomerization of alkanes (Lankhorst, Dejongste, and Ponec, 1982) to aromatization (Davis, Zaera, and Somorjai, 1984) and hydrodesulfurization (Topsoe, et al., 1981; Clausen et al., 1981) of various organic molecules. The developments of modern surface science techniques (Somorjai, 1981) and of solid state nuclear magnetic resonance (Thomas and Klinowski, 1985; Duncan and Dybowski, 1981) were primarily responsible for providing the researcher with the ability of molecular level scrutiny of catalytic systems. Table 1 lists the techniques that have been most frequently employed in catalysis science in recent years. Once the correlation between the kinetics (turnover rates, activation energies) and selectivity (product distribution) of a given catalytic system and its molecular properties of structure, oxidation states, and composition are established, a new generation of improved catalysts can be readily developed (Somorjai, 1985a). Next, the exploration of how to build an entirely new catalyst system using the principles that were uncovered by studies of the working system can commence.

Research and development in modern catalysis science was also aided greatly by the detection sensitivity of modern analytical techniques that include gas chromatography, mass spectroscopy, and ion scattering from surfaces. As a result there is diminished need for the use of high-surface area samples ($\sim 10^2$ m^2/g) to obtain a detectable signal. For catalytic reaction studies, surface areas in the range of 1 cm^2 are often adequate to permit more definitive investigations because of structural and chemical homogeneity and uniformity (Somorjai, 1984).

While molecular level scrutiny of catalyst systems has been possible only

TABLE 1. Frequently Used Techniques of Surface Science for Studies of Catalysts

1. Electron Scattering
    a. Electron spectroscopies (XPS, HREELS, AES)
    b. Low-energy electron diffraction (LEED)
    c. Electron Microscopy
2. Photon scattering (high and low intensities)
    a. Spectroscopies (IR, FTIR, Raman, Solid state NMR, ESR, EXAFS, NEXAFS, Laser techniques)
    b. Grazing angle x-ray diffraction
3. Molecule and ion scattering
    a. Molecular beam–surface interaction
    b. SIMS, ISS
4. Other techniques
    a. Radiotracer labeling
    b. Mössbauer spectroscopy
    c. Thermal Desorption

1. INTRODUCTION

TABLE 2. List of Surface-Science Studies of Catalyst Systems

1. Hydrogen–deuterium exchange by platinum
2. Oxidation of carbon monoxide by platinum
3. Hydrogenation of ethylene by platinum, rhenium
4. Ammonia synthesis by iron, rhenium
5. Hydrocarbon conversion by platinum (dehydrocyclization, isomerization, hydrogenolysis, hydrogenation dehydrogenation)
6. Hydrogenation of carbon monoxide by transition metals (Fe, Ni, Rh, Ru, Re, Co, Mo)
7. Partial oxidation of ethylene by silver and methanol by molybdenum oxide
8. Hydrodesulfurization of thiophene by molybdenum

in the last 10–15 years several reactions and associated catalysts have been investigated and are listed in Table 2. Out of the detailed studies came the recognition of patterns of behavior that suggests the existence of classes of catalytic reactions. First we will review some of the principles that govern catalytic studies that were learned from the large body of experimental observation in past decades (Section 1). In Section 2 we will attempt to summarize what has been gleaned from the molecular-level surface-science studies of catalytic systems and then try to classify them according to their behavior. Then we will discuss how these reaction types may be identified by experiments (Section 3). Next, some of the methods that could be employed to synthesize catalysts with desirable properties will be reviewed (Section 4). Finally, we will discuss important areas of catalysis science where acquiring better understanding is necessary in attaining our goals to build novel and improved catalytic systems (Section 5).

1.1. Catalytic vs. Stoichiometric Reactions

There are major differences in studies of catalytic vs. stoichiometric reactions. They are rarely emphasized during the educational process in chemistry. A molecule adsorbs on a catalyst's surface, it chemically rearranges while visiting several reaction sites by surface diffusion, and then it desorbs as the product to the gas phase. During the lifetime of a good catalyst, the reaction turns over a million times (10^6 product molecules per site or more). If the reactant forms strong chemical bonds upon adsorption on the surface, there is no catalysis, and the result is a stoichiometric reaction with a turnover of 1 (1 product molecule per site). Chemists are taught only about stoichiometric reactions during their formal training. Yet, many important life-sustaining reactions including photosynthesis, biological processes of our bodies, and the synthesis of ammonia are catalytic. During the catalytic process, therefore, a reactant cannot be strongly bound since that would poison the catalyst or would render it inactive. However, if the bonding were too weak, there would be no opportunity for chemical bond breaking, which is an integral part of any catalytic process. Thus bonding of intermediate strength between the catalyst and the surface is needed, and surface sites are required where both bond breaking and bond formation are possible within the residence time of the intermediates (Somorjai, 1985b).

This is well demonstrated by recent studies of the catalytic hydrodesulfurization of dibenzothiophene:

$$\text{dibenzothiophene} + 2H_2 \longrightarrow \text{biphenyl} + H_2S$$

The reaction was studied systematically by Chianelli and coworkers (Pecoraro and Chianelli, 1981) over a variety of transition metal sulfides of known structure and composition. While the third row of transition metal sulfide materials all exhibited low rates for the reaction, the fourth and fifth row sulfides generally exhibited high activity; this activity varied by over three orders of magnitude across the periodic table. When hydrodesulfurization activity (rates) were plotted as a function of the heat of formation of the metal sulfide, as shown in Figure 1, a clear volcano relationship appears which can be interpreted in terms of the principle of intermediate bonding (Chianelli et al., 1984); for example, for maximum hydrosulfurization rates, the metal–sulfur bonds strengths should be neither too strong nor too weak. The optimum enthalpies of sulfide formation can be estimated to be in the range of 35–45 kcal/g-atom of metal. A stronger metal–sulfur bond inhibits the release of sulfur from the catalyst's surface, whereas too weak a metal–sulfur bond provides little driving force for carbon–sulfur bond cleavage. The heats of formation of transition metal sulfide materials correlate linearly with the heats of adsorption for sulfur chemisorbed on clean transition metal surfaces.

1. INTRODUCTION

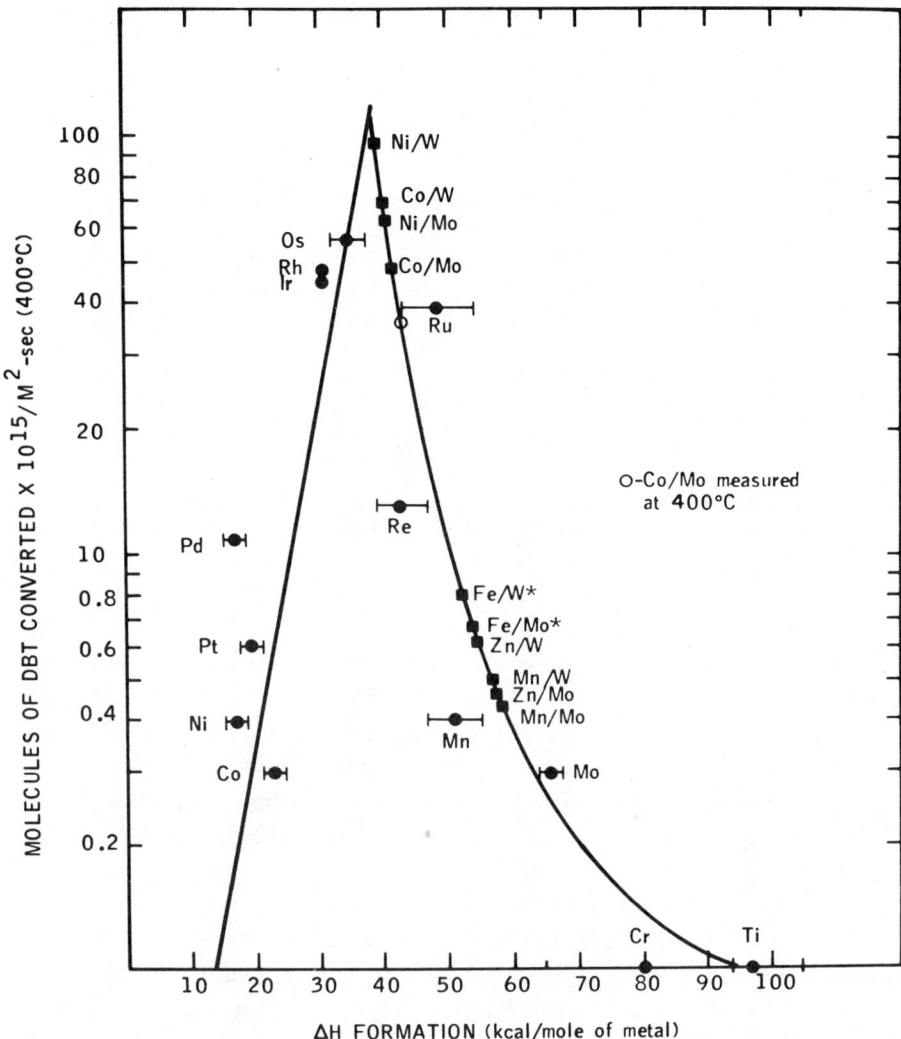

FIGURE 1. Volcano plot showing the correlation of hydrodesulfurization catalytic activity with heat of formation for mono- and bimetallic sulfides as reported by Chianelli et al. (1984).

1.2. Useful Turnover Rates and Activation Energies of Catalytic Reactions

As mentioned in Section 1.1, the reaction cannot be considered catalytic until it produces more than one product molecule per surface site. In order to determine the kinetic parameters for a reaction, one has to determine

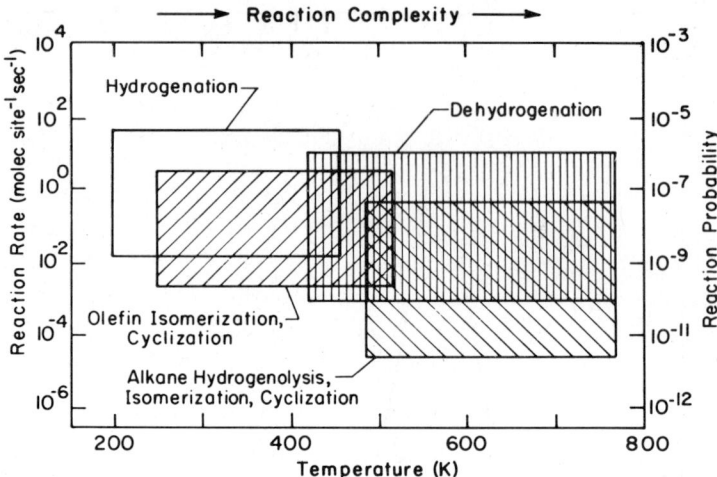

FIGURE 2. Block diagram of hydrocarbon conversion over platinum catalysts showing the approximate range of reaction rates and temperature ranges that are most commonly studied.

the surface area of the catalyst along with the rates of the reactions under steady-state conditions. By dividing the rates (product molecules per second) by the number of exposed surface sites, the specific turnover rate (product molecules per site per second) can be obtained. This type of analysis assumes that every surface site is active. It is likely, however, that only 10% or less of surface sites are active in any given reactions. In this case, the specific turnover rate is a conservative estimate of the real turnover.*

The turnover rates for hydrocarbon reactions are shown in Figure 2 (Davis and Somorjai, 1982). The turnover rates between 10^{-4} and 100 are used in the various technologies, and thus the temperature employed is adjusted to obtain the desired rates. The more complex isomerization, cyclization, dehydrocyclization, and hydrogenolysis reactions have activation energies in the range of 35–45 kcal/mole, and thus according to the Arrhenius expression, $k = A \exp(-\Delta E/RT)$, high temperatures are required to carry them out at the desired rates. Hydrogenation reactions have activation energies of 6–12 kcal/mole, and therefore may be performed at high rates at 300 K or below. Thus, there are at least two classes

---

*Another reaction rate parameter, the turnover frequency (product molecules per second), is often quoted. As defined, it does not assume knowledge of the surface area. However, the turnover frequency increases with increasing active surface area and as a result may be different from one catalyst to another for the same reaction.

1. INTRODUCTION

of reactions distinguishable by their very different activation energies that may be carried out at high and at low temperatures, respectively, under very different experimental conditions.

Let us consider a catalytic reaction where the desorption of the product molecules is the rate-limiting step (Somorjai, 1985a). In Figure 3, a plot of the turnover rate is presented as a function of the activation energy for the reaction at different temperatures. In obtaining these values we assume a preexponential factor, A, of 10^{12} sec^{-1}. Molecules that have desorption activation energies of less or equal to that represented by the solid line at the given temperature can participate in the catalytic reactions while those that have higher activation energies are too strongly bound to be able to turn over at the observed rates. Of course, those molecules that are bound so weakly as to have very small activation energies of desorption may not be able to react because their surface residence times are very short. Nevertheless, the surface coverages of weakly absorbed species may be controlled by the reactant pressures. High pressures favor catalytic reactions that involve weakly bound adsorbates. In Figure 3, the turnover rates and

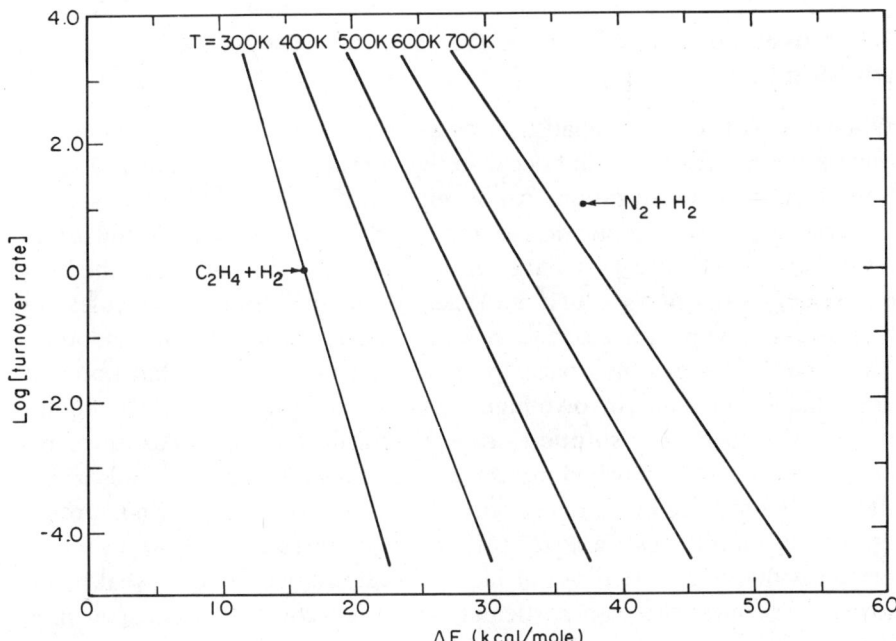

FIGURE 3. Turnover rates as a function of the activation energy for reactions at different temperatures. The preexponential factor of 10^{12} is assumed.

reaction activation energies of the hydrogenation of ethylene and ammonia synthesis are indicated to show two reactions that occur under widely different conditions.

1.3. Secondary Reactions

Most heterogeneous catalytic experiments are designed in such a way that both the reactants and the products have many opportunities for readsorption on the catalyst surface. The catalysts are usually high surface-area solids and the molecules collide with the surface many times before leaving the catalyst bed. Thus there is a finite probability that secondary reactions could occur. The likelihood of such reactions is enhanced if the products of the primary catalytic reactions are more reactive than the reactants. For example, ammonia is more reactive than nitrogen or hydrogen and methanol is more reactive than carbon monoxide and hydrogen. Especially when the conversion of a large fraction of the reactants to the products (over 20%) is desired, the product distribution includes the products of secondary reactions (Dwyer and Somorjai, 1978).

1.4. Coverage Dependence of the Heat of Adsorption and Bonding

When one plots the heat of adsorption as a function of adsorbate coverage, there are major changes that occur in this thermodynamic value as a function of surface concentration. An example of this is shown for the heat of adsorption of carbon monoxide as a function of coverage over a palladium crystal surface (Figure 4) (Conrad et al., 1974). At low coverages the heat of adsorption per mole is high, and these strongly adsorbed molecules are thus used in low-pressure surface-science studies because of their stability. These molecules can also participate in catalytic processes that occur at high temperatures and involve high activation energies. At high coverages the average heat of adsorption drops significantly mostly because of repulsion among the adsorbed molecules, or because the strongly adsorbing sites on the surface are already occupied. The overall heat of adsorption drops to around 10 kcal/mole for carbon monoxide (see Figure 4). Catalytic reactions that are carried out at high coverages involve these weakly adsorbed molecules that can participate in the reactions that occur at high turnover frequency and relatively low temperatures. It should be noted that homogeneous catalytic reactions and catalytic reactions that occur in

1. INTRODUCTION

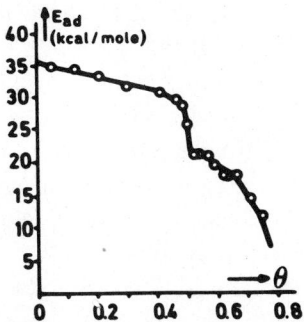

FIGURE 4. Isosteric heat of adsorption for CO on the Pd(111) crystal face as a function of CO coverage, θ (from Conrad et al., 1974).

biological systems occur at 300 K or below very frequently and by necessity must have low activation energies. In these circumstances, the reaction of only weakly bound species is implicated. Reactions that occur only at high temperatures and have high activation energies implicate the turnover of more strongly adsorbed chemical species.

1.5. Why Metals Are Good Catalysts

One of the important functions of transition metals in catalytic reactions is to atomize diatomic molecules and then to supply the atoms to other reactants and reaction intermediates (Somorjai, 1981). H_2, O_2, N_2, and CO are the diatomic molecules of importance, in order of increasing bond energy. The strength of bonding H, O, N, and C to transition metal surfaces provides the thermodynamic driving force for both the atomization and for the release of atoms for reactions with other molecules. It follows that dissociative adsorption of polyatomic molecules can also readily occur on metals since other chemical bonds (C–H, C–N, N–O, N–H, Cl–Cl, P–H, etc.) all fall in the bonding energy range represented by H_2 on the low bond-energy side and CO and N_2 on the high bond-energy side.

Metal surfaces also have other unique properties with regard to catalyzing a sequence of complex reactions that begins with dissociative adsorption followed by complex rearrangements through the formation and breaking of multiple bonds and finally desorption of the products. They have many binding sites where simultaneous bonding to 1, 2, 3, or 4 metal atoms is possible, counting only the nearest neighbors in the topmost atomic layer. If we include bonding possibilities to atoms in the second layer, and there are very many experimental indications that this occurs, there are many other possible bonding arrangements as well. All of these sites are available at close proximity, well within the range of distances the adsorbed intermediates can visit by diffusion within their residence times on the surface. The high density of reaction sites that are available in great variety is the unique feature of metal surfaces that makes them so active and versatile in catalyzing so many reactions.

The disadvantage of metal surfaces in catalytic reactions is caused by the same phenomena as the advantages. Because of the diversity of available sites, several competing reactions may all occur with similar probability, thereby reducing the selectivity. Often the desorption of the final product molecules can be rate limiting because of the strong metal–adsorbate bond that increases their surface residence times. In these circumstances selective blocking of certain metal sites and alteration of bonding at others by addition of suitable coadsorbates become part of the catalyst preparation and formulation.

1.6. Why d-Electron Transition Metals Are Such Good Catalysts

Recently a theory of metal catalysis has been proposed that considers those materials active in breaking and forming chemical bonds that have a high concentration of degenerate electronic states of low energy, permitting charge fluctuations (Somorjai and Falicov, 1985). The d-electron metals in which the d-band is mixed with the s and p electronic states to provide a large concentration of low-energy electronic states and electron vacancy states are ideal for catalysis because of the multiplicity of degenerate electronic states. Moreover, those sites where the degenerate electronic states have the highest concentrations are most active in breaking and forming chemical bonds. These electronic states have charge fluctuations, configurational fluctuation, and spin fluctuations (Somorjai and Falicov, 1985), and these are occurring at metal sites of high coordination. The density of hole states, which is a measure of the charge fluctuation probability, is shown in Figure 5 for various sites on a nickel surface on which six atoms are placed. The higher the atomic coordination of the site the higher the density of electron hole states, n. High coordination sites may be made available in catalytic reactions by using surfaces with open atomic structures so that atoms in the second layer become available to the incoming reactants or by using stepped and kinked surfaces. These will be discussed later.

NICKEL d-SHELL OCCUPIED BY 9.44 ELECTRONS

BULK NICKEL	Z(Ni)=12		n = 0.56
(111) SURFACE NICKEL ATOM	Z(Ni)= 9		n = 0.38
OUTSTEP NICKEL ATOM a (FIG. 1)	Z(Ni)= 7		n = 0.25
OUTSTEP NICKEL ATOM b (FIG. 1)	Z(Ni)= 7		n = 0.24
INSTEP NICKEL ATOM c (FIG. 1)	Z(Ni)=11		n = 0.52
INSTEP NICKEL ATOM d (FIG. 1)	Z(Ni)=10		n = 0.45
Ni UNDER A (111) Cu MONOLAYER	Z(Ni)= 9	Z(Cu)= 3	n = 0.52
Ni MONOLAYER ON Cu(111)	Z(Ni)= 3	Z(Cu)= 6	n = 0.46

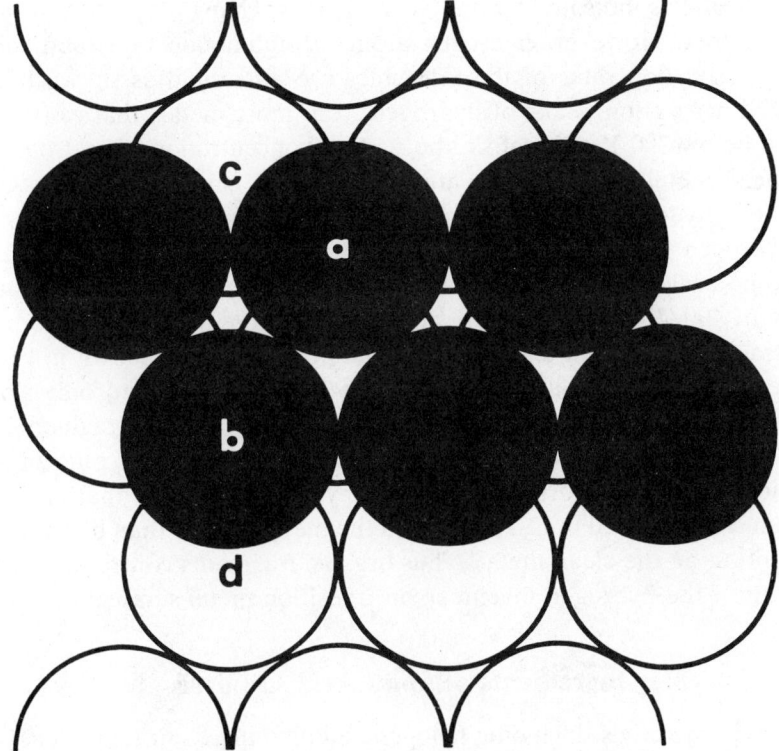

FIGURE 5. Model of the nickel (111) surface with six-atom cluster placed on top of it.

Thus the *d*-electron metals are most important for catalytic reactions because they provide a large concentration of degenerate low-energy electronic states within which charge fluctuations can occur involving electrons as well as electron vacancies. Those metals that have low density of states at the Fermi level even at the high coordination sites (e.g., gold) are not usually active in catalysis.

1.7. Hydrogen Inventory in Hydrocarbon Conversion Reactions

Most transition metals can readily dissociate the hydrogen molecules even below 300 K as indicated by the facile H_2/D_2 exchange reported by many investigators. However, hydrogen atom recombination rates and subsequent desorption rates of the molecules are also rapid as shown by the thermal desorption peaks of hydrogen from most metals that can be detected below 300 K. Therefore the surface concentration of hydrogen on the clean metal surface at elevated temperature is quite low unless high (>1 atm) hydrogen pressures are employed.

Hydrogen desorption from dissociating and strongly chemisorbed organic molecules from metal surfaces occur in sequence in the temperature range of 400–1000 K (Salmeron and Somorjai, 1982). Typical desorption spectra from adsorbed alkenes on platinum surfaces are shown in Figure 6. Because of the strong C–H bonds, hydrogen is desorbed only slowly from organic fragments. Yet hydrogen atoms from these fragments are available to coadsorbed reactants, as indicated by their rapid rates of deuteration. Thermal desorption spectroscopy studies indicate that more hydrogen can be stored in C–H bonds on the metal surface than by hydrogen adsorption on the clean metal. Thus organic fragments contribute a great amount of the hydrogen inventory on transition metal surfaces.

1.8. Molecular Ingredients of Solid Acid Catalysis

Solid acid materials, including halogenated aluminas, silica-aluminas, and crystalline zeolites, are essential components in virtually all commercial catalysts for converting heavy petroleum fractions into lighter feedstocks suitable for upgrading into gasoline, diesel, and jet fuels. Fluid catalytic cracking is by far the largest single catalytic process with worldwide throughput in the range of 2–3 billion barrels annually. Despite the wide application and importance of solid acid materials in isomerization, reforming, hydrocracking, and fluid cracking, until recently the nature of the surface acid

1. INTRODUCTION

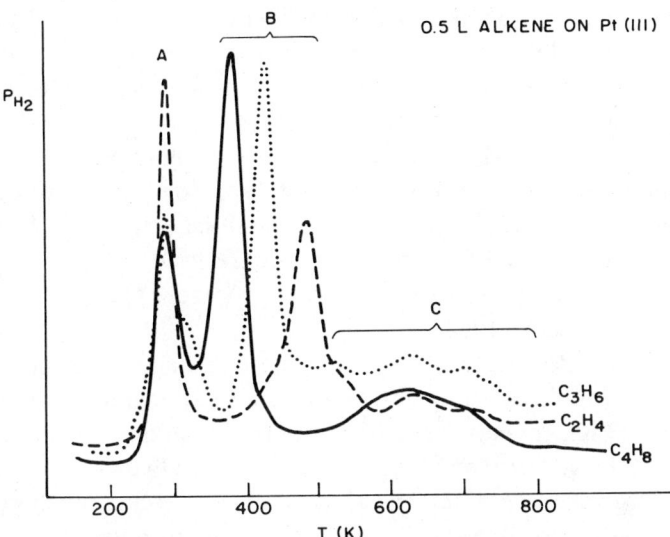

FIGURE 6. Hydrogen thermal desorption spectra demonstrating the sequential dehydrogenation of ethylene, propylene, and butene chemisorbed on Pt(111) at 120 K.

sites responsible for these reactions has remained largely obscure (Somorjai and Davis, 1985). Of particular importance are the acid site coordination environment, structure, composition, strength, and density.

The advent of magic angle spinning nuclear magnetic resonance (MAS–NMR) in combination with improved methods of application for infrared spectroscopy and base adsorption–thermal desorption has recently enabled catalytic chemists to scrutinize the distribution of acidic sites in solid acid materials with greatly improved effectiveness. One of the most important conclusions to clearly emerge from recent research on zeolitic catalysts (Haag, Lago, and Weisz, 1984) is that tetrahedral aluminum ions are the essential active sites for a variety of catalyzed hydrocarbon conversions. These sites are exceedingly reactive, indeed, even at low concentrations in the parts per million range. Turnover frequencies per tetrahedral aluminum atom for many hydrocarbon reactions catalyzed over zeolites frequently equal or exceed those for metals and the most efficient enzyme catalysts (Haag, Lago, and Weisz, 1984). There is a linear relationship between first-order rate constants for n-hexane cracking and aluminum content in HZM-5 zeolites (Haag, Lago, and Weisz, 1984; Olson, Haag, and Lago, 1980). In this case, MAS–NMR studies have demonstrated that all aluminums have tetrahedral structure (Kentgens, Scholle, and Veerman, 1983). In the dehydrated state this Al-structural environment is assymetric as a result of coordination of a single OH group that displays an OH stretching frequency near 3600 cm^{-1} (Jacobs and Von Ballmoos, 1982). Upon interaction with water or ammonia, the aluminum coordination becomes symmetrical, presumably as a result of proton transfer to the chemisorbed base. The linear dependence of catalytic activity on tetrahedral aluminum content is well followed over at least three orders of magnitude variation in the aluminum concentration. (Kentgens, Scholle, and Veerman, 1983; Fyfe et al., 1982).

The relative rate constants for n-hexene cracking, 1-hexene cracking, and 1-hexene double bond isomerization catalyzed near atmospheric pressure and at 727 K over catalysts with variable aluminum content have been determined (Haag, Lago, and Weisz, 1984). As with hydrocarbon reactions on metals, rearrangement and cracking of alkenes occurs much more rapidly as compared to alkanes. The difference in turnover frequency for hexene and hexane cracking is almost a factor of 1000. All reaction rates vary linearly with aluminum content, and the parallel lines tend to confirm that a common active site is operative in each reaction.

It has long been argued (Greensfelder, Voge, and Good, 1949; Thomas, 1949; Van Hook and Emmett, 1965) that hydrocarbon conversion over solid-acid materials involves carbonium-ion-like intermediates that result from the interaction of reactant hydrocarbons with surface hydroxyl groups that are nominally similar to common Brønsted acids. However, recent studies of small-alkane reactions catalyzed over amorphous solid acids and zeolites have revealed drastic differences in reaction selectivity which clearly emphasize the occurrence of different reaction pathways. Studies by McVicker and coworkers (McVicker, Kramer, and Ziemak, 1983), for example, revealed that amorphous solid acids, such as SiO_2/Al_2O_3, π-Al_2O_3, and halogenated Al_2O_3 produce isobutane-cracking products that are indicative of radical-like chemistry. Halogen addition strongly promotes the radical-like cracking pathway. Essentially no isomerization to n-butane could be detected over the amorphous solid acids. By contrast, ultrastable-Y zeolite produced reaction products requiring a combination of radical-like and carbonium-ion-like hydrogen abstraction processes as a prerequisite for subsequent carbonium-ion-like reactivity. Once initiated, intermolecular hydrogen transfer occurred readily, yielding mostly isomerization and alkylation products that are those expected for an acid-catalyzed reaction pathway. These studies demonstrate that the intrinsic "acidity" of solid-acid materials is considerably different than was previously thought, especially that of amorphous materials.

Solid-acid materials are usually nonconducting, and, as such, their investigation with common ultrahigh vacuum surface characterization tools poses special operation problems. Moreover, much of their chemical activity originates from internal surfaces of the microporous materials that are usually not accessible to electron or ion scattering. Nevertheless, many opportunities exist for systematic investigations coupling reactivity studies with surface-science investigations. It is our hope that such studies will be initiated in the near future.

2. CLASSIFICATION OF CATALYTIC REACTIONS INVOLVING METALS

The technologically important catalyst systems that have been subjected to combined surface-science and kinetic studies are listed in Table 2. The systems studied have been mostly metals. The results of these investigations suggest that there are three classes of catalytic reactions (Somorjai, 1985a):

1. Those that occur directly on the metal surface.
2. Reactions that occur on top of a strongly bound layer of adsorbates in the second layer.
3. Reactions that occur on coadsorbate-modified surfaces.

2.1. Catalytic Reactions on Metal Surfaces

These reactions usually involve strongly adsorbed intermediates and are surface *structure sensitive*. Atomically rough surfaces usually exhibit the highest turnover rates. The best example of this type of process is the synthesis of ammonia. Figure 7a shows the rates over three iron single-crystal surfaces (Spencer, Schoonmaker, and Somorjai, 1982). The (111) orientation crystal face is about 500 times as active as the (110) close-packed surface. Adsorption studies by Ertl (Ertl, 1980, 1981) show equally large differences in the sticking probability of N_2 on these surfaces, confirming that the dissociation of N_2 is the rate-limiting step in this reaction. The ammonia synthesis over rhenium crystal surfaces (Somorjai et al., 1986) shows even larger structure sensitivity as shown in Figure 7b. The

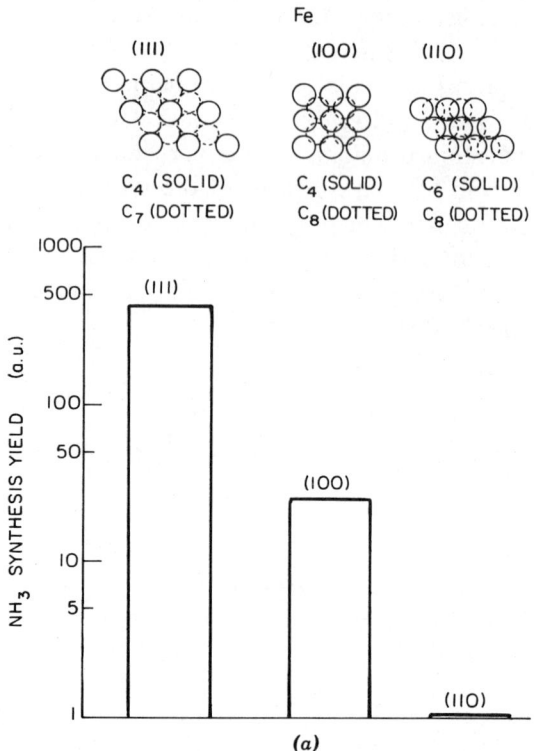

FIGURE 7. (*a*) Bar graph showing the profound influence of surface structure on the rates of ammonia synthesis over iron single-crystal surfaces at 20 atm and 700 K. (*b*) Rates of ammonia synthesis over rhenium single-crystal surfaces at 20 atm and 700 K. (*c*) Metal surfaces that are most active for ammonia synthesis.

(1120) and the (1121) crystal faces of this hexagonal close-packed metal are over 1000 times more active than the close-packed (0001) crystal face. Figure 7c shows the three metal surfaces that are most active for ammonia synthesis. These surfaces are rough on the atomic scale and have open structures that expose high coordination sites especially in the second layer. These atoms with large numbers of nearest neighbors are accessible to the incoming N_2 molecules and appear to be most active for their dissociation.

The ammonia synthesis may be viewed as an example of a reaction that obeys Langmuir–Hinshelwood kinetics, that is, dominated by the reactions between adsorbed species on the metal surface.

High coordination sites may be made available in catalytic reactions by

2. CLASSIFICATION OF CATALYTIC REACTIONS INVOLVING METALS

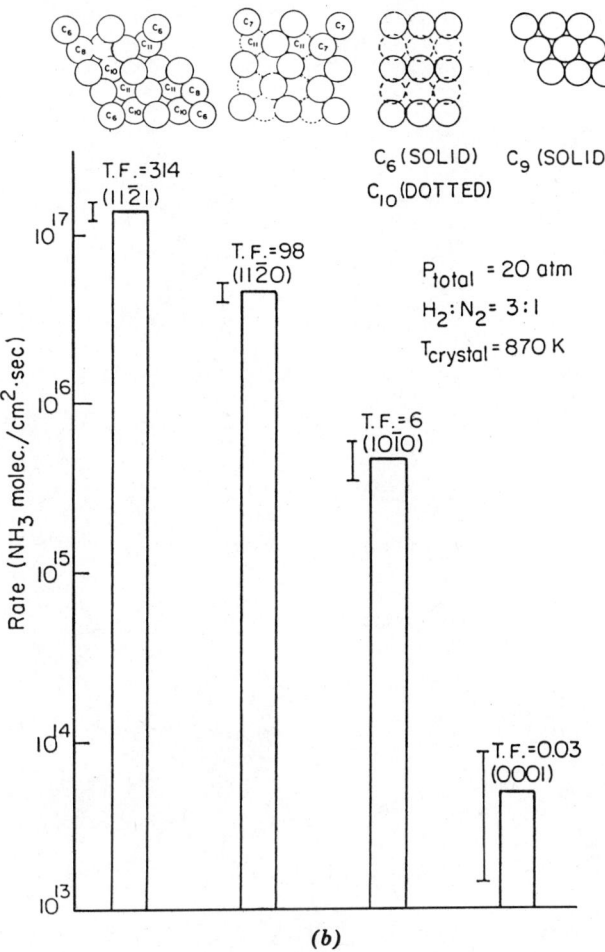

FIGURE 7. Continued.

using surfaces with open atomic structures so that atoms in the second layer become available to the incoming reactants. This is clearly the situation during ammonia synthesis where these open, high Miller index surfaces that are most active for this reaction are stabilized in the presence of nitrogen despite their high surface energy as compared to the close-packed low Miller index surfaces, bcc(110) and hcp(0001). Another more frequent method of generating rough surfaces is through the creation of atomic steps and kinks (Somorjai, 1983). Figure 8a illustrates different

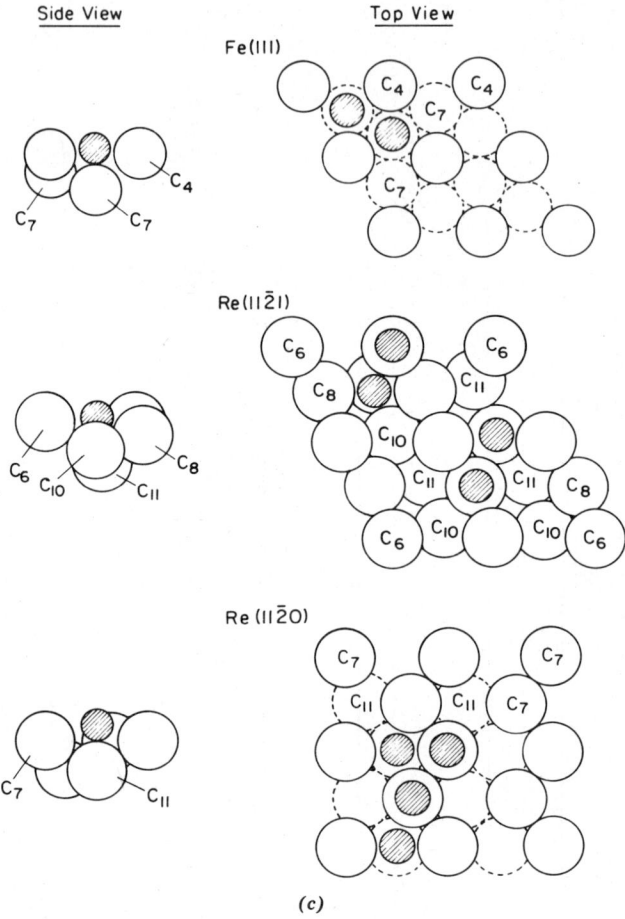

(c)

FIGURE 7. Continued.

close-packed surfaces with variable concentrations of steps. These are usually one atom in height separated by terraces of low Miller index orientations. There may be kinks in the steps (Figure 8b). The relative concentration of atoms at kinks, steps, and in terraces depends on the conditions of preparation. For single crystals, the direction of cut determines the surface structure. For small particles, the particle size and shape determine the concentrations of the various irregularities, steps, and kinks. Usually the smaller the particle, the higher the concentration of kinks and steps.

The bottom of step or kink exposes atoms of high coordination while

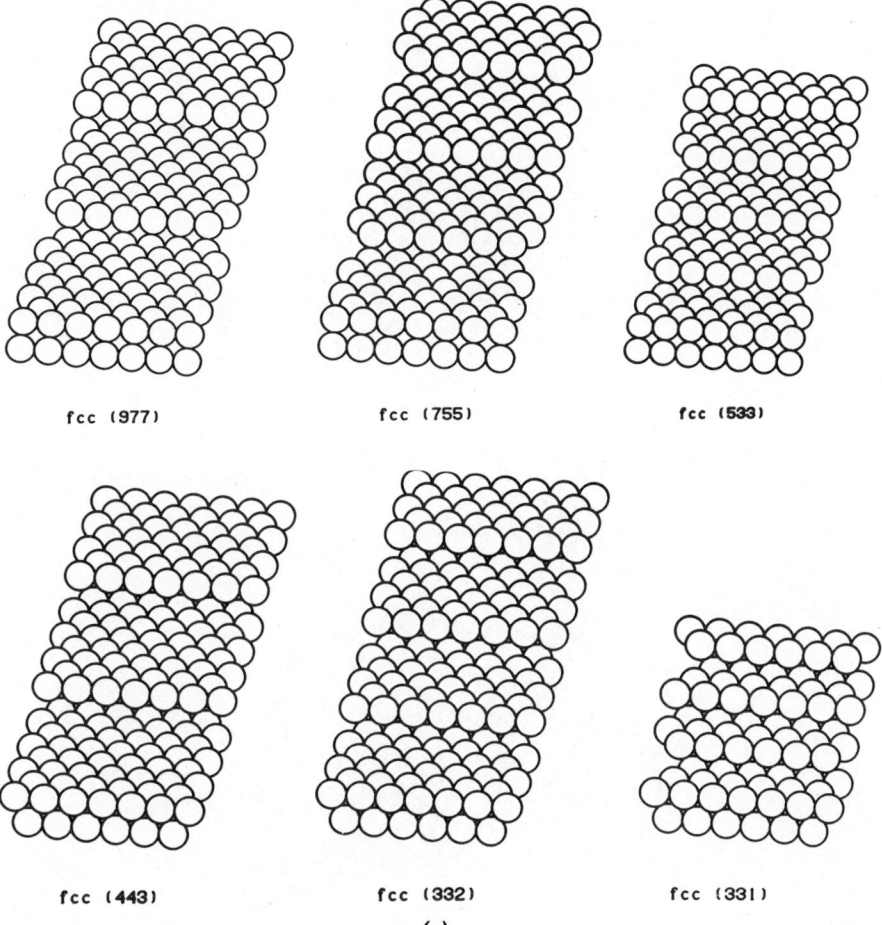

FIGURE 8. (*a*) Structure of several high Miller index stepped surfaces with different terrace widths and step orientations. (*b*) Surface structures of several high Miller index surfaces with deferring kink concentrations in the steps.

the top consists of atoms with smaller number of nearest neighbors. Kink sites in platinum single-crystal studies proved to be centers of strong hydrogenolysis (C–C bond breaking) activity as shown in Figure 9 (Davis, Zaera, and Somorjai, 1984). The rates of hydrogenolysis of light molecules or *n*-heptane correlates well with increasing kink concentrations.

There is increasing evidence that while high coordination sites enhance

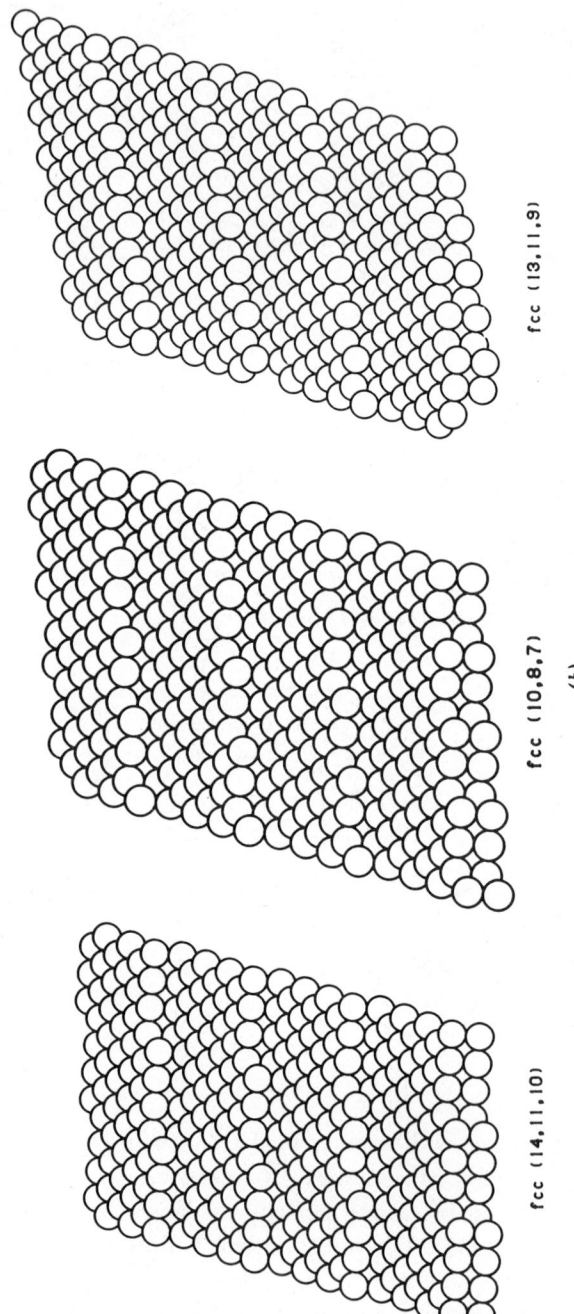

fcc (13,11,9)

fcc (10,8,7)

(b)

fcc (14,11,10)

FIGURE 8. Continued.

2. CLASSIFICATION OF CATALYTIC REACTIONS INVOLVING METALS

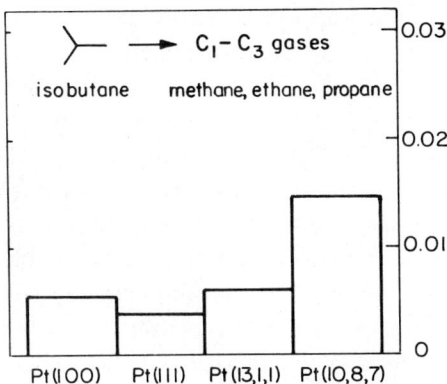

FIGURE 9. Reaction rates shown as a function of surface structure for isobutane hydrogenolysis catalyzed at 570 K, at atmospheric pressure over four platinum surfaces. Hydrogenolysis rates are maximized when kink sites are present at high concentrations as in the platinum (10,8,7) crystal surface.

the catalytic reaction rates for structure-sensitive reactions, low coordination sites adsorb molecules strongly and hold them tenaciously. On a stepped or kinked surface, both high coordination and low coordination sites are present side by side. Therefore sites of increased turnover have neighboring sites where strongly adsorbed species exist with resident times that are over an order of magnitude greater than the turnover times for the catalytic reaction. Studies using ^{14}C and ^{35}S isotope labeling of reactants clearly indicate the presence of a strongly adsorbed overlayer that partially covers a catalytically active metal surface. It is one of the ironies of metal catalysis that the creation of active sites of high turnover brings about the production of other sites of strong bonding and thus minimal turnover in equal numbers.

2.2. Catalysis on Top of a Strongly Adsorbed Overlayer

Reactions of this type do not occur directly on the metal surface and therefore are usually *structure insensitive* (Somorjai, 1985a). In fact, the role of the metal is reduced to providing atoms, hydrogen for example, via the dissociation of diatomic molecules. The metal is usually covered by strongly adsorbed overlayers and thus the incoming reactants (other than hydrogen) cannot form strong metal–adsorbate bonds. An example of this type of reaction is the hydrogenation of ethylene (Somorjai, et al.,

1985). This facile reaction occurs at 300 K and at atmospheric pressures on many transition metal surfaces. It has been the subject of investigations of many researchers including Farkas and Farkas (Farkas, Farkas, and Rideal, 1934), Eley (Eley and Tuck, 1936), Beek (Beek, 1950), Twigg and Rideal (Twigg and Rideal, 1939), Horiuti and Polanyi (Horiuti and Polanyi, 1934), Roberts (Roberts, 1963), and many others (Horiuti and Miyahara, 1968). I shall restrict my comments to platinum and rhodium, which are among the most active catalysts for this process. Table 3 shows that hydrogenation occurs equally well on platinum crystals, films, foils, and supported particles, indicating that the reaction is structure insensitive (Somorjai and Zaera, 1984). When the clean metal surfaces are exposed to ethylene, a strongly adsorbed overlayer of ethylidyne (C_2H_3) forms. This molecule, shown in Figure 10 along with its vibration spectrum, is obtained by high-resolution electron energy loss spectroscopy HREELS (Somorjai, Koestner, and Van Hove, 1982). The kinetics of ethylene hydrogenation and those of ethylidyne have been studied extensively over the (111) faces of rhodium and platinum, and the rates of these processes are displayed in Figure 11. Ethylene hydrogenation occurs at a rate six orders of magnitude higher than the rehydrogenation of the strongly adsorbed ethylidyne (Somorjai et al., 1985). Even the deuteration of the methyl group of ethylidyne occurs very slowly. Studies using ^{14}C labeling of ethylidyne and vibrational spectroscopy confirm these findings.

The (111) faces of platinum and rhodium are covered with a monolayer of ethylidyne during ethylene hydrogenation since reaction rates are nearly identical over initially clean surfaces and surfaces precovered with ethylene. Vibrational spectroscopy studies confirm that the adsorbed monolayer structure on these surfaces after hydrogenation is ethylidyne. Thus ethylene hydrogenation occurs rapidly on the C_2H_3 covered surfaces. The packing of the ethylidyne on the overlayer does not permit C_2H_4 adsorption directly on the metal surface, as proven by exchange studies with C_2H_4 on C_2D_4. On the other hand, thermal desorption studies show that H_2 (D_2) can be dissociated and readsorbed on the ethylidyne-covered surfaces up to about one-fourth of a monolayer coverage.

A reaction model that explains these results is shown in Figure 12 (Somorjai et al., 1985). The hydrogen atom is transferred to the ethylene molecule that is weakly adsorbed on top of the ethylidyne and in the second layer perhaps by forming an ethylidene intermediate. This model of hydrogen transfer from hydrocarbons to ethylene was first proposed by Thomson and Webb (1976) and our studies corroborate their findings. Our mech-

TABLE 3. Comparison of Ethylene Hydrogenation Rates and Kinetic Parameters for Platinum Catalysts with Different Surface Morphologies[a]

Catalyst	Log Rate[b]	Orders, in Partial Pressure		E_a (kcal/mole)
		Ethylene	Hydrogen	
Platinized foil	1.9	−0.8	1.3	10
Platinum evaporated film	2.7	0	1.0	10.7
1% Pt/Al$_2$O$_3$	—	−0.5	1.2	9.9
Platinum wire	0.6	−0.5	1.2	10
3% Pt/SiO$_2$	1.0	—	—	10.5
0.05% Pt/SiO$_2$	1.0	0	—	9.1
Pt (111)	1.4	−0.6	1.3	10.8

[a] G. A. Somorjai (1985).
[b] Rate in molecules per Pt atom · sec, corrected for the following conditions: $T = 323$ K, $P_{C_2H_4} = 20$ torr, $P_{H_2} = 100$ torr.

Different ethylidyne species: bond distances and angles
(r_C = carbon covalent radius; r_M = bulk metal atomic radius)

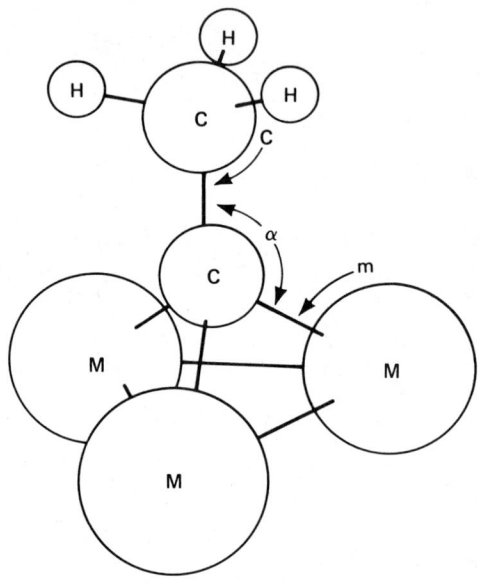

	C [Å]	m	r_M	r_C	α [°]
$Co_3(CO)_9 CCH_3$	1.53 (3)	1.90 (2)	1.25	0.65	131.3
$H_3 Ru_3(CO)_9 CCH_3$	1.51 (2)	2.08 (1)	1.34	0.74	128.1
$H_3 Os_3(CO)_9 CCH_3$	1.51 (2)	2.08 (1)	1.35	0.73	128.1
Pt (111) + (2 × 2) CCH_3	1.50	2.00	1.39	0.61	127.0
Rh (111) + (2 × 2) CCH_3	1.45 (10)	2.03 (7)	1.34	0.69	130.2
$H_3C - CH_3$	1.54			0.77	109.5
$H_2C = CH_2$	1.33			0.68	122.3
$HC \equiv CH$	1.20			0.60	180.0

(a)

FIGURE 10. (a) Surface structure of ethylidyne. Bond distances and bond angles are compared with several trinuclear metal cluster compounds of similar structure. (b) High-resolution electron-energy-loss vibrational spectrum of ethylidyne and deuterated ethylidyne of rhodium (111).

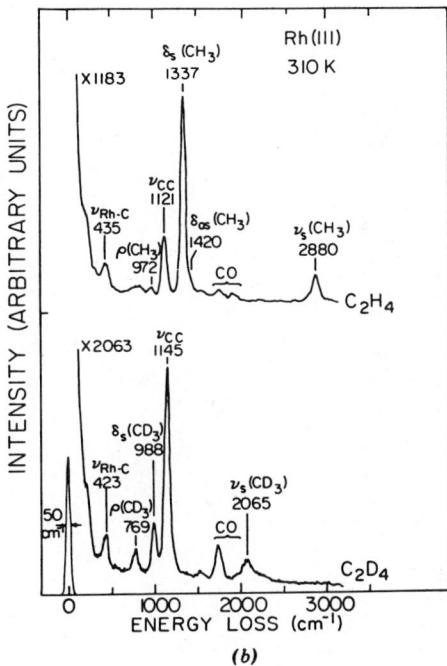

FIGURE 10. Continued.

anism is of the Eley–Rideal type and is characterized by low activation energy and structure insensitivity.

However, there are other mechanisms of C_2H_4 for hydrogenation that our studies and those of others have uncovered (Somorjai et al., 1985). At higher temperatures, the rate of rehydrogenation of C_2H_3 is significant and the bare metal becomes available, in part, for C_2H_4 hydrogenation. During the electrochemical hydrogenation of C_2H_4, the platinum surface is covered with a layer of hydrogen atoms (hydride) that react rapidly with the approaching C_2H_4 and do not permit the formation of ethylidyne. The complexity of surface reactions cannot be underestimated.

Nevertheless, ethylene hydrogenation provides an example of reactions of weakly adsorbed molecules in the second layer, an important class of catalytic reactions that could occur at low temperatures or high pressures. The hydrogenation of CO over certain transition metals that exhibit positive-order pressure dependence on both H_2 and CO pressures are thought to occur in this way.

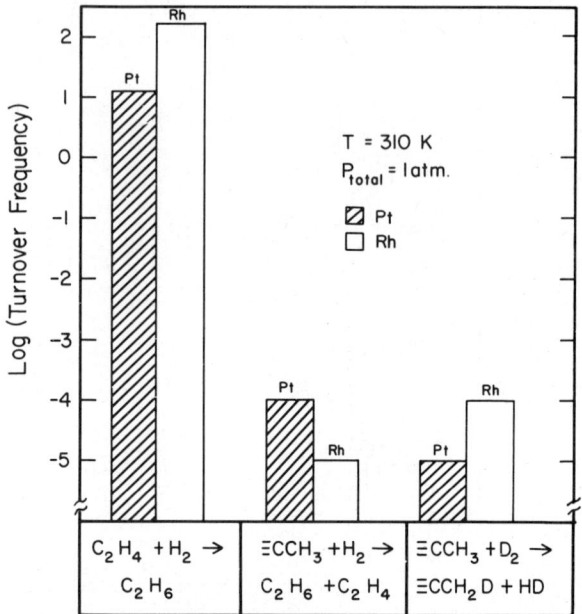

FIGURE 11. Turnover rates for ethylene hydrogenation, the rehydrogenation of ethylidyne, and the deuteration of the methyl group of ethylidyne on platinum and rhodium crystal surfaces.

These types of structure-insensitive reactions may be compared with homogeneous catalytic reactions that are facile, occurring at lower temperatures, and include hydrogenation or hydroformylation. Since the metal plays secondary roles in this process, high coordination sites are not needed to carry out the reaction. It is hoped that future studies will reveal the

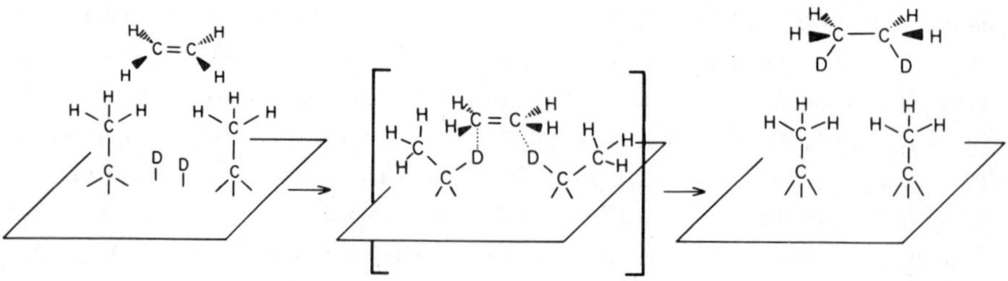

FIGURE 12. Proposed mechanism for the rapid, structure-insensitive hydrogenation of ethylene.

2. CLASSIFICATION OF CATALYTIC REACTIONS INVOLVING METALS

possible correlation between homogeneous catalytic reactions and heterogeneous reactions of this type.

It is interesting to speculate on the mechanism of hydrogen transfer when it is mediated by the strongly adsorbed organic overlayer and on the nature of bonding of the incoming reacting molecules in the presence of this overlayer. Rapid hydrogen transfer and weak bonding of the reactants are indicated by the low activation energy and the rapid turnover of these processes. Many homogeneous catalytic processes, including enzyme-catalyzed reactions, also display these properties. If the mechanisms of these reactions can be correlated with the mechanism of ethylene hydrogenation, it brings into focus the key role of the strongly adsorbed overlayer, both for hydrogen transfer and for acting as a template to the reactant molecules. Is it possible that the direct bonding of reactants to metal sites is not necessary for the reaction to occur? Our data is consistent with such a model; this should force a reevaluation of many of the reaction mechanisms that are suggested for homogeneous catalytic processes.

The organic overlayer may also serve as a template to orient or align the reactants. LEED surface crystallography and HREELS studies of the structure of these monolayers indicate that their structural integrity is preserved at temperatures as high as 400 K; thus their presence only allows us to carry out various specific reactions below this temperature. Above 400 K, fragmentation to small organic CH and C_2H groups occurs (Figure 13 shows this process). While at low temperatures (Koel et al., 1986), benzene and ethylene maintain their molecular identity on the platinum and rhodium crystal surfaces, above 400 K the fragments are the same small organic moieties. Thus enzymelike catalysis that requires a template to line up the reactant molecules can be carried out only below 400 K.

2.3. Reactions over Modified Surfaces

Let us consider how the location and bonding of an adsorbed molecule is altered when another molecule or atom coadsorbs with it. The relatively weak interactions of benzene and carbon monoxide when coadsorbed on platinum or rhodium demonstrate this (Somorjai and Mate, 1985). Benzene forms a disordered monolayer over platinum in the absence of CO. When CO is introduced, several ordered structures form that are detectable by low-energy electron diffraction, LEED. These structures change depending on the CO:benzene ratio on the Pt or Rh surfaces, which can be monitored

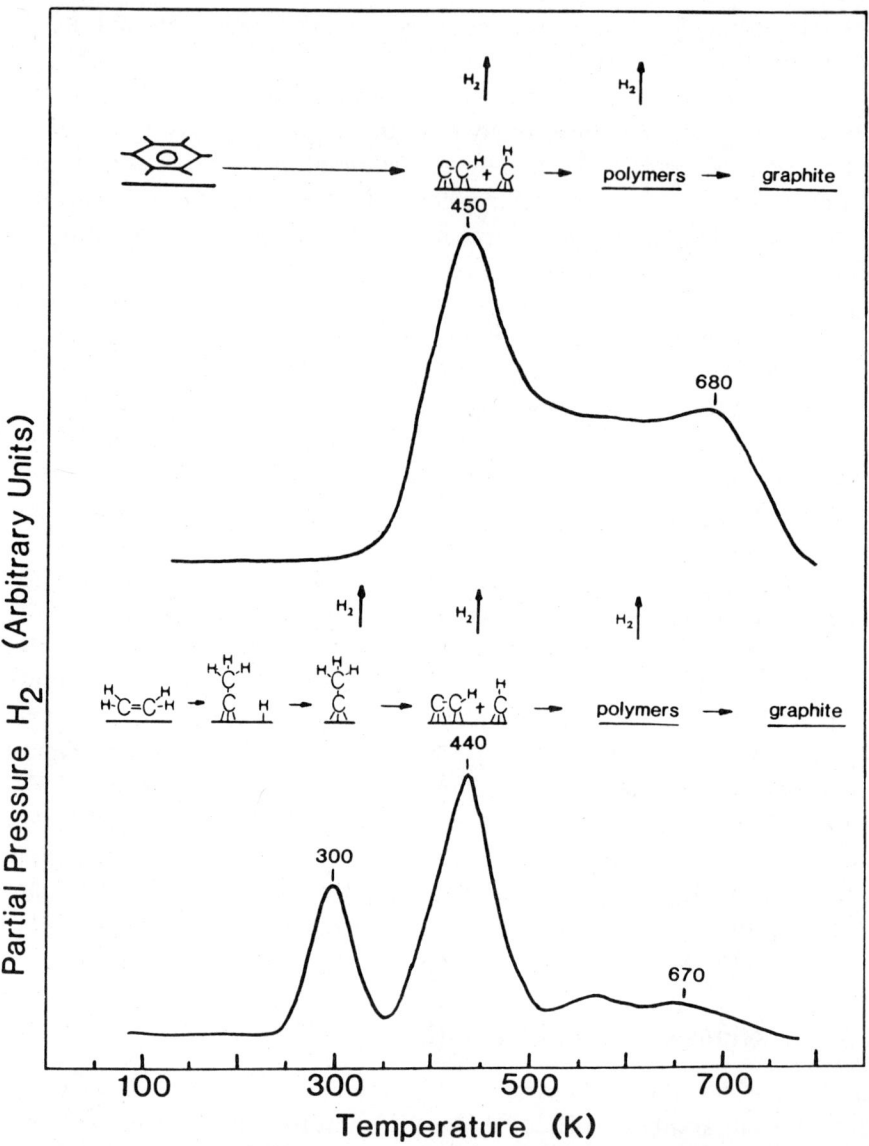

FIGURE 13. Fragmentation of ethylene and benzene adsorbed on the rhodium (111) crystal surface as a function of temperature as determined by high-resolution electron-energy-loss spectroscopy studies.

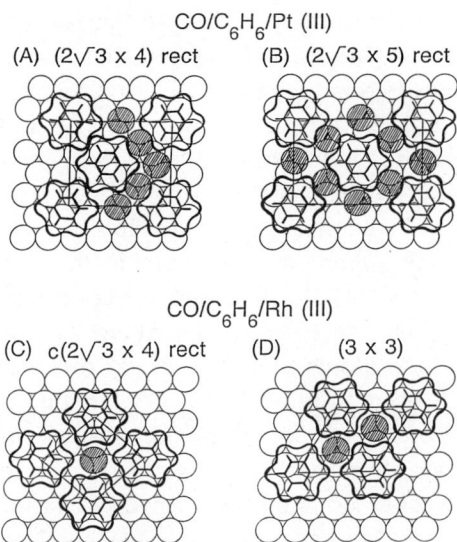

FIGURE 14. Coadsorbed surface structures of carbon monoxide and benzene on the platinum (111) and rhodium (111) single-crystal surfaces.

by HREELS and thermal desorption spectroscopy, TDS. Figure 14 shows the various ordered structures that consist of mixed CO–benzene layers with one, two, or three CO molecules in the unit cell. The ordering of benzene is facilitated by the weak attractive interaction with CO that also blocks certain alternative adsorption sites. It should be noted that CO is located in a threefold site (as determined by LEED surface crystallography and HREELS) that it would never occupy in the absence of benzene (Anderson and Pendry, 1978; Behm et al., 1979).

The coadsorbed atoms or molecules may be viewed as surface modifiers and they have indeed profound influence on the structure and distribution of the bonding sites and also on the nature of the chemical bond that the reactants form with the catalyst surface. We shall arbitrarily divide the surface modifiers into two groups: structure modifiers and bonding modifiers. It should be noted that often the modifier influences both the surface structure and the bonding of the reactants.

2.3.1. Structure Modifiers

a. SITE BLOCKING BY SULFUR. Let us consider the interaction of coadsorbed sulfur with thiophene which occurs during the hydrodesulfurization of thiophene on the molybdenum (100) crystal surface (Somorjai, Gellman, and Neiman, to be published). This gentle reaction removes the sulfur from the molecule as H_2S in the presence of hydrogen, leaving behind the C_4 species that readily hydrogenates to butadiene, butenes, and butane without fragmentation. Molybdenum metal strongly adsorbs and decomposes thiophene and butenes as shown by surface studies, and thus, the clean surface cannot be an active catalyst. MoS_2 is a layer compound and its basal plane holds thiophene so weakly that its thermal desorption occurs at 165 K (Somorjai et al., 1984). Thus this surface is not chemically active. The active molybdenum surface contains about one-half monolayer of strongly adsorbed sulfur. These atoms block the metal sites where thiophene decomposition would occur. Studies using ^{35}S labeling indicate that these sulfur atoms remain permanently on the metal surface during the catalytic reactions. The sulfur atom that is removed from the thiophene molecule occupies sites of weaker bonding where hydrogenation to H_2S and subsequent desorption occurs while the C_4 species becomes partly hydrogenated and desorbs.

Thus the blockage of certain adsorption sites on the surface of early transition metals attenuates the strong bonding and permits the catalytic reaction to occur.

The hydrogenolysis of organic molecules is frequently an undesirable reaction that leads to the production of lower molecular weight products. Kink sites on transition metal surfaces are especially active for the C–C bond-breaking reaction (Somorjai et al., 1981). While their surface concentration is no more than about 5% of the total number of metal sites, they may account for 90% of hydrogenolysis activity. These hydrogenolysis sites can often be poisoned by the chemisorption of controlled amounts of sulfur (produced by H_2S decomposition) that binds more strongly to kink sites as compared to terrace sites. In this way, the hydrogenolysis reaction can be poisoned selectively as the kink sites are blocked and rendered inactive.

b. ENSEMBLE EFFECT IN ALLOY CATALYSIS AND THE CREATION OF NEW SITES BY ALLOYS. As compared to pure platinum, bimetallic alloys such as platinum-rhenium and platinum-gold frequently exhibit superior activity, selectivity, and deactivation resistance while catalyzing reforming reactions. The influence of gold on hydrocarbon conversion catalysis by platinum was recently studied by evaporating gold into platinum single-crystal surfaces (Sachtler and Somorjai, 1983). At low temperatures, gold forms epitaxial overlayers on platinum but upon heating it dissolves to form an alloy in the near surface region. This Pt–Au alloy displays markedly different activity and selectivity for the conversion of n-hexane as shown in Figure 15a. Isomerization activities increase substantially as compared to that for clean platinum while the aromatization and hydrogenolysis rates decrease exponentially with increasing gold surface concentration. This

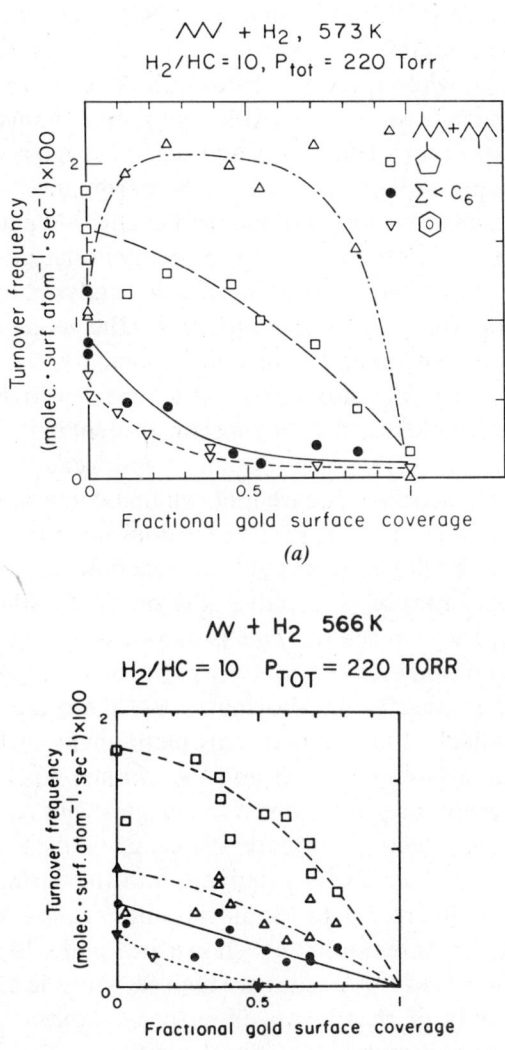

FIGURE 15. Rates of formation of various products from n-hexane conversion as a function of fractional gold coverage for gold-platinum alloys that were prepared by vaporizing gold onto (a) platinum (111) and (b) platinum (100) crystal surfaces, respectively.

remarkable change in catalytic behavior can be explained by a change in the geometric distribution of platinum sites that are present in the (111) alloy surface. Substitution of gold atoms dilutes the surface platinum atoms such that the high-coordination threefold platinum sites are eliminated much faster than the twofold bridge and single-atom top sites. This change in the size distribution of the available reaction sites is frequently called the ensemble effect (Sachtler and Somorjai, 1983). As a result of this effect, catalyzed reactions that involve adsorption and rearrangement at threefold sites are eliminated, while reactions that require one or two atom sites are attenuated to a much lesser extent. Although minor changes in electronic structure may also occur at the alloy surface sites, most of the reaction results can be simply explained by this high-coordination-site elimination model. Similar results revealing pronounced changes in catalytic behavior with alloy composition were recently reviewed by Ponec (1983) and Sinfelt (1983). For a variety of hydrocarbon reactions, catalyzed over metal films and high-area-supported catalysts, in most cases the geometrical ensemble effect is decisive in controlling the reaction selectivity.

The effect of alloying is also surface-structure sensitive, as shown by recent studies where gold was the alloying constitutent in the Pt(100) crystal face instead of the Pt(111) surface (Somorjai and Yeates, to be published 1987). The (100) surface has a square unit cell that contains fourfold bridge and top sites and unlike the (111) surface it does not have threefold sites. When this surface is alloyed with gold, all reaction rates decline in proportion to the concentration of inactive gold on the Pt(100) surface when n-hexane was used as a reactant. This is shown in Figure 15b. Thus the enhancement of the isomerization activity requires a presence of threefold sites. When gold is used as an alloying agent, there are three types of threefold sites available. One contains only platinum atoms while the other two mixed Pt–Au sites contain one and two atoms, respectively (Figure 16). Thus alloying produces new mixed metal sites with catalytic behavior that can modify the selectivity. Figure 15a clearly indicates that the high isomerization rate of n-hexane is sustained until the surface was covered up to two-thirds monolayer gold (Somorjai and Yeates, to be published 1987). Thus all three threefold sites, shown in Figure 16, are active for isomerization. The mixed Pt–Au sites are then responsible for the enhanced isomerization activity of the Pt–Au alloy that exhibits markedly higher rates than the pure platinum (111) crystal surface.

Boudart and coworkers have shown fiftyfold increase in the rate of H_2/O_2 reaction to produce water over Pd–Au alloys (Lain, Criodo, and Bou-

2. CLASSIFICATION OF CATALYTIC REACTIONS INVOLVING METALS

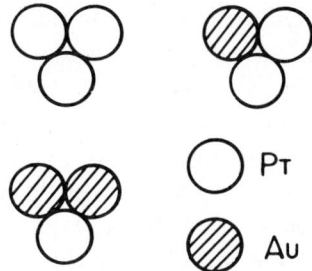

FIGURE 16. Platinum and mixed platinum–gold threefold sites that are active for hexane isomerization.

dart, 1977). Such large effects cannot be explained by site-blocking ensemble effects. The new sites that are created by alloying have unique structure and bonding. In fact, a new catalyst is created with structural and bonding properties that are not derived from the structural and bonding properties of the pure alloy constituents.

c. MODIFIERS THAT CAUSE SURFACE RESTRUCTURING. We have previously shown that the iron (111) crystal face is about 500 times more active for the synthesis of ammonia from nitrogen and hydrogen than the iron (110) crystal face. However, when alumina is present on the metal, which is then heated in a nitrogen or ammonia atmosphere, massive restructuring occurs. The iron (110) face recrystallizes into (111) orientation facets that become the predominant surface orientation after only 1 hr of heat treatment at 500°C in 1 atm of nitrogen (Strongin, Bare, and Somorjai, 1986). The restructured iron is now as active for ammonia synthesis as is the clean iron (111) crystal face. Al_2O_3 is called a structural promoter of iron catalysts for good reason. The surface restructuring must occur through the formation of an iron aluminate–iron nitride phase as both the presence of alumina and nitrogen are needed for it to take place. While the close-packed (110) iron surface has a lower surface free energy than the (111) face when clean, in the presence of nitrogen, the (111) crystal face appears to be more stable as it forms stronger metal–nitrogen bonds. This provides the thermodynamic driving force for restructuring, and the presence of alumina is needed for compound formation that accelerates the kinetics of restructuring that would otherwise be much too slow to occur under the available experimental conditions.

Nickel, an excellent hydrogenation catalyst, can be poisoned by sulfur that is present on the metal surface in concentrations amounting to a few percent of the monolayer. Surface-science studies revealed that the adsorbed sulfur aids the reconstruction of the (111) crystal face to form the (100) face. Sulfur, which is responsible for this restructuring, produces an ordered overlayer on this reconstructed surface (Goodman, 1984a). Again the stronger metal–sulfur bond on the (100) face as compared to the (111) face provides the thermodynamic driving force for this restructuring. In this circumstance, however, the reconstruction leads to poisoning of the catalytic activity and to chemical passivation of the metal surface.

Surface restructuring is important not only for structural promotion or poisoning, but also for catalyst aging and regeneration. A growth of catalyst particles often occurs in chemical environments that promote restructuring leading to a drastic loss of active surface area. Regeneration of catalysts using active redispersing agents that alter the surface structure and the surface free energies of the particles is an important area of catalysis science. It is hoped that correlations between changes of bonding and surface structure and the kinetics of restructuring will be established with beneficial effects on both catalyst life and chemical stability.

d. REACTIONS REQUIRING BOTH METAL SITES OF HIGH COORDINATION AND STRONGLY ADSORBED OVERLAYERS. DEHYDROCYCLIZATION, ISOMERIZATION, AND HYDROGENOLYSIS. There are many hydrocarbon conversion reactions that need strong metal–adsorbate bond-breaking structure-sensitive reaction steps that also require hydrogen transfer mediated by a strongly adsorbed organic overlayer. These reactions occur at high temperatures and have high activation energies. Dehydrocyclization, hydrogenolysis, and isomerization are important examples. These processes occur exceptionally well over platinum during reforming of low-octane linear alkanes such as n-hexane or n-heptane to aromatics or branched isomers of high octane number. The complexity of hydrocarbon conversion reactions require both structure-sensitive bond scissions and rearrangements and hydrogen transfer and desorption via weak adsorption sites over the strongly adsorbed overlayer. Therefore, while the structure sensitivity is clearly present, it is less pronounced than in the case of ammonia synthesis (a factor of 5 instead of 500 differences in rates from crystal face to crystal face) (Somorjai, Davis, and Zaera, 1982). As a consequence of surface-structure sensitivity and strong metal–carbon bonding with reaction intermediates, these processes have high activation energies and therefore are carried out at high temperatures to obtain suitable turnover rates. Nevertheless, the active surface must also be covered by organic fragments that contain hydrogen for the necessary hydrogen transfer and desorption reaction steps. The upper temperature limit of these reactions is determined by the complete loss of hydrogen from these fragments, leading to graphitization of the organic overlayer that renders the catalyst inactive.

Figure 17 illustrates the rates of benzene production from n-hexane near atmospheric pressure and at 573 K over the various platinum single-crystal surfaces that were shown earlier. Also shown are turnover rates for a competing cyclization reaction that produces methylcyclopentane. Platinum surfaces with hexagonal (111) terrace structure display superior activity and selectivity for the more important aromatization reaction. In this case, the hexagonal surface is 3–5 times more active than the square (100) surface. Similar results were also obtained for n-heptane aromatization (Gillespie et al., 1981). Steps and kinks promote the aromatization reaction

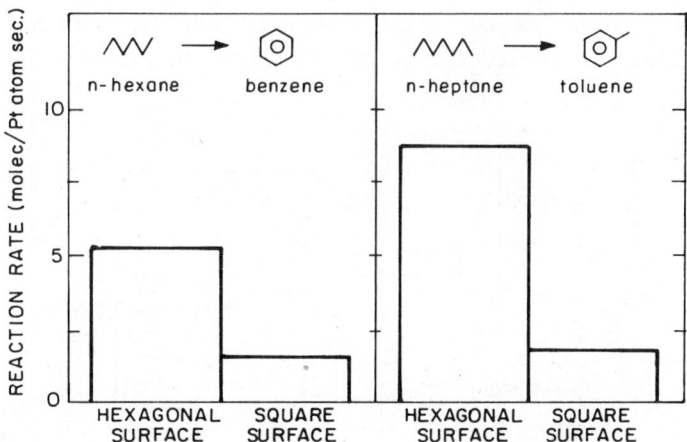

FIGURE 17. Dehydrocyclization of alkanes to aromatic hydrocarbons is one of the most important petroleum reforming reactions. The bar graphs shown here compare the reaction rates for n-hexane and n-heptane aromatization catalyzed at 573 K and atmospheric pressures over the two flat platinum single-crystal faces with different atomic structure. The platinum surface with the hexagonal atomic arrangement is several times more active than the surface with a square unit cell over a wide range of reaction conditions.

(Gillespie et al., 1981), although their influence appears to be less important than the terrace structure. By contrast, the rate of methylcyclopentane formation appears to show little dependence on platinum surface structure.

The production of both methylcylopentane and benzene over platinum catalysts illustrates that cyclization may occur by two mechanisms involving either 1,5- or 1,6-ring closure. Our studies show that 1,5-ring closure predominates under mild conditions, whereas the structure-sensitive aromatization reaction via 1,6-ring closure becomes strongly favored at high temperatures. The temperature range where this change in selectivity occurs increases with increasing hydrogen pressure (Davis, Zaera, and Somorjai, 1984). Figure 18 summarizes the selectivities for producing benzene over methylcyclopentane that were measured as a function of temperature. It can be seen that the hexagonal platinum surfaces exhibit superior aromatization selectivity over a wide range of conditions.

Conversely, the flat (100) platinum surfaces with square atomic arrangement are often much better isomerization catalysts as compared to the hexagonal platinum surfaces (Davis, Zaera, and Somorjai, 1982). This is shown in Figure 19, where reaction rates are compared for isobutane isomerization and hydrogenolysis catalyzed over four platinum surfaces with different atomic structures. Depending on the catalyst preparation (structure), it is clear that one may obtain superior aromatization or isomerization activity. This is also well documented in the literature for practical catalysts

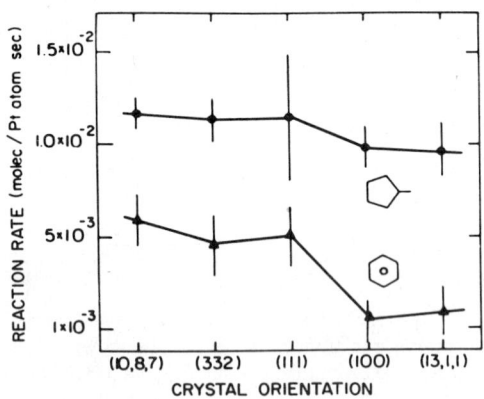

FIGURE 18. Reaction rates for n-hexane dehydrocyclization to benzene and methylcyclopentane catalyzed at 573 K (H_2/HC = 10, P_{tot} = 220 torr) over platinum single-crystal surfaces with different atomic structure.

2. CLASSIFICATION OF CATALYTIC REACTIONS INVOLVING METALS

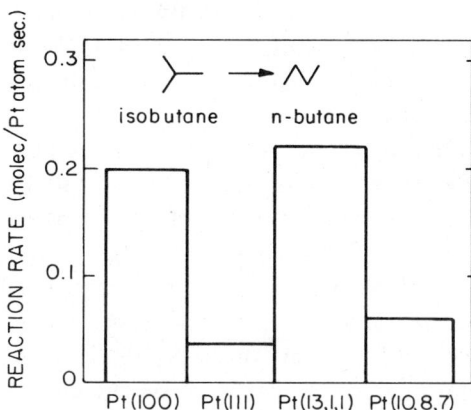

FIGURE 19. Catalytic acitivity for isobutane isomerization at 573 K as a function of surface structure for four platinum surfaces. The (100) and (13,1,1) surfaces with square atomic arrangement exhibit maximum isomerization rates.

(Lankhorst, Dejongste, and Ponec, 1982; Amir-Ebrahami et al., 1979; Dartigues et al., 1979). The hydrogenolysis reaction illustrated in Figure 19 proceeds with maximum rates on surfaces that contain large concentrations of kink sites during reforming. It is frequently desirable to poison the hydrogenolysis sites by chemisorbing controlled amounts of H_2S or other strongly bound additives that bind more strongly at step and kink sites as compared to the terrace sites. In this way, the hyperactive kink sites can be selectively poisoned and rendered inactive for the undesirable hydrogenolysis reaction. Because isomerization accounts for over 90% of the isobutane conversion over clean platinum (Davis, Zaera, and Somorjai, 1982a), it is clear from Figure 19 that the (100) surface is several times more active than Pt(111) for this small-alkane reaction.

Similar results revealing marked structure sensitivity for simpler-hydrocarbon reactions catalyzed over clean metal surfaces were also obtained by Goodman (1982, 1984b) in studies of ethane and butane hydrogenolysis catalyzed over Ni(100) and Ni(111). Near atmospheric pressure and at 450–650 K, the (100) surface with square atomic arrangement was several times more active than the hexagonal Ni(111) surface. The turnover rates, energetics, and product distributions for alkane hydrogenolysis catalyzed over both platinum and nickel model catalysts were very similar to those frequently reported for high-area-supported catalysts (Davis and Somorjai, 1985). Thus, it appears clear that the square (100) atomic arrangement leads to the enhancement of chemical activity for small-hydrocarbon conversion reactions over both platinum and nickel. With platinum, skeletal rearrangement is favored over hydrogenolysis; with nickel, hydrogenolysis is the exclusive reaction pathway.

In addition to altered reaction rates, the distribution of hydrogenolysis products also varies sharply over the (111) and (100) platinum surfaces. The hexagonal surface displays high selectivity for scission of the terminal C–C bond, whereas the (100) surface with the square unit cell prefers cleavage of C–C bonds located near the center of the reactant molecule as indicated by studies (Davis, Zaera, and Somorjai, 1984; Davis, Zaera, and Somorjai, 1982b) of a variety of C_4–C_7 alkane reactions. Consecutive rearrangements during a single residence on the catalyst also occur more readily on the Pt(100) surface (Davis, Zaera, and Somorjai, 1982a).

The Carbonaceous Deposit. Perhaps the most general feature of hydrocarbon catalysis over platinum is the rapid and unavoidable buildup of one or more monolayer equivalents of strongly bound carbonaceous deposit. The formation of this deposit, its rehydrogenation dynamics, and the nature of its participation in the catalytic cycle for hydrocarbon skeletal rearrangement was revealed in detailed studies (Davis, Zaera, and Somorjai, 1982a) correlating the rates and selectivities with the deposit surface coverage, composition, and structure. Carbon-14-labeled deposits were used to ascertain the residence time of the strongly bound carbon, while thermal desorption of hydrogen and CO were applied to monitor the deposit composition and the concentration of platinum sites that remain uncovered during the catalytic reactions, respectively. As a result of these studies a molecular model of the working platinum reforming catalyst (Davis, Zaera, and Somorjai, 1982a) was developed (see Figure 20). During skeletal rearrangement, most of the surface remains continuously covered by a disordered polymeric residue. The amount of deposit increases with increasing time and temperature, and the morphology gradually changes from two-dimensional at low temperatures to three-dimensional at temperatures above about 625 K. These changes are accompanied by a decrease in the hydrogen content of the deposit with increasing temperature. This is shown in Figure 21, where hydrogen thermal desorption results illustrate the sequential dehydrogenation of deposits prepared at several temperatures (Davis, Zaera, and Somorjai, 1982a). Also shown is the temperature dependence of the amount of deposit and its average (H/C) composition. As long as the reaction temperature is below about 800 K, the deposit

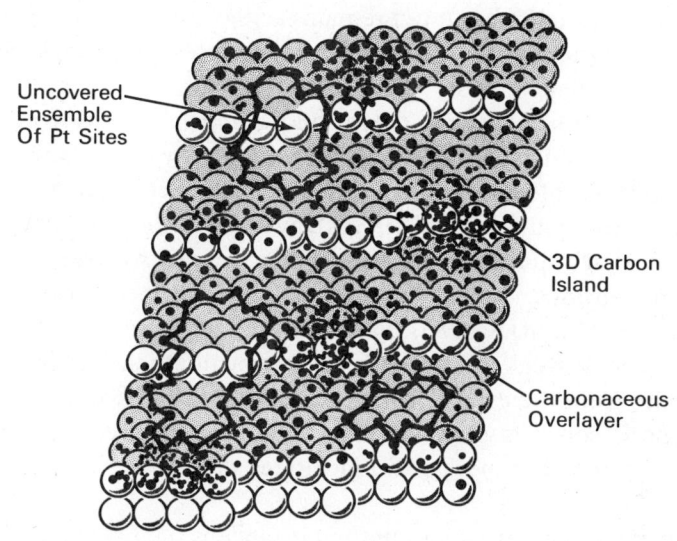

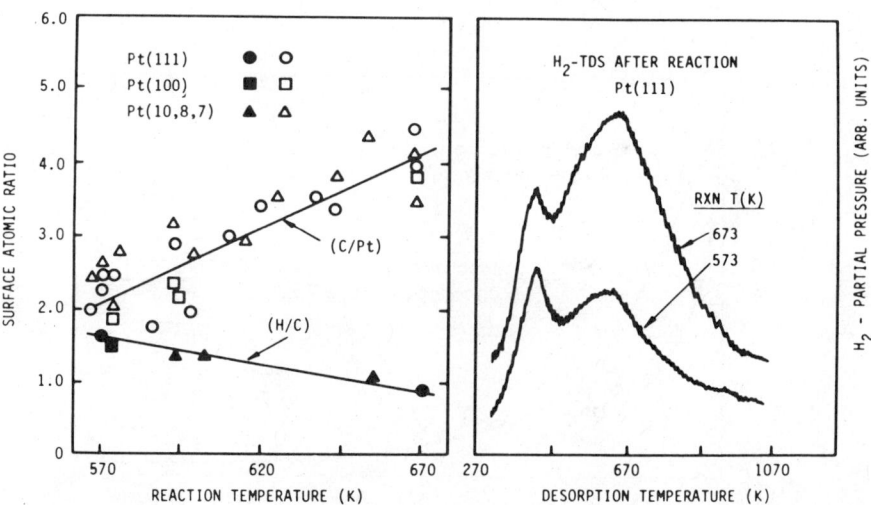

FIGURE 21. Important properties of the carbonaceous deposits produced during n-hexane reforming over platinum single-crystal surfaces shown as a function of reaction temperature. The left frame shows the amount of deposit expressed as carbon atoms per surface platinum atom as measured by AES and the H/C stoichoimetry of the deposit as measured by H_2 TDS. The right frame illustrates the sequential dehydrogenation of deposits produced during n-hexane reaction studies using platinum (111) at the indicated temperatures.

contains hydrogen and the catalyst remains active. At higher temperatures, the deposit completely dehydrogenates, condenses into graphitic structures, and all catalytic activity vanishes.

Radiotracer studies (Davis, Zaera, and Somorjai, 1984) showed that under typical reforming conditions, hydrogen transfer between this deposit and reacting molecules occurs easily. Deuterium exchange also occurred readily (Salmeron and Somorjai, 1982; Davis and Somorjai, 1983; Koel et al., 1984) although the residence time of the deposit was very long compared to the turnover period for skeletal rearrangement (Davis, et al., 1985; Zaera and Somorjai, 1983). Furthermore, the presence of this deposit appears to provide weak adsorption sites that facilitate the release of product molecules from the catalyst surface. Thus, the presence of the carbonaceous deposit converts platinum into a catalyst that readily exchanges hydrogen with reacting species and facilitates the release of product molecules.

Another ingredient of paramount importance is the presence of a small (5–30%) concentration of uncovered platinum surface sites that always persist in the presence of the active carbonaceous deposit. These sites mostly exist as patches or ensembles that contain several contiguous surface

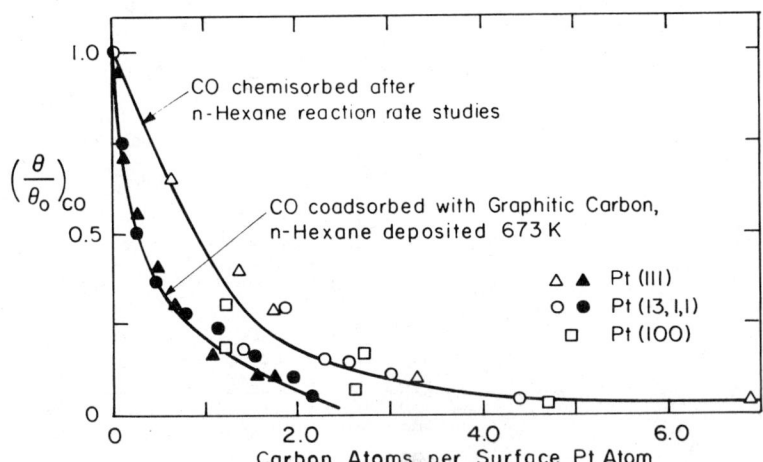

FIGURE 22. Fractional concentration of uncovered platinum surface sites determined by CO adsorption and desorption as a function of surface carbon coverage on the (100), (111), and (13,1,1) platinum crystal surfaces. A comparison is made between the CO uptake determined following n-hexane reaction studies and CO uptake determined when CO was coadsorbed with graphite surface carbon.

atoms (Davis, Zaera, and Somorjai, 1982b). In the steady state, these sites are responsible for the initial dissociative chemisorption and subsequent rearrangement of incident reactant molecules. The presence of these sites accounts for the structure sensitivities frequently observed for important aromatization and isomerization reactions. Figure 22 shows the dependence of the vacant site concentration on reaction temperature as measured by CO adsorption–thermal desorption following n-hexane reaction studies over the (100), (111), and (13,1,1) platinum surfaces (Davis, Zaera, and Somorjai, 1982a). The concentration of vacant sites approaches zero only at high temperatures (~800 K) where complete dehydrogenation also takes place. Also shown is the ratio of the steady-state reaction rate to the initial rate over clean platinum as a function of vacant site concentration. The deviation from first-order behavior appears to result from the participation of the carbonaceous deposit as a hydrogen transfer agent.

2.3.2. Bonding Modifiers

a. THE ROLE OF POTASSIUM DURING AMMONIA SYNTHESIS AND CO HYDROGENATION. Potassium is frequently used as a promoter in many catalytic reactions. The hydrogenation of CO and the synthesis of NH_3 are perhaps the best known examples of potassium promotion. Potassium has a very high heat of adsorption at low coverages, about 60 kcal/mole on most transition metal surfaces, indicating complete ionization of the atom. At high coverages, however, mutual dipolarization of the charged potassium species leads to neutralization. When about 50% coverage is reached, the heat of adsorption equals the heat of sublimation of metallic potassium, about 23 kcal/mole, indicating that the adsorbed atoms are no longer ionized (Somorjai and Garfunkel, 1982).

During CO hydrogenation over transition metal surfaces, the presence of potassium usually increases the rates and also selectivities for C_2^+ hydrocarbon production as expected if the CO dissociation rate is increased. As noted earlier, potassium adsorbed on transition metals exists in largely ionic states; this results from the transfer of valence charge density into the metal d-band, which reduces the metal work function. This charge transfer has a profound influence on the adsorption behavior of CO as revealed by thermal desorption and vibrational spectroscopy studies using platinum, nickel, and ruthenium single-crystal surfaces. In all cases, the CO desorption temperature is increased by 100–200 K in the presence of potassium (Somorjai, Garfunkel, and Crowell, 1982) reflecting a 5–12 kcal/

mole increase in the heat of molecular CO chemisorption. In addition, the CO bond is weakened substantially as compared to adsorption on clean metal surfaces. Figure 23 illustrates the HREELS spectra for CO coadsorbed with potassium at several coverages on the hexagonal (111) platinum surface (Somorjai, Crowell, and Garfunkel, 1982). With increasing coverage, there is a continued shift of the CO stretching frequencies from 1875 and 2120 cm^{-1} to 1565 cm^{-1}. These shifts correlate with a change in bonding from mostly top sites to bridge sites and a decrease in CO bond order from 2.0 to 1.5. This dramatic bond weakening reflects enhanced population of the CO $2\pi^*$ antibonding orbital as a result of the increased density of metal electronic states in the presence of potassium. One should anticipate that the weakened CO bond and strengthened metal–carbon bond would facilitate CO dissociation. This was demonstrated by Goodman

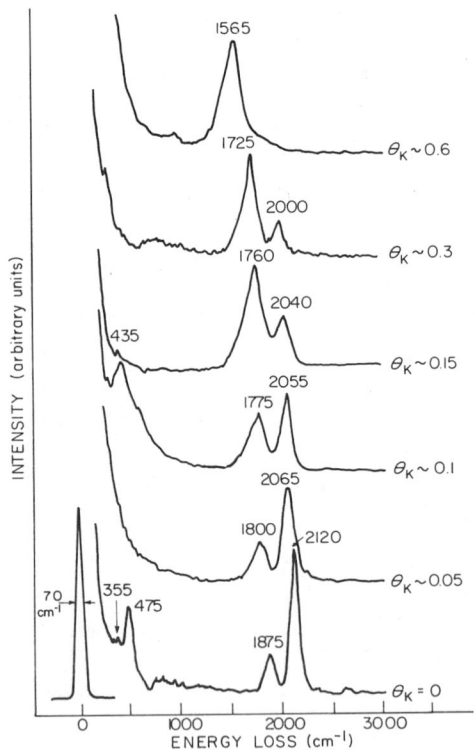

FIGURE 23. Vibrational spectra of CO at the saturation coverage when chemisorbed on Pt(111) at 300 K as a function of preadsorbed potassium coverage.

2. CLASSIFICATION OF CATALYTIC REACTIONS INVOLVING METALS 53

(1984a) using nickel (100) where the CO dissociation rate was increased fourfold at potassium coverage of 10% of a monolayer. The activation energy for CO dissociation was also lowered from about 23 to 10 kcal/mole.

It is now clearly established that potassium chemisorbed on transition metals functions as an unusually powerful donor. This increases the density of surface electron states available for back-bonding with certain adsorbates, if they possess orbitals with energy and symmetry that correlate near the Fermi energy of the metal. Examples of such orbitals would be the $2\pi^*$ of CO and N_2. The important general consequences of this interaction include increased heat of molecular adsorption and increased dissociation probability. Commercial fused iron catalysts normally contain about 1.8% potassium per mole, which greatly accelerates the rate of ammonia synthesis over this high surface area material. The role of potassium in promoting the N_2 dissociation was revealed by Ertl (1980,1981) in studies of N_2 chemisorption on the low activity (100) and (110) iron surfaces. With low concentrations of potassium promoter, the rate of N_2 dissociation was increased by up to two orders of magnitude. The same studies revealed that the heat of molecular N_2 chemisorption was increased by about 4 kcal/mole while the activation energy for N_2 dissociation was reduced to almost zero.

On the clean rhodium (111) surface, CO stays molecularly adsorbed at low pressures while it dissociates in the presence of potassium (Somorjai, Crowell, and Tysoe, 1985). This can be studied by the adsorption of a mixture of $^{12}C^{18}O$ and $^{13}C^{16}O$ and detecting $^{13}C^{18}O$ and $^{12}C^{16}O$, the products of scrambling, which clearly identify the dissociation of molecular CO on the metal surface. In Figure 24, we show that three CO molecules may dissociate per potassium atom at a potassium coverage where maximum charge transfer to the transition metal occurs.

b. THE ROLE OF OXIDE SUPPORTS TO STABILIZE DIFFERENT OXIDATION STATES OF TRANSITION METAL CATALYSTS. Studies of carbon monoxide hydrogenation over initially clean rhodium foil and rhodium (111) single crystals at 6 atm and at 500–600 K showed high selectivities for methane production along with small yields of ethylene and propylene (Somorjai and Sexton, 1977; Castner and Blackadar, 1980). By contrast, when rhodium oxide (Rh_2O_3) was investigated for synthesis activity, high yields of C_2- and C_3- oxygenated hydrocarbons were produced. Ethylene addition to the CO/H_2 mixture produced propionaldehyde with high selectivity, indicating the exceptional carbonylation activity of oxidized rhodium species (Somorjai and Watson, 1981, 1982). Under similar conditions, carbonylation activity was totally absent with rhodium metal. Thus it appears that higher oxidation state rhodium ions, Rh^+ and Rh^{3+}, are essential for chemical activity in reactions that produce oxygenated products. Photoemission studies confirm that a combination of rhodium metal and rhodium cations is probably needed for maximum activity.

Unfortunately, rhodium oxide was gradually reduced under typical synthesis conditions, resulting in unfavorable changes in reactions selectivity. Higher oxidation state rhodium ions were stabilized under synthesis conditions by reacting Rh_2O_3 with La_2O_3 (Somorjai and Watson, 1981, 1982) in the presence of oxygen to produce the perovskite compound lanthanum rhodate $LaRhO_3$. With this material, catalytic activity for producing oxygenated products was stable over long periods. Figure 25 shows the dependence of the synthesis selectivity on reaction temperature at 6 atm total pressure. Remarkably the selectivity for oxygenate synthesis was greater than 80 mole % over a wide range of conditions. At low temperatures, methanol was formed with maximum selectivity and at low activation en-

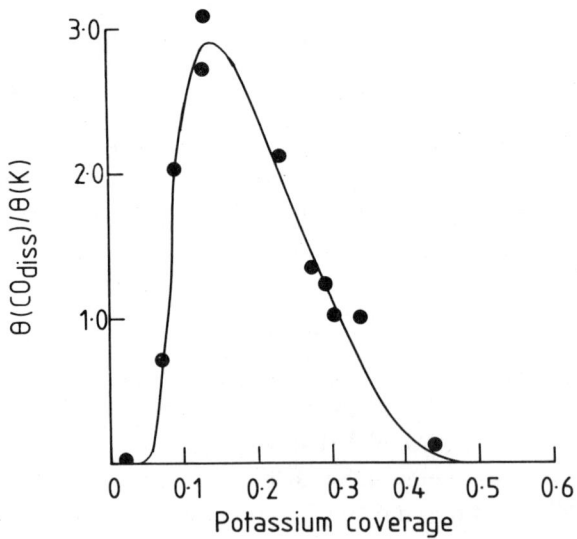

FIGURE 24. Amount of dissociated carbon monoxide per potassium atom as a function of potassium coverage on the rhodium (111) single-crystal surface.

2. CLASSIFICATION OF CATALYTIC REACTIONS INVOLVING METALS

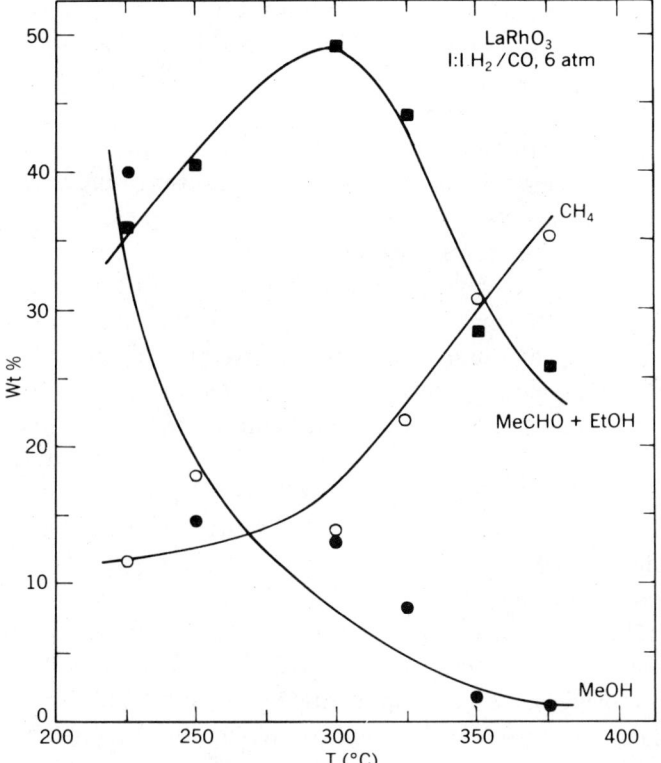

FIGURE 25. Temperature dependence and selectivities for producing methane, methanol, and C_2-oxygenated products over $LaRhO_3$ at 6 atm and 1:1 H_2/CO ratios.

ergy of 16 kcal/mole. At intermediate temperatures of 520–600 K, carbonylation activity was accentuated, resulting in high selectivities for the synthesis of C_2 and C_3 aldehydes and alcohols. Under these conditions, it appears that activities for CO dissociation, CO hydrogenation, and hydrogenation of carbon fragments as well as carbonylations all become comparable.

The unique selectivity of oxidized rhodium species for producing oxygenated products results largely from the altered strength of CO and H_2 chemisorption as the metal oxidation state is varied. Thermal desorption studies showed that the CO adsorption strength is decreased whereas the heat of hydrogen chemisorption is increased by 3–8 kcal/mole upon oxidation. These changes can be correlated with the elimination of high coordination bonding sites and the formation of surface hydroxyl groups, respectively. The combination of metallic and ionic sites provides the catalyst with high hydrogen storage capacity and bifunctional activity for carbonylation and hydrogenation.

The importance of higher oxidation state metal ions during hydrocarbon synthesis was further delineated in model studies using iron catalysts (Dwyer and Hardenbergh, 1984). In addition to linear alkanes, olefins, and small alcohols, the water–gas shift reaction also accounts for a significant fraction of the CO conversion. The evolution of the active surface composition was investigated in extensive studies by Dwyer and coworkers using single-crystals, foils, and medium area powders (Dwyer and Hardenbergh, 1984). At low CO conversions, photoelectron spectroscopy studies clearly showed that carbides were formed in the near surface region with compositions similar to bulk iron carbides. As the carbides formed, catalytic activity increased markedly, small alcohols were formed selectively, and the hydrocarbon products became heavier and more olefinic. These results are in good agreement with earlier Mössbauer emission spectroscopy studies by Raup and Delgass (1979), which showed that catalytic activity correlates well with the extent of carbide formation. With small iron particles, Fe_2C carbide was formed whereas Fe_5C_2 carbide was favored with larger particles. When dispersed on carbon or MgO supports the larger particles give rise to maximum turnover frequencies. While the catalytic activity of carbided iron powders was stable for many hours, low area foils and single crystals deactivated quickly due to the formation of an unreactive surface carbon phase with properties similar to graphite.

c. DUAL AND MULTIPLE FUNCTIONAL CATALYTIC SYSTEMS. For complex molecular rearrangements or for simultaneous reactions of a mixture of reactants, a system of catalysts must be used that is capable of carrying out the variety of desired reactions within the time of contact of the reactants with the catalyst. Usually a system of dispersed metal particles on a high surface area oxide already exhibits at least dual functionality. For example, in the platinum–alumina system, platinum alone or in combination with other metals provides for hydrogenation, dehydrogenation, and for the isomerization and aromatization of longer-chain alkanes. The halogenated high area oxide provides acidity for olefin and methylcyclopentane isomerization. Well-documented synergism (Heinemann, 1981) between the metal and acid functions yields reforming catalysts that are often far more active than either one of the individual components.

Such a synergism by consecutive or parallel reactions that occur on transition metal and oxide surfaces is the result of the long surface residence times, t, that can be estimated from the turnover frequencies (proportional to $1/t$) and the heats of adsorption of reactants and products [$t = t_0 \exp(\Delta H/RT)$]. It is in the range of 10^{-3}–10 sec for most heterogeneous catalytic processes in the temperature range of interest. Assuming $t_0 = 10^{-12}$ sec, the adsorbed reaction intermediates have sufficient time to diffuse all over the metal catalyst particles, spill over, and diffuse further on the oxide before desorption; this provides ample interaction with different active catalytic sites for sequential reactions to occur. The simplest example is perhaps the spillover of hydrogen or oxygen. These diatomic molecules can readily atomize on metal surfaces while dissociation would not occur on the oxide surfaces. However, when the metal is dispersed on an oxide, hydrogen or oxygen atoms formed at metal sites diffuse over to the oxide sites, where they can participate in a variety of oxide-catalyzed reactions. Thus multiple reaction steps that occur on both metal and oxide catalysts can take place upon a single adsorption event. It is, in fact, questionable whether we could talk about a single active site for a reaction where several active sites may be necessary such as in the case of complex organic rearrangements to produce a desired product molecule.

This view of long residence time, sequential multistep catalytic reaction events brought about in recent years the development of multifunctional catalysts that consist of an oxide and one or more metals. The oxide is usually a molecular sieve (aluminosilicate) or other crystalline microporous solid. These materials have well-defined one- or two-dimensional channel structures with channels of molecular dimensions. Thus, only those molecules that are small enough to fit into the channels can undergo catalytic reactions. The ordered channel structures bring about shape-selective catalysis (Weisz, 1980), which usually includes selective alkylation and isomerization. The metal particles that are placed inside the molecular sieve or physically mixed with it atomize hydrogen and perform catalytic chemistry that the oxide cannot carry out. For example, methanol can be converted to aromatic molecules or to straight-chain alkanes in the appropriate boiling range to produce high-octane gasoline or jet fuel on a well-chosen molecular sieve. In the presence of metals, the same product mixture may be obtained when carbon monoxide and hydrogen are used as reactants instead of methanol.

Recently, a new aromatization catalyst has been reported containing platinum dispersed on alkali-exchanged L-zeolite (Somorjai et al., 1985).

The oxide support has no acid character and its structure is composed of one-dimensional channels where the channel diameter varies between 7 and 13 Å. Its activity and selectivity is far superior to other dehydrocyclization catalysts. One reason for this appears to be the presence of special sites in the L-zeolite for the nucleation of platinum particles with (111) surface structure that provide the best configuration for this structure-sensitive reaction (Bernard, 1980; Barthomeuf, 1984).

d. THE STRONG METAL-SUPPORT INTERACTION. EVIDENCE FOR SURFACE COMPOUND FORMATION. The properties of titania-supported metal catalysts have been the subject of many studies since it was first reported that such catalysts can exhibit specific activities for CO hydrogenation which are substantially larger than those for silica- or alumina-supported metals (Vannice and Garten, 1979; Meriaudeau et al., 1982; Ko and Garten, 1981; Vannice, Twu, and Moon, 1983). More recently, similar increases in activity have been observed for NO reduction by CO and H_2 over TiO_2-supported Rh, Pd, and Pt (Solymosi, Volgyesi, and Sarkany, 1978; Solymosi, Volgyesi, and Rasko, 1980; Nakamura et al., 1981a,b; Pande and Bell, in press). Studies of H_2 and CO adsorption have demonstrated that the chemisorption of these gases declines with increasing catalyst reduction temperature (Tauster and Fung, 1978a,b). TEM observations (Baker, Prestridge, and Garten, 1978,1979) have established that this loss of adsorption capacity is not due to sintering and that the only change occurring upon high-temperature reduction is a flattening of the particle shape due to an improved wetting of the support surface. Evidence for an interaction between the metal and the support has also been inferred from XPS observations. It has been proposed (Horsley, 1979; Bahl, Tsai, and Chung, 1980; Fung, 1982; Sexton, Hughes, and Foger, 1982) that a shift in the core-level binding energy of a metal support on TiO_2 relative to that for bulk metal is partly due to charge transfer from the support to the metal. The fact that the effects of using TiO_2 as the support can be evidenced for large (~50 Å) metal particles as well as very small particles (~20 Å) has led to the suggestion that the observed modifications in adsorption and catalytic properties may be a result of the decoration of the metal particle surface by TiO_x moieties (Santos, Phillips, and Dumesic, 1983; Resasco and Haller, 1984; Sadeghi and Henrich, 1984; Belton, Sun, and White, 1984).

The idea that the metal-support effects may be a result of the decoration of the metal surface by TiO_x moieties has generated an interest in studying

2. CLASSIFICATION OF CATALYTIC REACTIONS INVOLVING METALS

the effect of TiO_x deposited on metal foils and single crystals. Chung and coworkers (Chung, Xiong, and Kao, 1984) have observed that the deposition of TiO_x on Ni (111) results in a maximum methanation rate at a Ti coverage of 1.5×10^{14} cm^{-2}. The effects of TiO_x surface species on the chemisorption of CO and H_2 on polycrystalline Ni have been examined by Raupp and Dumesic (1984). These researchers noted that TiO_x moieties caused a significant reduction in the activation energy for CO desorption relative to clean Ni (Rives-Arnau and Munuera, 1980).

In studies by Levin et al. (1986), overlayers of TiO_x on Rh foil have been grown by Ti evaporation in vacuo followed by oxidation in 100 torr of O_2 at 420 K. Auger electron spectra taken as a function of Ti dosing show a two-dimensional growth of TiO_x until completion of one monolayer; thereafter, growth is three-dimensional. The decrease in CO chemisorption capacity of the foil is substantially greater than that expected for physical blockage of adsorption sites by TiO_x. This behavior is enhanced by H_2 reduction at 770 K. Temperature-programmed desorption of CO shows a weakening in the strength of CO adsorption on Rh for TiO_x

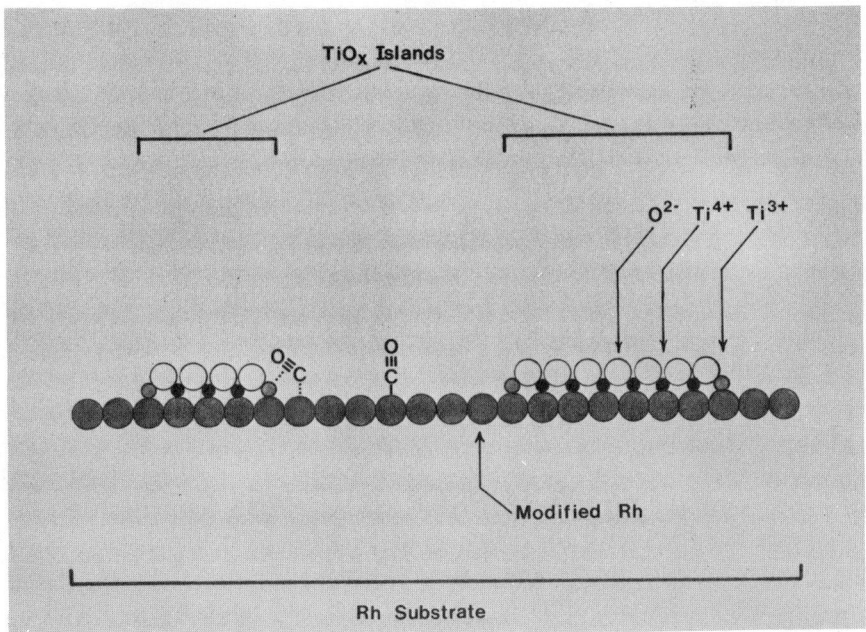

FIGURE 26. Model of the titanium oxide–rhodium (oxide–metal) interface system CO chemisorption showing inhibition at rhodium sites adjacent to TiO_x species, in addition to rhodium sites covered by TiO_x.

coverages exceeding 0.5 monolayers. The data can be interpreted in terms of a model in which CO chemisorption is inhibited at Rh sites adjacent to TiO_x species in addition to the Rh sites covered by TiO_x. This model is shown in Figure 26.

Oxide–metal interfaces show unique chemical composition and charge distribution that can have marked effects on the catalytic and chemisorption properties. Indeed, we may view these interfaces as surface compounds because there do not appear to be bulk, three-dimensional compounds with similar properties. Bimetallic clusters (Sinfelt, 1983) that form at high dispersions are another example of surface compounds since the miscibility of metals is only exhibited at high dispersions when most of the metal atoms are on the surface. Thus, the surface environment promotes unique interactions at the oxide–metal interface and for bimetallic systems that can greatly modify the catalytic and chemisorption properties.

3. IDENTIFICATION OF TYPE OF CATALYZED REACTION

Building of new selective catalyst systems that perform superior activity requires detailed knowledge, on the molecular level, of the existing catalyst system. Several of the surface analytical techniques that are listed in Table 1 provide means to determine the chemical composition of the catalyst surface. This should be determined when the catalyst is clean as well as when the catalyst is covered with the reactant mixture. The chemical analytical techniques are of two types: those that can be used with microporous high internal surface area catalysts and those that can be used only with the external surfaces of model catalysts. *X-ray photoelectron spectroscopy* can be used for surface chemical analysis on both types of catalyst. In addition, this technique can determine the oxidation states of surface atoms. *Solid state nuclear magnetic resonance* is extremely useful for determining the chemical composition of microporous solids but it is not sensitive at present for surface areas of less than 1 m^2. *Ion scattering techniques* are most sensitive for the chemical analysis of external surfaces but can also be utilized for the analysis of internal surfaces.

Surface area determination is one of the most important and well-developed techniques. Total areas may be determined by physical adsorption of nitrogen or argon at low temperatures. Metal surface areas can be established by chemisorption of hydrogen, carbon monoxide, or oxygen

3. IDENTIFICATION OF TYPE OF CATALYZED REACTION

molecules that would not readily chemisorb on the oxide supports under conditions of the adsorption experiments. Without knowledge of the surface area, one cannot determine specific reaction rates and other kinetic parameters. Variations of active surface areas are often responsible for catalyst degradation or regeneration. Thus, it is important to monitor catalyst surface areas under all conditions of the experiments.

The *structure sensitivity* of the reaction is determined by studies of the variations of specific reaction rates (molecules per area per second) with particle size or dispersion (concentration of surface atoms divided by concentration of total number of atoms) for high surface area catalyst systems. The reaction is considered structure sensitive if the specific turnover rates vary with particle size by more than the uncertainty of the other experimental variables (surface area, temperature, etc). Conversely, the reaction is structure insensitive if its turnover rate shows no variation with particle size. Single-crystal model catalyst studies provide excellent evidence of structure sensitivity if they exhibit variation of turnover rates and selectivities from crystal plane to crystal plane.

The interpretation of these studies is made difficult by structure-sensitive side reactions that can give rise to apparent structure sensitivity. For example, the hydrogenolysis of organic molecules, a structure-sensitive reaction, can often produce a carbonaceous deposit that renders part of the catalyst surface inactive. Thus, structure sensitivity of turnover rates may be caused by the varying amounts of inactive deposits with alteration of particle size. The active surface area, therefore, must be determined by selective chemisorption or some other appropriate techniques for precise determination of turnover rates.

It is important to determine the *atomic surface structure* of small particles that exhibit structure-sensitive behavior in catalytic reactions in order to identify the surface sites that are responsible for observed structure sensitivity. Unfortunately, this is an arduous and difficult task. Electron microscopy and extended x-ray absorption fine structure (EXAFS) appear, at present, to be the most promising methods for obtaining this information.

Temperature-programmed thermal desorption measurements provide significant information about the bonding of molecules adsorbed on the catalyst under reaction conditions. Upon heating with a heating rate of 1–10°/sec the adsorbed molecules desorb intact or undergo sequential bond breaking with increasing temperature to release hydrogen or other gaseous molecules. Desorption peaks appear at well-defined temperatures as detected by gas chromatography mass spectroscopy where the adsorbed species have maximum desorption rates. Since the adsorbate bond depends on the surface structure and the chemical environment on the surface in the proximity of the adsorption site, the thermal desorption spectrum yields a fingerprint of the catalyst surface, namely its structure and bonding characteristics. While the interpretation of the thermal desorption spectrum on the molecular level needs complementary studies using other techniques, the simplicity of thermal desorption experiments can be readily used to compare catalysts that are prepared differently or show different chemical behavior for the same reaction.

Pressure dependence measurements that determine how the rates and product distributions are changed with the partial pressures of the reactants permit the establishment of macroscopic rate equations in a wide range of applicable, experimental circumstances. This type of study is clearly necessary to predict how the catalyst is to behave with changing experimental conditions. The wider the pressure range of the investigation, the better, because the reaction mechanism often changes with pressure. As the pressure is increased, weakly adsorbed molecules are stabilized on the catalyst surface and may participate in chemical reactions of different types that nevertheless have high turnovers. *Temperature dependence* studies of the total rates and the rates of each product lead to the determination of the apparent activation energies of the reactions. While the interpretation of its value, in terms of the rate-determining elementary reaction steps, is difficult, it is often possible to give sound physical interpretation by comparing these values with the results of theoretical calculations.

Identification of *secondary reactions* requires running the catalytic process both at low and high conversions. As long as the product concentration is low, secondary reactions are less likely to be important. This may be accomplished by variation of the flow rates that control the contact time of the reactants with the catalyst or by injection of the product along with the reactant into the catalytic reactor.

Determination of the *surface residence* times of reaction intermediates often can be accomplished by isotopic labeling of the reacting molecules. ^{14}C and ^{35}S are excellent long-life beta-emitter isotopes whose concentrations can be monitored readily in the catalyst bed by the use of sensitive semiconductor detectors that have 1% of a monolayer sensitivity. *Tritium* labeling can also be used although hydrogen-deuterium exchange studies are more frequently employed to monitor hydrogen exchange processes.

Pulse studies that perturb the surface steady state have been useful in obtaining valuable kinetic information about the catalytic process including surface residence times or retention times of molecules in the catalyst bed. A pulse of labeled reactant is introduced into the catalytic reactor and its reappearance at the outlet is monitored as a function of time along with the product distribution. If the labeled species stay on the surface much longer than the turnover time, their role in the catalytic reaction is indirect, whereas they are participating directly if their rate of reappearance is similar to the rate of catalytic reaction.

The *role of the catalyst support* must be determined by appropriate experiments. The support is usually a high surface area microporous oxide. It may act as a dispersing agent of the active metal catalyst, nothing more. In this circumstance, exchanging it with another support or removing it altogether from the reaction mixture should have minimal impact on the kinetics of the reaction. The support may be an oxidation state modifier or, in fact, it can be a catalyst component. In these circumstances, it is a vital catalyst ingredient and its removal would greatly alter the reaction rates and the product distribution. It is then important to determine (a) the type of reaction that cannot be carried out without the presence of the support and (b) the conditions of catalyst preparation that with a solid state reaction between oxide and metal, will optimize the reaction conditions.

4. BUILDING OF IMPROVED CATALYSTS

The task ahead is to utilize our knowledge of the necessary molecular ingredients of the complex catalytic system that are presently employed and the information available on the thermodynamics and the kinetics of the reactions to be catalyzed to design a superior catalyst system. One of the first challenges is the *synthesis of microporous catalysts* and materials of high surface area that remain stable under reaction conditions. Most of the synthetic methods, at present, use solutions and solid precipitation from a colloid phase. There are other synthetic routes that may be explored. One could utilize the controlled decomposition of vapor phase molecules, for example. Both the active catalyst and the catalyst support must be available in high surface area forms. The use of crystalline microporous solids as supports has the advantage of structural uniformity. This, in turn, allows a better definition of how the atomic structure influences the catalytic chemistry, since the ordered structure is available to scrutiny by diffraction experiments and determination by crystallography. For structure-sensitive reactions, such a support can provide a template to produce the desired catalyst surface structure. The *control of the catalyst structure* is often the key to the actual selectivity. Where metal sites are deposited on oxide supports, the conditions of reduction often determine the surface structure of metal particles that form. When the rate of reduction is slow, surface structures of low surface energy are produced. These are the (111) faces of face-centered cubic metals, the (110) face of body-centered cubic metals and the (0001) face for hexagonal close-packed materials. When the rate of reduction is rapid, higher Miller index, rougher, or more open surfaces will form. Additives that are *structural promoters* can change the surface structure. Just as aluminum oxide facilitates the restructuring of iron in the presence of nitrogen, other additives might play the same role for other catalyst materials. Alloy components that may not participate in the reaction chemistry, but modify the structure and site distribution on the catalyst surface, should be explored utilizing the knowledge accumulated from the working alloy systems.

Site blocking could improve selectivity as has been proven for many working catalyst systems. Sulfur or silicon or other strongly adsorbed atoms that seek out the active site that is responsible for the undesirable side reaction could improve the selectivity. Well-chosen alloy constituents can act the same way, blocking low coordination sites, kinks, or steps that may be desirable in some circumstances, just as in the case of sulfur on platinum. Blocking of high coordination sites may profoundly alter the selectivity as in the case of sulfur on the nickel (100) face or gold in platinum (111).

Bonding modifiers may be employed to weaken or strengthen the chemical bonds. Strong electron donors, like potassium, or electron acceptors, like chlorine, may be used for this purpose. *Alloying may create new active sites* in addition to providing an ensemble effect that can greatly modify reactivity and selectivity. Many more alloy systems must be explored systematically to be able to predict the effects of alloying on catalytic behavior.

Perhaps the most promising area of new catalyst preparation lies in the *combined use of metal and ion (acid) catalysis* by building a system that contains both the oxide and metal. Multifunctional catalysts may lead to entirely new reaction schemes as shown by the production of gasoline from carbon monoxide and hydrogen over mixed metal–zeolite catalysts. *Surface compounds* that exhibit unique chemistry can form at the metal–oxide interface. Spillover of reactants and reaction intermediates from the metal through the metal–oxide compound will interface to the oxide and lead to consecutive reactions that will occur with high selectivity and lead to the formation of complex products. Crystalline micropores of molecular dimensions can lead to *shape selectivity* that can enhance the formation of stereospecific products.

Our knowledge of the possibilities in catalyst design provides only guidelines to the research path that has proven successful in the past. There is no substitute for the dreams and imagination of creative researchers who try *new ideas that have not been explored before.* The use of vibrationally or electronically excited molecules, of new materials, carbides, nitrides, sulfides, unique reaction conditions, that is, low-temperature liquid-phase enzymelike catalysis, and the exploration of catalysis of new reactions will keep the field of catalysis an exciting frontier area of research for many generations of scientists to come. *New instrumentation* will be discovered that permits closer scrutiny of the catalysts and the molecules adsorbed on the catalyst surface on the molecular level. *Theory* is providing more information that is useful to the practical catalyst researcher. Not only bonding information is becoming available, but calculations of the reaction path are possible that can help to at least rule out some of the many reaction mechanisms.

5. AREAS OF CATALYSIS SCIENCE WHERE IMPROVED UNDERSTANDING IS ESSENTIAL

The rate of catalytic reactions (product molecules per second) is proportional to the surface area. Thus practical heterogeneous catalysis will always be with materials of high surface area (i.e., microporous solids). These are usually metastable in the thermodynamic sense, with respect to low surface area solids of the same chemical composition, because the positive surface free energy must be minimized. Nevertheless, the transformation to low surface area dense materials can be arrested indefinitely if the kinetics of sintering remains very slow. As long as strong chemical bonds must be broken in order to destroy the porous structure, high surface area catalysts can readily be stabilized under most conditions of catalytic reactions. It should be remembered that the human body is a microporous high surface area complex catalytic system and is very stable, capable of long life although it is metastable with respect to its low surface area constituents, organic and inorganic.

We should gain understanding of the phase diagram of the microporous solids and the roles of chemical bonding that control their formation. New synthetic methods should be developed to produce them from the vapor phase or from solution. While most of the materials research in catalyst synthesis has been concentrating in recent years on microporous crystalline aluminosilicates (zeolites), it is clear that other materials (elements and compounds) could also be produced in high surface area forms. The development of new methods of synthesis of microporous materials and major efforts of characterizing the kinetics of their formation, their thermodynamic properties, structure, and bonding should lead to improved understanding of catalytic materials and to the discovery of new materials.

Catalyst poisoning, short-term and long-term deterioration of catalyst reactivity, and catalyst selectivity demand more basic research attention. The sintering and growth of small particles and the selective blocking of active sites have not been subjected to systematic surface-science studies as yet. There are often irreversible solid state reactions that lead to alteration of surface chemical composition to volume change that plugs the micropores. Organic fragments undergo temperature-dependent variation of bonding that leads to coking and catalyst deactivation. Systematic and broad studies should be carried out to gain better understanding of catalyst deactivation.

Catalysts could be regenerated by a variety of solid state or solid–vapor reactions. Redispersion of large particles to smaller ones produces higher and active catalyst areas. Selective removal of strongly adsorbed inhibitors reactivates poisoned catalyst sites. The regeneration of catalysts is another area of catalysis science where systematic research can greatly enhance our ability to develop longer-life, active, and selective catalysts.

REFERENCES

V. Amir-Ebrahami, F. Garin, F. Weisang, and F. G. Gault, *Nouv. J. Chim.* **3,** 529 (1979).
S. Anderson and J. B. Pendry, *Surf. Sci.* **71,** 75 (1978).
M. K. Bahl, S. C. Tsai, and Y. W. Chung, *Phys. Rev. B* **21,** 1344 (1980).
R. T. K. Baker, E. B. Prestridge, and R. L. Garten, *J. Catal.* **59,** 390 (1978).
R. T. K. Baker, E. B. Prestridge, and R. L. Garten, *J. Catal.* **59,** 293 (1979).
D. Barthomeuf, *J. Phys. Chem.* **88,** 42 (1984).
O. Beek, *Discuss. Faraday Soc.* **8,** 118 (1950).
H. J. Behm, D. Christmann, G. Ertl, and M. A. Van Hove, *Surf. Sci.* **88,** L59 (1979).
D. N. Belton, Y. M. Sun, and J. M. White, *J. Phys. Chem.* **88,** 5172 (1984).
J. R. Bernard, *Proceedings of the 5th International Conference on Zeolites, Naples, 1980*. L. V. C., Rees, Ed., p. 686, Heyden, London, 1980.

REFERENCES

D. G. Castner and R. L. Blackadar, *J. Catal.* **66,** 257 (1980).

R. R. Chianelli, T. A. Pecoraro, T. R. Halpert, W. H. Pan, and E. I. Stiefel, *J. Catal.* **86,** 226 (1984).

Y. M. Chung, G. Xiong, and C. C. Kao, *J. Catal.* **85,** 237 (1984).

B. S. Clausen, H. Topsoe, R. Candia, J. Villadsen, B. Lengeler, I. A. Nielsen, and F. Christensen, *J. Phys. Chem.* **85,** 3068 (1981).

H. Conrad, G. Ertl, J. Koch, and E. E. Latta, *Surf. Sci.* **43,** 462 (1974).

J. M. Dartigues, A. Chambellan, S. Corroleur, and F. G. Gault, *Nouv. J. Chim.* **3,** 591 (1979).

S. M. Davis and G. A. Somorjai, *The Chemical Physics of Solid Surfaces and Heterogeneous Catalysis,* Elsevier, Amsterdam, 1982.

S. M. Davis and G. A. Somorjai, *J. Phys. Chem.* **87,** 1545 (1983).

S. M. Davis and G. A. Somorjai, *Bull. Soc. Chem.* **3,** 271 (1985).

S. M. Davis, F. Zaera, and G. A. Somorjai, *J. Am. Chem. Soc.* **104,** 7453 (1982a).

S. M. Davis, F. Zaera, and G. A. Somorjai, *J. Catal.* **77,** 439 (1982b).

S. M. Davis, F. Zaera, and G. A. Somorjai, *J. Catal.* **85,** 206 (1984).

S. M. Davis, F. Zaera, B. E. Gordon, and G. A. Somorjai, *J. Catal.* **92**(2), 240 (1985).

J. A. Dumesic, H. Topsoe, and M. Boudart, *J. Catal.* **37,** 513 (1975).

T. M. Duncan and C. Dybowski, *Surf. Sci. Reports* **1,** 157 (1981).

D. J. Dwyer and J. H. Hardenbergh, *J. Catal.* **87,** 66 (1984).

D. J. Dwyer and G. A. Somorjai, *J. Catal.* **52,** 291 (1978).

D. D. Eley and J. L. Tuck, *Trans. Faraday Soc.* **32,** 1425 (1936).

P. H. Emmett, *The Physical Basis of Heterogeneous Catalysis,* Plenum, New York, 1975.

G. Ertl, *Catal. Rev. Sci.* **21,** 201 (1980).

G. Ertl, *Proc. Welch Conf.* **25,** 179 (1981).

A. Farkas, L. Farkas, and E. K. Rideal, *Proc. Roy. Soc.* **A146,** 630 (1934).

C. A. Fyfe, G. C. Gobbi, J. Klinowski, J. M. Thomas, and S. Ramdas, *Nature* **296,** 530 (1982).

W. D. Gillespie, R. K. Herz, E. E. Petersen, and G. A. Somorjai, *J. Catal.* **70,** 147 (1981).

D. W. Goodman, *Surf. Sci.* **123,** 6679 (1982).

D. W. Goodman, *Acc. Chem. Res.* **17,** 194 (1984a).

D. W. Goodman, *Proceedings of the 8th International Congress on Catalysis, Berlin,* Vol. 4, p. 3, Verlag Chemie, Berlin, 1984b.

B. S. Greensfelder, H. H. Voge, and G. Good, *Ind. Eng. Chem.* **41,** 2573 (1949).

W. O. Haag, R. M. Lago, and P. B. Weisz, *Nature* **309,** 589 (1984).

H. Heinemann, *Catalysis Science and Technology,* J. Anderson and M. Boudart, Eds., p. 16, Springer Verlag, Berlin, 1981.

J. Horiuti and K. Miyahara, *Nat. Stand. Ref. Data Ser. Nat. Bur. Stand.* **13,** (1968).

J. Horiuti and M. Polanyi, *Trans. Faraday Soc.* **30,** 1164 (1934).

J. Horsley, *J. Am. Chem. Soc.* **101,** 2870 (1979).

P. A. Jacobs and R. J. Von Ballmoos, *J. Phys. Chem.* **86,** 3050 (1982).

A. P. M. Kentgens, K. F. M. Scholle, and W. S. Veerman, *J. Phys. Chem.* **87,** 4357 (1983).

E. I. Ko and R. L. Garten, *J. Catal.* **68,** 233 (1981).

B. E. Koel, J. Crowell, B. Bent, and G. A. Somorjai, *Surf. Sci.* **146,** 211 (1984).
B. E. Koel, J. Crowell, B. Bent, and G. A. Somorjai, *J. Phys. Chem.* **90,** 2949 (1986).
Y. Lain, J. Criodo, and M. Boudart, *Nouv. J. Chim.* **1,** 461 (1977).
P. P. Lankhorst, H. C. Dejongste, and V. Ponec, *Catalyst Deactivation,* Elsevier, Amsterdam, 1982, 34.
M. Levin, M. Salmeron, A. T. Bell, and G. A. Somorjai, *Surf. Sci.* **169,** 123 (1986).
G. B. McVicker, G. M. Kramer, and J. J. Ziemak, *J. Catal.* **83,** 286 (1983).
P. Meriaudeau, O. H. Ellestad, M. Dufaux, and C. Naccache, *J. Catal.* **75,** 243 (1982).
R. Makamura, S. Nakai, K. Sugiyama, and E. Echigoya, *Bull. Chem. Soc. Jpn.* **54,** 1950 (1981a).
R. Makamura, K. Yamagami, S. Nishiyama, and E. Echigoya, *Chem. Lett.* **2,** 275 (1981b).
D. H. Olson, W. O. Haag, and R. M. Lago, *J. Catal.* **61,** 390 (1980).
N. K. Pande and A. T. Bell, Appl. Catal. (in press).
T. A. Pecoraro and R. R. Chianelli, *J. Catal.* **67,** 430 (1981).
V. Ponec, *Adv. Catal.* **32,** 1 (1983).
G. B. Raupp and W. N. Delgass, *J. Catal.* **58,** 316 (1979).
G. B. Raupp and J. A. Dumesic, *J. Phys. Chem.* **88,** 660 (1984).
D. E. Resasco and G. L. Haller, *J. Catal.* **82,** 279 (1984).
V. Rives-Arnau and G. Munuera, *Appl. Surf. Sci.* **6,** 122 (1980).
R. W. Roberts, *J. Phys. Chem.* **67,** 2035 (1963).
J. W. A. Sachtler and G. A. Somorjai, *J. Catal.* **81,** 77 (1983).
H. R. Sadeghi and V. E. Henrich, *J. Catal.* **87,** 279 (1984).
M. Salmeron and G. A. Somorjai, *J. Phys. Chem.* **86,** 341 (1982).
J. Santos, J. Phillips, and J. A. Dumesic, *J. Catal.* **81,** 147 (1983).
B. A. Sexton, A. E. Hughes, and K. Foger, *J. Catal.* **77,** 85 (1982).
J. H. Sinfelt, *Bimetallic Catalysts,* John Wiley & Sons, New York, 1983.
F. Solymosi, L. Volgyesi, and J. Sarkany, *J. Catal.* **54,** 336 (1978).
F. Solymosi, L. Volgyesi, and J. Rasko, *Z. Phys. Chem. N. F.* **120,** 79 (1980).
G. A. Somorjai, *Chemistry in Two Dimensions: Surfaces,* Cornell University Press, New York, 1981.
G. A. Somorjai, *Heterogen. Catal.* **25,** 83 (1983).
G. A. Somorjai, *8th Int. Cong. Catal.* **1,** 113 (1984).
G. A. Somorjai, Discussion Meeting of Royal Society, Philosophical Transactions of the Royal Society, London (1985a).
G. A. Somorjai, *Science* **227,** 902 (1985b).
G. A. Somorjai and S. M. Davis, *J. Chim.,* 271 (1985).
G. A. Somorjai and L. M. Falicov, *Proc. Nat. Acad. Sci.* **82,** 2207 (1985).
G. A. Somorjai and E. L. Garfunkel, *Surf. Sci.* **115,** 441 (1982).
G. A. Somorjai and C. M. Mate, *Surf. Sci.* **160**(2), 542 (1985).
G. A. Somorjai and B. A. Sexton, *J. Catal.* **46,** 167 (1977).
G. A. Somorjai and P. R. Watson, *J. Catal.* **74,** 282 (1982).
G. A. Somorjai and R. Yeates, *J. Catal.* (to be published 1987).

REFERENCES

G. A. Somorjai and F. Zaera, *J. Am. Chem. Soc.* **106**(8), 2288 (1984).
G. A. Somorjai, J. Crowell, and E. L. Garfunkel, *Surf. Sci.* **121,** 303 (1982).
G. A. Somorjai, J. Crowell, and W. T. Tysoe, *J. Phys. Chem.* **89,** 1598 (1985).
G. A. Somerjai, S. M. Davis, and F. Zaera, *J. Catal.* **77,** 439 (1982).
G. A. Somorjai, E. L. Garfunkel, and J. E. Crowell, *J. Phys. Chem.* **86,** 310 (1982).
G. A. Somorjai, A. J. Gellman, and D. Neiman, *J. Catal.* (to be published).
G. A. Somorjai, R. J. Koestner, and M. A. Van Hove, *Surf. Sci.* **121,** 321 (1982).
G. A. Somorjai, M. H. Farias, A. J. Gellman, R. R. Chianelli, and K. S. Liang, *Surf. Sci.* **140,** 181 (1984).
G. A. Somorjai, A. Wieckowski, S. Rosasco, G. Salaita, A. Hubbard, B. Bent, and F. Zaera, *J. Am. Chem. Soc.* **107,** 5910 (1985).
G. A. Somorjai, M. Asscher, J. Carrazza, M. Khan, and K. Lewis, *J. Catal.* **98,** 277 (1986).
G. A. Somorjai, W. D. Gillespie, R. K. Herz, and E. E. Petersen, *J. Catal.* **70,** 147 (1981).
N. D. Spencer, R. C. Schoonmaker, and G. A. Somorjai, *J. Catal.* **74,** 129 (1982).
D. Strongin, S. Bare, and G. A. Somorjai, *J. Catal.* (to be published 1986).
S. J. Tauster and S. C. Fung, *J. Am. Chem. Soc.* **100,** 170 (1978a).
S. J. Tauster and S. C. Fung, *J. Catal.* **55,** 29 (1978b).
C. L. Thomas, *Ind. Eng. Chem.* **41,** 2573 (1949).
J. M. Thomas and J. Klinowski, *Adv. Catal.* **33,** 200 (1985).
S. J. Thomson and G. Webb, *J. Chem. Soc. Chem. Comm.,* 526 (1976).
H. Topsoe, B. S. Clausen, R. Candia, D. Wivel, and S. Morup, *J. Catal.* **68,** 433 (1981).
G. H. Twigg and E. K. Rideal, *Proc. Roy. Soc.* **A171,** 55 (1939).
W. A. Van Hook and P. H. Emmett, *J. Am. Chem. Soc.* **87,** 939 (1965).
M. A. Vannice and R. L. Garten, *J. Catal.* **56,** 235 (1979).
M. A. Vannice, C. C. Twu, and S. H. Moon, *J. Catal,* **79,** 70 (1983).
P. B. Weisz, *Pure Appl. Chem.,* **52,** 2091 (1980).
F. Zaera and G. A. Somorjai, *J. Catal.* **84,** 375 (1983).

3
ORGANOMETALLIC CHEMISTRY:
Basis for the Design of Supported Catalysts

B. C. GATES, *Center for Catalytic Science and Technology, Department of Chemical Engineering, University of Delaware, Newark, Delaware*

1. INTRODUCTION

1.1. Molecular Organometallic Chemistry: Catalysis by Transition Metal Complexes and Clusters

Organometallic chemistry is central to almost all of catalysis: Most catalysts are metals, metal oxides, or metal sulfides; most reactants are organic; and most intermediates are organometallic, incorporating reactant-derived ligands bonded to metal centers offered by the catalyst. Whether the metal center is part of a molecule in solution or part of a surface, the underlying principles of organometallic chemistry pertain.

Organometallic chemistry has developed explosively in the past several decades, being driven by recent commercial successes and prospects of new ones in homogeneous organometallic catalysis. Prominent milestones in the history of organometallic chemistry include the discoveries of Ziegler's catalyst for olefin polymerization in 1955, Wilkinson's catalyst for olefin

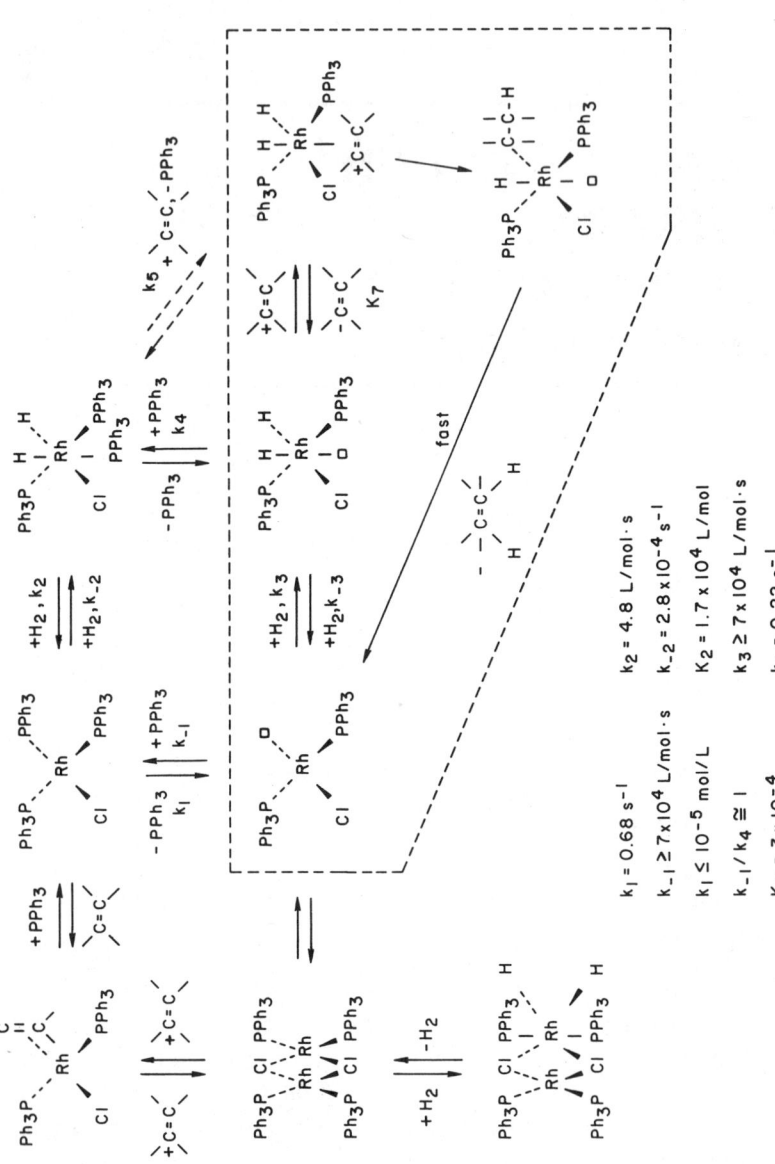

FIGURE 1. Catalytic cycle for the Wilkinson hydrogenation in solution. The cycle (adapted from that shown by Collman and Hegedus, 1980) was determined by Halpern and coworkers (as cited by Collman and Hegedus, 1980).

1. INTRODUCTION

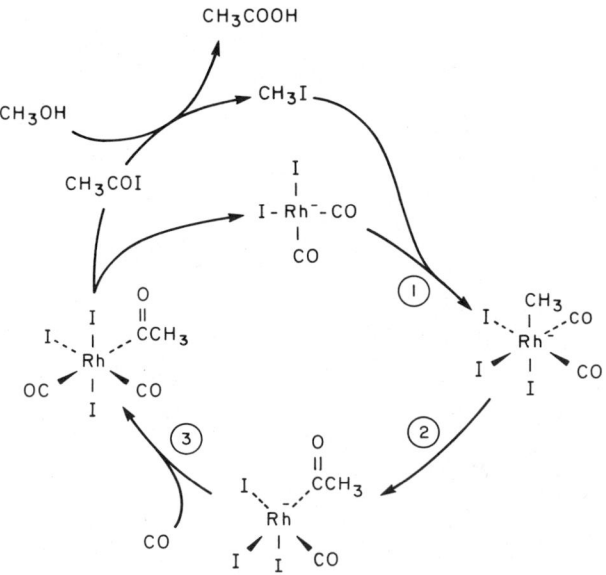

FIGURE 2. Catalytic cycle for methanol carbonylation in solution (adapted from Forster, 1976).

hydrogenation in 1965, and catalysts for olefin hydroformylation and methanol carbonylation soon thereafter (Collman and Hegedus, 1980). Ziegler and Wilkinson were awarded Nobel Prizes for their work in organometallic chemistry.

Organometallic chemistry is the chemistry of transition metals and the compounds formed by bonding organic groups (ligands) to them; organometallic compounds are characterized by metal–carbon bonds. There is now a massive literature on the synthesis, structure, bonding, and reactivity of these compounds (Collman and Hegedus, 1980), and there are some well-defined catalytic cycles, such as those of the Wilkinson hydrogenation (Figure 1) and the methanol carbonylation (Figure 2). The cycle for olefin hydroformylation is less well known, but that shown in Figure 3 is regarded as a good approximation. These cycles illustrate the delicate dynamic balance required for catalysis.*

In these and other examples of organometallic catalysis, the transition

---

*Catalytic cycles and kinetics and their place in catalyst design are discussed by Boudart in Chapter 5 of this book.

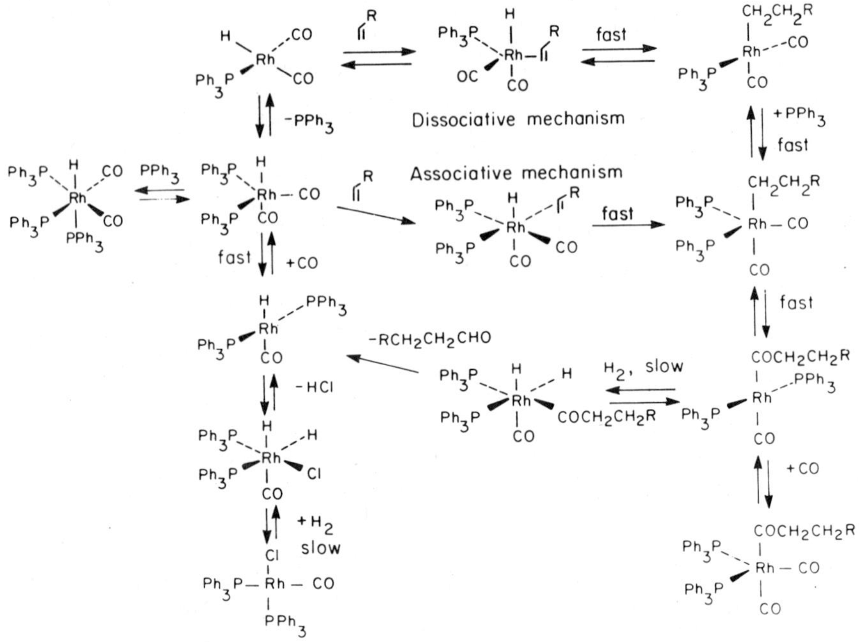

FIGURE 3. Catalytic cycle for olefin hydroformylation in solution, adapted from the work of Wilkinson et al., as cited by Pruett (1979) (Gates, Katzer, and Schuit, 1979).

metal ion or atom provides a template where the organic ligands react to be converted into products. Consider the example of the commercially important methanol carbonylation (Figure 2). The reactants are CO and methanol, and the product is acetic acid. The elementary step labeled ① involves a complex of rhodium in the $+1$ oxidation state [Rh(I)] with two iodide ligands and two carbonyl ligands; this complex is an anion and is soluble in polar solvents. It reacts in a rate-determining step with a co-catalyst (or promoter), methyl iodide; the step is called an *oxidative addition* because the rhodium formally is oxidized from the $+1$ to the $+3$ oxidation state as two ligands are added, one a methyl and one an iodide. The next step, labeled ②, is an *insertion* of the carbonyl ligand between the metal and the methyl group; this is more precisely described as a methyl migration reaction. The next step (labeled ③) is a *ligand association* of the reactant CO at the open bonding site. The final step in the cycle is a *reductive elimination;* it closes the loop, so that we now have catalysis, and it generates acetyl iodide, which in complicated chemistry (involving

1. INTRODUCTION

unknown steps) is converted into product, regenerating the methyl iodide cocatalyst and the rhodium complex mentioned initially. It is evident that open sites (or sites of *coordinative unsaturation*) on the rhodium are needed for this reactivity; the crucial chemistry takes place within the coordination sphere of the rhodium. All the elementary steps illustrated in Figures 1 and 2 are now understood quite well; these kinds of steps account for most of the known reactions of organometallic compounds.

The transition metal complexes mentioned above are all *mononuclear*, each containing a single metal atom or ion. There are also many *polynuclear* transition metal complexes, known as *transition metal clusters;* these have two or more metal atoms and metal–metal bonds. Metal clusters have been found to be catalysts for a number of reactions in solution; however, there are only a few well-defined catalytic cycles involving intact metal clusters (Markó and Vizi-Orosz, 1986; Muetterties and Krause, 1983), since most metal clusters are fragile, reacting to give mononuclear metal complexes and/or metal aggregates under catalytic conditions. Nonetheless, the clusters are attractive in prospect because their neighboring metal centers offer opportunities for ligand–metal bonding and reactivity not offered by mononuclear metal complexes. The clusters are akin to metal particles in having neighboring metal centers; the larger cluster carbonyls, for example, $[Os_{10}C(CO)_{24}]^{2-}$, which has the following structure (where the CO ligands are omitted for clarity),

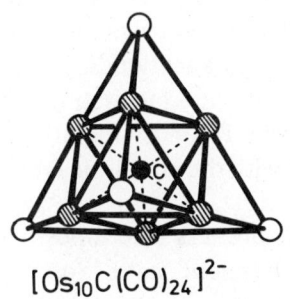

$[Os_{10}C(CO)_{24}]^{2-}$

are very much like small metal particles covered by CO ligands.

There is one example of catalysis that apparently involves metal clusters (rhodium carbonyls) and an industrially important reaction (ethylene glycol synthesis from $CO + H_2$); pressures of hundreds of atmospheres are required, and the process is not applied commercially.

1.2. Design Variables Affecting Reactivity and Catalytic Activity

At the risk of oversimplification, we can identify the following principal variables for design of reactivity in organometallic complexes:

1. The metal(s) in the complex.
2. The nuclearity (number of metal atoms).
3. The oxidation state of the metal.
4. The ligands.

Organometallic chemistry provides some guidelines for prediction of reactivities of transition metal complexes, but much remains to be learned, and only a few useful generalizations can be stated, as follows (Collman and Hegedus, 1980):

The second- and third-row metals usually have stronger metal–hydrogen, metal–carbon, and metal–metal bonds than the first-row metals. The third-row complexes usually exhibit lower reactivity than the second- or first-row elements, and the most active catalysts are usually from the second or first row.

The metals on the left-hand side of the periodic table (groups IV, V, and VI) are the more electropositive and tend to form stronger bonds to hard (i.e., small and weakly polarizable) electron donor atoms such as oxygen. These elements have relatively few d electrons and therefore high coordination numbers (numbers of ligands). The π-acid ligands (such as CO and olefins) that undergo backbonding with transition metals (weakening $C\equiv O$ and $C=C$ bonds and facilitating reaction of these ligands) do not bond as strongly to group IV and V metals,, and the migratory insertion reactions are often facile, with the equilibria lying to the side of the insertion products.

The group VIII metals are less electropositive than the others and are reduced easily to the zerovalent state. Soft ligands such as iodide and phosphine form strong bonds with these metals. Many of the complexes readily undergo oxidative addition and reductive elimination reactions. The group VIII metals, especially in the second and third rows, often have low coordination numbers and relatively stable coordinatively unsaturated complexes. These metals often occur in complexes with strongly bonded π-acid ligands such as CO and unsaturated hydrocarbons.

1.3. Illustration of the Design Variables with Catalytic Cycles

We can easily illustrate some of the ideas stated above by reference to the catalytic cycles shown in Figures 1–3. For example, the cycle for hydroformylation of olefins in the presence of rhodium complexes (Figure 3) demonstrates—after the fact of the discovery of the catalyst—why rhodium is an appropriate metal: it is a transition metal that coordinates all of the reactants, H, CO, and olefins, undergoing backbonding to the latter two. It is a second-row metal; the element below it in the periodic table (Ir) is also applicable (but the cycle is not so rapid); the element above it in the periodic table (Co) is also applicable and is used on a large scale in industrial hydroformylation processes [but these processes require pressures of hundreds of atmospheres (in contrast to about 10 atm for rhodium) because of the lack of stability of the cobalt carbonyl complexes]. Rhodium has stable complexes with the metal in the +1 and +3 oxidation states, which implies that elementary steps requiring a change of 2 in the oxidation state (oxidative addition or reductive elimination) can occur. Further, this group VIII metal provides the necessary stability of coordinatively unsaturated metal complexes required for ligand association and oxidative addition reactions.

The triphenylphosphine ligands on the rhodium complexes were chosen because they are good electron donors (helping to activate the reactant ligands) and because they are bulky; the attendant steric hindrance is believed to suppress the undesired reaction giving branched aldehydes, thereby increasing the selectivity to the desired straight-chain aldehydes. The process for propylene hydroformylation is carried out with liquid-phase reactants and catalyst in the presence of a large excess of triphenylphosphine to maximize the selectivity (Pruett, 1979).

The complexes in the catalytic cycle are all mononuclear; in the processes carried out with cobalt catalysts, cobalt carbonyl clusters [e.g., $Co_2(CO)_8$, $Co_4(CO)_{12}$, and others] can form; apparently, these form increasingly as the CO partial pressure is reduced, and ultimately cobalt metal is plated out on the reactor wall. In this example, the clusters are apparently an undesired and negligible complication during normal operation. In other processes (e.g., the synthesis of ethylene glycol from CO and H_2 at high pressures in the presence of rhodium complexes), metal clusters may be the key catalytic species.

This example of hydroformylation illustrates the importance of each of the "design variables" listed above; two points of clarification deserve emphasis: (1) The catalysts were found as a result of research that could be described as exploratory in nature; they were not the result of any rational design—although insights into the detailed chemistry were undoubtedly of great value in the research. (2) The "design variables" cannot be manipulated straightforwardly and independently of each other. All that can be set independently are the obvious variables of the metal, solvent, temperature, pressure, contacting scheme, and so on; taken together, these determine the chemistry.

A second example, methanol carbonylation (Figure 2), illustrates many of the same points and also the role of a promoter. Again, cobalt, rhodium, and iridium all give catalytically active complexes, with rhodium complexes being the most active and selective catalysts. The pressure (about 15 atm) is about the same as that required for hydroformylation with rhodium, and, again, cobalt complexes require pressures of hundreds of atmospheres. The methyl iodide promoter serves the purpose of introducing the methyl ligand into the stable, coordinatively unsaturated rhodium complex $[Rh(CO)_2I_2]^-$ in an oxidative addition step.

These examples (and the Wilkinson hydrogenation, Figure 1) begin to demonstrate why rhodium complexes find such wide application in organometallic catalysis; numerous other examples, including the stereospecific L-dopa synthesis, could be added.

2. "MOLECULAR" ORGANOMETALLIC CATALYSTS ON SURFACES

2.1. Introduction

Surface science and surface catalysis are increasingly being influenced by the concepts of organometallic chemistry; structure and bonding in organometallic compounds closely parallel structure and bonding of organic ligands on metal surfaces, but the analogy does not extend to reactivity and catalytic activity (Ertl, 1986).

We now focus the discussion on simple organometallics bonded to surfaces and having structures that have close molecular analogues; in this special case there are close analogies even in reactivity and catalytic activity between soluble catalysts and surface catalytic species, and there has been substantial progress in catalyst design. The designs, in essence, involve translation of the solution chemistry onto surfaces.

Why would one want to put an organometallic species onto a surface to do catalysis? The obvious answer is to gain the advantages of organometallics in solution (primarily, this means high selectivity associated with the simplicity of the catalytic cycle) combined with the advantages of solid catalysts, the classic ones being ease of separation from products, lack of corrosiveness, and one would hope—and this is a major issue— the robustness and stability that are characteristic of the useful solid catalysts.

The catalysts described here are referred to as "molecular" organometallics on supports; they are just some of the kinds of catalysts intermediate between the classical limiting cases involving molecular catalysts in solution ("homogeneous catalysis") and surfaces ("heterogeneous catalysis"). All the intermediate classes of catalysts (Figure 4) have some of the characteristics of molecules in a homogeneous medium (e.g., uniformity of environment and structure) and some of the characteristics of solid surfaces (e.g., being part of a phase separate from that holding the bulk of the reactants). The uniformity of the environment of catalytic species is often associated with high selectivity—and this, at least in prospect, is a major advantage of all the intermediate catalysts (Figure 4) that can be described as "molecular." It should not be forgotten, however, that a major virtue of classical surface catalysts is often their very heterogeneity—especially for complex reactant mixtures (such as petroleum frac-

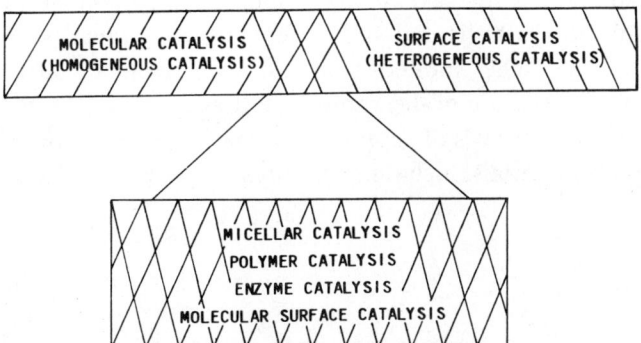

FIGURE 4. Classes of catalysis, with emphasis on the intermediate classes between the limiting cases of homogeneous catalysis and surface catalysis.

tions): The goal may be to catalyze simultaneously a set of reactions of a set of reactants, and a catalyst like Pt–Re/Al_2O_3 offers such a multiplicity of catalytic sites as to be able to provide pathways for many desired conversions of the hydrocarbons in petroleum naphtha (Boudart, 1982).

The supported molecular organometallic catalysts of interest here are to be contrasted to traditional supported metal catalysts, exemplified by Pt–Re/Al_2O_3. The former incorporate supported structures analogous to molecular organometallic species; the latter incorporate metal particles (or crystallites or aggregates, sometimes referred to as "clusters"*). The metal particles in conventional supported metal catalysts are nonuniform in size, shape, surface structure, and catalytic activity.

2.2. Synthesis

Organometallic species anchored to solid supports have been prepared by a variety of routes inferred from organometallic solution chemistry combined with the functional group chemistry of the support. The emphasis here is on materials with demonstrated catalytic activity which are evidently structurally unique in the sense of having "molecular" structures of a single type as determined spectroscopically. The supports include metal oxides as well as polymers and metal oxides functionalized with ligands analogous to those common in organometallic complexes (e.g., phosphines). The synthesis reactions, illustrated in the following sections, include simple ligand exchange, deprotonation of hydrido complexes on basic supports, oxidative addition involving surface OH groups, and multistep syntheses mimicking solution syntheses, among others. Often, the materials are air-sensitive and the synthesis and subsequent handling must be done in an inert atmosphere. The subtleties of surface chemistry often prevent straightforward extension of the known solution chemistry to the supports. Most attempts at synthesis of supported organometallics have led to structurally complex materials, including mixtures of metal complexes and metal aggregates.

2.3. Characterization

The characterization techniques of greatest value in the determination of structures of surface-bound organometallics are closely comparable to those

---

*Here we restrict the term "cluster" to its usage in organometallic chemistry.

used commonly in conventional organometallic chemistry (Table 1). In the following paragraphs, results of a few of the most important characterization methods are illustrated and the techniques are evaluated briefly. The spectra are interpreted on the basis of comparisons of spectra of supported species and those of their molecular analogues; consequently, precise structural inferences are often attainable.

Infrared (IR) spectroscopy is the method most easily and widely applied to characterize surface-bound organometallics. Spectra in the carbonyl stretching region provide a basis for identification of many surface structures. This sensitive and easily applied technique is capable of providing characterization of ligands other than carbonyls, but there are few reported examples.

Laser Raman spectroscopy provides complementary data, but this technique is difficult to apply successfully because of the lack of sensitivity and the complication of fluorescence of the surface and destruction of the sample by the incident laser radiation. Raman spectroscopy provides evidence of metal–metal and metal–oxygen bonds, but there are few reported examples. A Raman spectrum of the cluster anion $[H_2Re_3(CO)_{12}]^-$ on MgO (Figure 5) provides this kind of information.

Inelastic electron tunneling spectroscopy is another vibrational spectroscopy that has been used to characterize surface-bound organometallics. The technique requires coating the surface of the sample with a conducting layer of a metal such as lead, and the spectra are typically determined at liquid helium temperature so that the electron tunneling process is rapid; the resolution is much less than that of infrared and Raman spectros-

TABLE 1. Spectroscopic Methods Used for Characterization of Supported Molecular Organometallic Complexes

Spectroscopic Method for Supported Organometallic Complexes	Comparable Method used in Conventional Organometallic Chemistry
IR spectroscopy	IR spectroscopy
^{1}H and ^{13}C NMR spectroscopy	^{1}H and ^{13}C NMR spectroscopy
EXAFS	X-ray diffraction crystallography
Raman spectroscopy	Raman spectroscopy
Inelastic electron tunneling spectroscopy	—
X-ray photoelectron spectroscopy	X-ray photoelectron spectroscopy

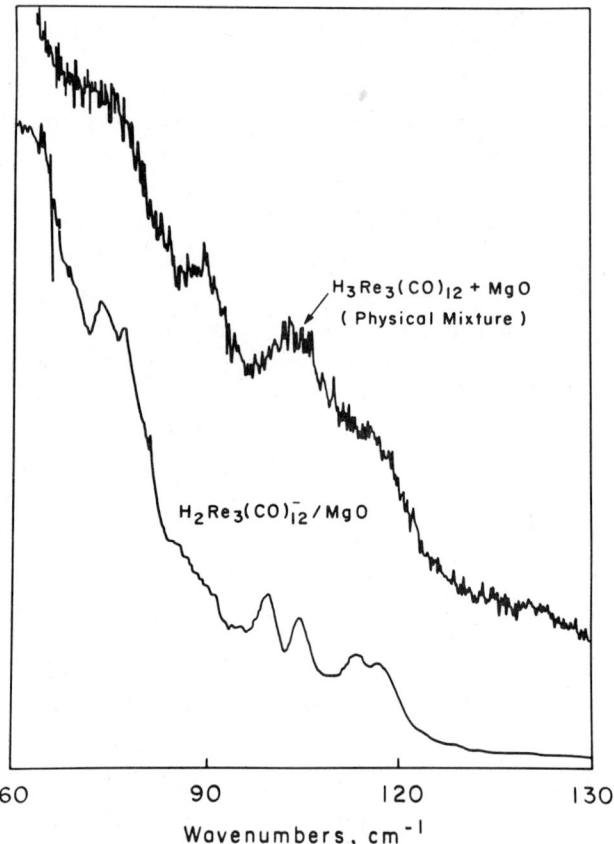

FIGURE 5. Laser Raman spectra in the metal–metal stretching and metal–carbon bending region of $H_3Re_3(CO)_{12}$ + MgO (physical mixture) and $H_2Re_3(CO)_{12}^-$ adsorbed on MgO (Kirlin et al., 1986).

copies, but the sensitivity is high. Results are illustrated for the anion $[Ru_6C(CO)_{17}]^{2-}$ on Al_2O_3 (Figure 6).

EXAFS (extended x-ray absorption fine structure) spectroscopy is in prospect the most powerful method for determining the structures of surface-bound organometallics. Spectra of well-characterized standards are essential for the proper interpretation of data; the most appropriate standards are the molecular analogues of the surface species themselves, and it is best to use structural parameters for the standards determined from x-ray crystallography as a basis for evaluation of the EXAFS data.

2. "MOLECULAR" ORGANOMETALLIC CATALYSTS ON SURFACES

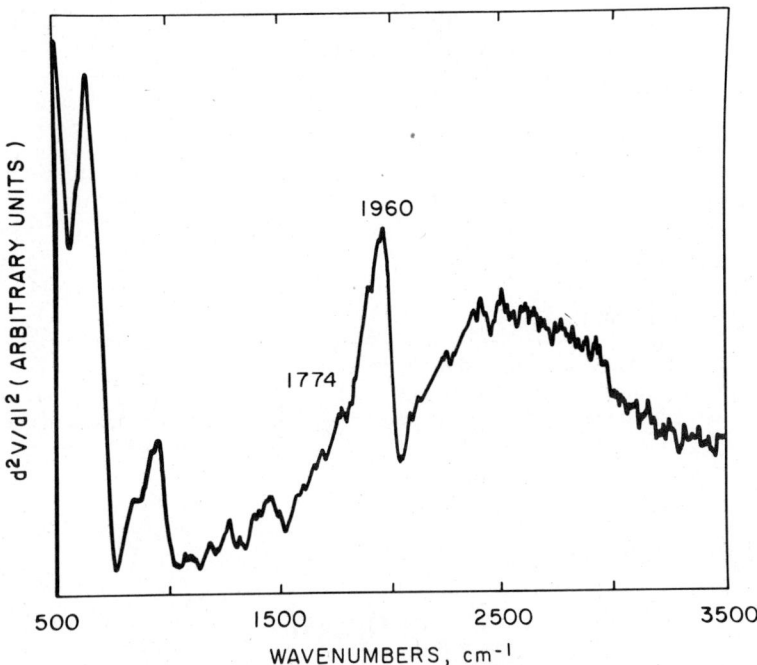

FIGURE 6. Inelastic electron tunneling spectrum of $[Ru_6C(CO)_{17}]^{2-}$ adsorbed on Al_2O_3 at liquid helium temperature (DeThomas et al., to be published).

The method has been applied to the γ-Al_2O_3-supported triosmium cluster:

Structural parameters characterizing the surface species are summarized in Table 2 (Duivenvoorden et al., 1986). The results provide a strong confirmation of the structure of the surface species, which is closely anal-

TABLE 2. Characterization of γ-Al$_2$O$_3$-Supported Osmium Carbonyls by EXAFS (Duivenvoorden et al., 1986)

Sample	Os$_3$(CO)$_{12}$ [a]	HOs$_3$(CO)$_{10}${OAl}	Os(CO)$_x${OAl}$_3$
Os–Os distance, Å	2.88[b]	2.88[c]	—
Os–Os coordination number	2[b]	2[c]	—
C:Os atomic ratio	4	3.35[c]	2.8
Os–O support distance, Å		2.12	2.19
Os–support coordination number	—	0.65	3

[a] Reference compound, characterized in the crystalline state.
[b] Value determined by x-ray crystallography.
[c] Value determined on the basis of data characterizing the reference compound and the surface species.

ogous to those of the following compounds:

$$\text{(CO)}_3\text{Os} \underset{\underset{\underset{R}{|}}{O}}{\overset{\overset{\text{Os(CO)}_4}{}}{\overset{H}{\diagup\diagdown}}} \text{Os(CO)}_3$$

[where R is H, alkyl, or SiEt$_3$ (D'Ornelas et al., 1985)]. The EXAFS data confirm the surface structure that had been inferred from the infrared spectra (Psaro et al., 1981; Deeba and Gates, 1981), Raman spectra (Deeba et al., 1981), inelastic electron tunneling spectra (Hilliard and Gold, 1985), and the stoichiometry of the surface synthesis (Psaro et al., 1981). This is regarded as one of the best-defined oxide-bound organometallic structures.

Ultraviolet (UV)-visible reflectance spectroscopy offers the advantage of ease of application, but it has been of minor value in the characterization of surface-bound organometallics. The method provides evidence of metal–metal bonds (Collier et al., 1983), but since it does not provide highly specific structural information, it is best used to complement the other methods. In a few instances, the UV–visible spectrum provides a good fingerprint for identification of a surface-bound organometallic; an example (Figure 7) is [Os$_{10}$C(CO)$_{24}$]$^{2-}$ on MgO; excellent agreement is observed between the spectrum of the [Et$_4$N]$^+$ salt deposited on MgO and that of the

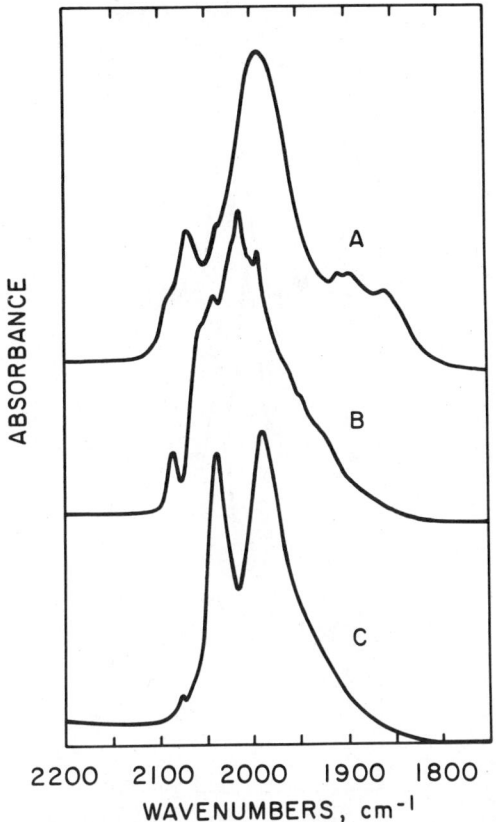

FIGURE 7. Infrared spectra of (A) unused catalyst incorporating HOs(CO)$_4^-$ prepared from H$_2$Os(CO)$_4$ and MgO; (B) sample after use as a CO hydrogenation catalyst; and (C) solid after extraction, incorporating Os$_{10}$C(CO)$_{24}^{2-}$ (Lamb and Gates, 1986). Reproduced from *Journal of the American Chemical Society*. Copyright 1986, American Chemical Society.

surface-bound dianion formed on MgO during CO hydrogenation (Lamb and Gates, 1986).

Nuclear magnetic resonance (NMR) spectroscopy has found only little application to surface-bound organometallics (Foley et al., 1983; Hanson et al., 1984; Lamb et al., to be published), but with modern high-field instrumentation, a greatly increased attention to this method is expected. ^{1}H NMR (without magic angle spinning) has been used to characterize supported mononuclear complexes, for example, rhodium allyl on SiO$_2$ (Foley et al., 1983). Changes in the spectrum (Figure 8) provide evidence

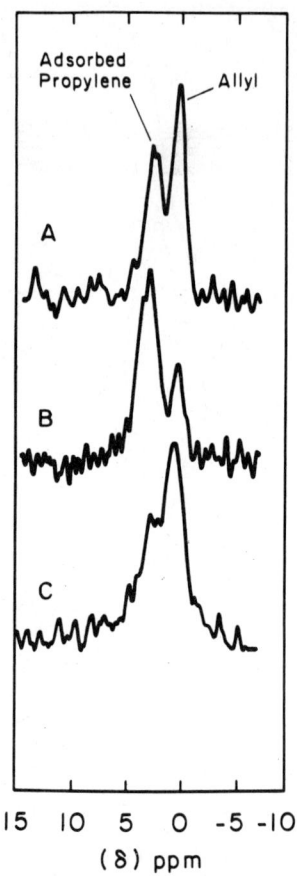

FIGURE 8. ^{1}H NMR spectra of {SiO}Rh(allyl)$_2$ (0.75 wt% Rh) during reaction with D$_2$: (A) after reaction with flowing D$_2$ for 30 min (760 torr, 25°C; 500 coadditions); (B) after reaction with D$_2$ for 48 hr (760 torr, 25°C; 500 coadditions); (C) subsequent evacuation of sample (1 × 10^{-3} torr, 25°C; 2500 coadditions, intensity multiplied by 2.25 vs. spectra A and B) (Foley et al., 1983). Reproduced from *Journal of the American Chemical Society*. Copyright 1983, American Chemical Society.

of conversion of the allyl ligands into propylene and propane as the rhodium (initially in the +3 oxidation state) was reduced at room temperature in H$_2$ to the zerovalent state, accompanied by formation of Rh aggregates on the SiO$_2$ surface.

^{13}C NMR spectroscopy with magic angle spinning and cross polarization has just begun to be used to characterize organometallics on oxide surfaces;

2. "MOLECULAR" ORGANOMETALLIC CATALYSTS ON SURFACES

mononuclear molybdenum carbonyl complexes on γ-Al_2O_3 were found to have spectra consistent with $Mo(CO)_x$ subcarbonyls (Figure 9) (Hanson et al., 1984), and a mononuclear osmium complex on MgO was found to have a spectrum consistent with the structure of the anion $HOs(CO)_4^-$ (Lamb et al., to be published); the latter has been characterized by several other techniques, and the surface chemistry is described in Section 3.4.

Metal NMR has not yet been reported for supported organometallics, but the prospects are excellent. High-field spectrometers are required, and these are now commercially available.

X-ray photoelectron spectroscopy (XPS) provides indications of oxidation states of metals in supported catalysts, but often the determinations are inexact and in need of confirmation by other methods. XPS is especially useful for detection of changes in oxidation states. The technique requires ultrahigh vacuum, and lack of stability of the samples is often a complication.

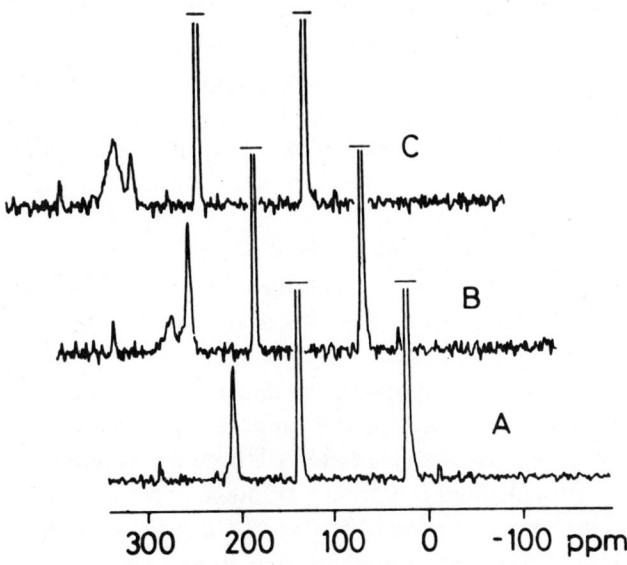

FIGURE 9. CP MAS ^{13}C NMR spectra of $Mo(CO)_6$ on γ-alumina. Spectrum A was obtained after activation at 30°C for 1 hr. Spectrum B was obtained on a separate sample heated to 100°C for 1 hr. Spectrum C was obtained on a separate sample heated to 100°C for 3 hr. In each spectrum, two very large peaks appear; these are due to hexamethylbenzene, which was added to the samples for magic angle and chemical shift calibration (Hanson et al., 1984). Reproduced from *Inorganic Chemistry*. Copyright 1984, American Chemical Society.

Temperature-programmed decomposition of supported organometallics provides quantitative information about the ligands bonded to the surface species. In this experiment, the temperature of a sample in an inert carrier gas stream is ramped at a known (and usually constant) rate, and the gaseous effluent is analyzed to provide a quantitative profile of what has desorbed or formed by chemical reaction; often the chemical reactions involve OH groups of the surface, which may oxidize the metal (Brenner, 1986). A variation of this technique involves use of a reactive gas stream such as H_2 in place of the inert carrier gas, and complementary data are obtained.

Transmission electron microscopy has yielded some striking results characterizing metal clusters on supports (Schwank et al., 1983; Iijima and Ichikawa, 1985). Individual metal atoms are discernible in some samples, but nearly the best available technology for high resolution microscopy is required (Long et al., in press).

Other surface characterization methods of potential value for supported metal complexes include electron spin resonance spectroscopy and magnetic susceptibility measurements; Mössbauer spectroscopy, which can be expected to be of greatest value for iron-containing samples; and UV photoelectron spectroscopy and electron-energy-loss spectroscopy, which ultimately are expected to offer interesting prospects as the techniques are developed for the supported species. The well-defined supported organometallics offer good prospects for extending the range of the aforementioned methods.

In summary, supported organometallics are among the best characterized supported metals. These samples have provided the opportunities for demonstrating the power of a range of surface characterization techniques, including EXAFS, laser Raman spectroscopy, and inelastic electron tunneling spectroscopy. It is emphasized that the spectroscopic methods of greatest value in characterizing the supported molecular organometallics are the techniques most closely related to those used commonly in organometallic chemistry (Table 1). A great advantage of many of these techniques is that they can be used for catalyst samples in reactive atmospheres, even at high pressures. Infrared spectroscopy is the method used most successfully with samples in the presence of reactive atmospheres (even at several hundred degrees Celsius and pressures of tens of atmospheres, as will be discussed later); Raman spectroscopy and EXAFS can also be used under such conditions. The potential exists for NMR as well, but this may be much more difficult when magic angle spinning of the sample is required.

2.4. Reactivity

Reactivities of supported organometallics have not been well demonstrated, but the available data are generally consistent with the statement that the chemistry of the supported species is often nearly the same as that of their analogues in solution.

Supports impart some unique reactivity to the complexes. For example, anchoring of complexes of titanium to a rigid polymer support, as follows,

$$\text{Ⓟ—C}_6\text{H}_4\text{—CH}_2\text{—Cp}^- \text{Li}^+ \xrightarrow{\text{TiCl}_3} \text{Ⓟ—C}_6\text{H}_4\text{—CH}_2\text{—Cp—Ti(Cl)}_2\text{Cp} + \text{LiCl} \quad (1)$$

(where Ⓟ represents the backbone of crosslinked polystyrene), followed by reduction, presumably gives the following coordinatively unsaturated structure (Bonds et al., 1975):

$$\text{Ⓟ—C}_6\text{H}_4\text{—CH}_2\text{—Cp—Ti—Cp}$$

Such a complex is not stable in solution because it reacts with itself, leading to the formation of metal–metal bonds. The solid support preserves coordinative unsaturation in the complex by being a rigid ligand that holds the organometallic groups apart from each other. Consequently, the supported complex is catalytically active for reactions such as olefin hydrogenation, and the soluble analogue is not. Surfaces of solid catalysts in general have the virtue of presenting stable sites of coordinative unsaturation. This is one of the principal reasons why most practical catalysts are solids.

If a polymer support is flexible, having a gel- or solutionlike character, it does not provide this stabilization by site isolation. However, the flexible (lightly crosslinked) polymers are useful when it is desirable to have the support act as a large chelating ligand. For example, when phosphine groups are incorporated into lightly crosslinked polystyrene to give Ⓟ-PPh$_2$, the resulting solid is able to strongly bond and chelate rhodium complexes similar to those shown in the catalytic cycle for olefin hydroformylation in solution (Figure 3). The high ratio of phosphine groups to rhodium complexes favors the selectivity to the desired straight-chain aldehyde product in the hydroformylation reaction catalyzed by the polymer (Pittman and Hanes, 1976).

The intrinsic reactivity of a support such as phosphine-functionalized crosslinked polystyrene is attributed almost entirely to the functional group rather than the nearly inert hydrocarbon backbone. Metal oxide supports, on the other hand, are not nearly so inert, and complicated patterns of reactivity of oxides with organometallics are the rule, not the exception.

For example, supported triosmium clusters have been anchored to silica by the following oxidative addition reaction (Psaro et al., 1981; Deeba and Gates, 1981):

$$Os_3(CO)_{12} + \underset{\pi\overset{Al}{\underset{/\!/}{}}\pi}{\overset{H}{\underset{|}{\overset{|}{O}}}} \longrightarrow (CO)_3Os\overset{Os(CO)_4}{\underset{\underset{\pi\overset{Al}{\underset{/\!/}{}}\pi}{\overset{|}{O}}}{\overset{H}{-\!\!\!-}}}Os(CO)_3 + 2\ CO \qquad (2)$$

Similar chemistry occurs on γ-Al$_2$O$_3$ and other metal oxide surfaces. Tetraosmium clusters are formed from H$_4$Os$_4$(CO)$_{12}$ on γ-Al$_2$O$_3$ and MgO in a simple deprotonation reaction giving [H$_3$Os$_4$(CO)$_{12}$]$^-$ ions on the surface (Krause et al., 1985).

The surface-bound triosmium and tetraosmium clusters show similar patterns of reactivity. The clusters are oxidized by surface OH groups. At temperatures greater than about 100°C, the clusters break apart, giving mononuclear Os^{2+} carbonyl complexes (Psaro et al., 1981; Knözinger and Zhao, 1981; Deeba and Gates, 1981; Krause et al., 1985). The stoichiometry of the surface reaction of the triosmium cluster has been determined from measurement of CO and H_2 evolution (Psaro et al., 1981). The resulting supported mononuclear complex on γ-Al_2O_3, having two or three carbonyl ligands, has been characterized precisely by IR spectroscopy and EXAFS (even the metal-support bonding distances are known, Table 2). The Al_2O_3-supported sample with the broken-up cluster, as expected, no longer gave Raman evidence of Os–Os bonds (Deeba et al., 1981). The resulting mononuclear complexes on γ-Al_2O_3 are surprisingly resistant to reduction in H_2—temperatures of 300–400°C are required for formation of Os metal, present in the form of aggregates (Psaro et al., 1981, Deeba and Gates, 1981; Knözinger and Zhao, 1981).

Organometallic complexes on metal oxide surfaces, especially those of group VIII metals, may be easily reduced to give aggregates of the zerovalent metal (Psaro and Ugo, 1986). Examples include rhodium allyl complexes on silica, which are reduced in H_2 at room temperature (Foley et al., 1983). In some instances, bimetallic clusters give aggregates that are probably alloylike, especially when group VIII metals are used (Anderson et al., 1977; Guczi, 1986), and in other instances the metals originally present in the cluster segregate, with the formation of aggregates of one metal and separate mononuclear complexes of the other. For example, $FeOs_3$ clusters on SiO_2 evidently give Fe aggregates and Os(II) complexes (Choplin et al., 1983); $RuOs_3$ clusters on γ-Al_2O_3 give Ru aggregates and mononuclear Ru and Os complexes (Budge et al., 1984), and $RhOs_3$ clusters on γ-Al_2O_3 give Rh aggregates and Os(II) complexes (Budge et al., 1985).

These reactions involving reduction of metal and formation of metal–metal bonds are roughly the reverse of the reactions described earlier as cluster breakup. Occasionally these reduction/aggregation processes have been observed to lead to formation of supported molecular metal clusters rather than structurally indistinct aggregates. For example, the $Os^{2+}(CO)_2\{OAl\}_3$ or $Os^{2+}(CO)_3\{OAl\}_3$ complexes formed by breakup of the above-mentioned surface-bound triosmium carbonyl clusters have been partially reconverted into the original supported clusters, as indicated by IR spectra (Psaro et al., 1981). More thorough evidence of molecular osmium carbonyl cluster formation has been reported for MgO-supported samples initially incorporating the anion $HOs(CO)_4^-$ (inferred to be present on the surface in an ion pair), which has the following structure (Lamb and Gates, 1986):

$$\begin{array}{c} OC \diagdown \diagup CO \\ H-Os-CO \\ | \\ C \\ \| \\ O \\ \downarrow \\ /\!/\!/ \, Mg^{2+} /\!/\!/\!/\!/ \end{array} \begin{pmatrix} H \\ | \\ O \end{pmatrix}^{-}_{/\!/\!/}$$

This was formed by deprotonation of $H_2Os(CO)_4$ on the basic MgO surface.

When the sample was heated with a 3:1 (molar) H_2:CO mixture at 275°C and 10 atm for about 20 hr (the catalytic CO hydrogenation reaction took place, as will be described later), a condensation reaction occurred, giving $[H_3Os_4(CO)_{12}]^-$ [which was extracted from the surface with $(CH_3)_4NCl$ in isopropanol]. The remaining solid had an IR spectrum characteristic of $[Os_{10}C(CO)_{24}]^{2-}$ (Figure 7); following the extraction, the solid was red; the UV–visible spectrum of the used catalyst (Figure 5) confirms the structural inference. The surface chemistry of the osmium carbonyls on the basic MgO closely parallels the solution chemistry of the osmium carbonyls: In the presence of CO and H_2, the tetranuclear cluster is relatively stable in solution (Nicholls et al., 1982) and the synthesis of $[Os_{10}C(CO)_{24}]^{2-}$ takes place in high yield in the presence of Na (a reducing agent) at about 260°C (Hayward and Shapley, 1982). It is inferred that the cluster anions observed on the MgO surface are the thermodynamically favored structures in the presence of CO + H_2. There was no evidence of Os aggregates on this surface (even though they form readily on γ-Al_2O_3 and SiO_2); the combination of ligands (provided by the basic MgO support and the reactive gas atmosphere) was crucial to the stabilization of the molecular clusters at high temperature. Evidently the osmium carbonyls are sufficiently mobile on the surface for the cluster formation reaction to occur.

In summary, the known solution chemistry provides the best guide to surface reactivity of organometallics, but it is insufficient, and the details of the surface chemistry of the support are often crucial. The properties of the support* need to be added to the list of "design variables;" the important properties include the concentrations and reactivities of func-

---

*The properties of catalyst supports are considered in detail by Bell in Chapter 4 of this book.

3. SUPPORTED ORGANOMETALLIC CATALYSTS

tional groups such as OH groups and Lewis acid centers (e.g., coordinatively unsaturated Al^{3+} and Mg^{2+}), the geometry of the surface, and physical properties such as rigidity and pore size.

3. SUPPORTED ORGANOMETALLIC CATALYSTS

To illustrate the design of supported molecular organometallic catalysts, we consider a set of examples, chosen to emphasize each of the "design variables." Since these "design variables" cannot be unraveled, the approach is an oversimplification and could easily be misconstrued. It is emphasized that most of the catalysts described in the following sections did not result from processes of rational design; rather, they resulted from research guided by the principles of organometallic and surface chemistry. We begin with the simplest examples, for which the term catalyst design is quite appropriate, and proceed to those that are more complex.

3.1. Polymer-Supported Rhodium Complex Catalysts for Olefin Hydroformylation

The choice of rhodium as the metal for a supported hydroformylation catalyst follows from the ideas described above and, quite straightforwardly, from the knowledge of how rhodium complexes act as soluble hydroformylation catalysts (Figure 3) (Haag and Whitehurst, 1973). The ligands are crucial, and the choice of phosphines also follows from the solution catalysis. Since high phosphine:rhodium ratios in solution give high selectivity to the straight-chain aldehyde product, a flexible polymer (crosslinked polystyrene functionalized with PPh_2 groups) is preferred and has been found to give high selectivities (Pittman and Hanes, 1976).

Catalysts of this type, presumably incorporating structures such as

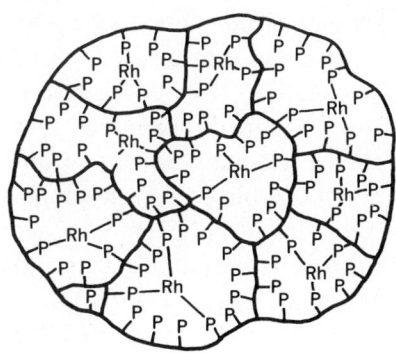

have undergone pilot-plant testing and evaluation for prospective industrial application (Lang et al., 1977). The catalysts are active and selective, but evidently not sufficiently stable, slowly losing rhodium in the product liquid flowing from the trickle-bed reactor. The lack of stability is associated with ligand association–ligand dissociation processes (Lang et al., 1977). Even small losses of the expensive rhodium would be enough to prevent economical application of such a catalyst.

3.2. Polymer-Supported Bifunctional Catalysts for Methanol Carbonylation

This example also emphasizes the choice of the metal, the ligands, and the physical properties of the support. The choice of rhodium is obvious on the basis of the foregoing discussion. The need for a cocatalyst (a second catalyst function) is evident from the cycle shown in Figure 2. Anchoring iodide stably to a support is not possible; therefore the need for a group with analogous reactivity is clear; compounds such as the "pseudohalides," for example

$$\text{Cl}_5\text{C}_6-\text{SCH}_3$$

an analogue of CH_3I, meet the need, being promoters of the catalytic reaction in solution (Webber, Gates, and Drenth, 1977).

A polymer-supported catalyst incorporating pseudohalide groups has been found to form complexes with rhodium and thereby to become catalytically active for methanol carbonylation (Webber, Gates, and Drenth, 1977/78). Presumably, a structure analogous to $Rh(CO)_2I_2^-$ in Figure 2 is formed, undergoing an oxidative addition reaction with the anchored analogue of CH_3I (as shown in Figure 10), whereby a CO ligand is positioned next to a CH_3 ligand on the rhodium. Since two bulky groups interact in this step, it is evident that a flexible polymeric support is appropriate. The catalyst supported on crosslinked polystyrene is active for the methanol carbonylation reaction, but it is unstable, losing rhodium to the reactant solution (Webber, Gates, and Drenth, 1977/78).

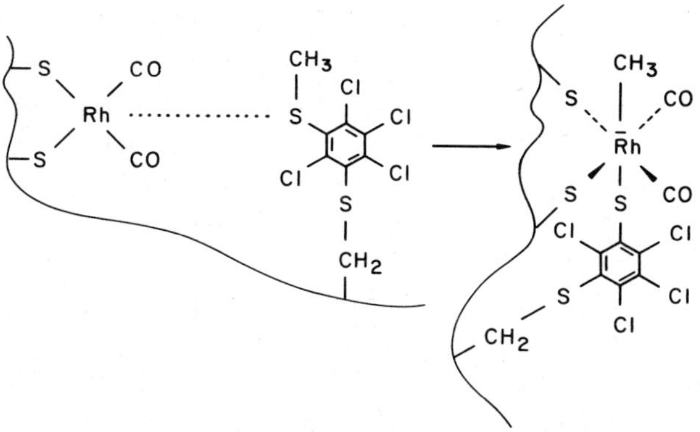

3. SUPPORTED ORGANOMETALLIC CATALYSTS

FIGURE 10. Suggested oxidative addition step occurring in a bifunctional polymer-supported catalyst for methanol carbonylation (Webber, Gates, and Drenth, 1977/78).

The foregoing examples emphasize the choice of metal, ligands, and physical properties of supports; the catalyst designs are straightforwardly built upon the catalytic cycles of the solution reactions. Usually we are not afforded the luxury of such detailed solution chemistry, however, and the design is more challenging, as illustrated in the following examples.

3.3. Oxide-Supported Molybdenum Catalysts for Hydrocarbon Conversion

Mononuclear complexes of many metals (Ti, Zr, Hf, Nb, Cr, Mo, W, Re, Rh, Ni, Pd, and Pt) have been anchored to metal oxide surfaces by reactions such as the following, giving anchored cationic complexes (Yermakov, Kuznetsov, and Zakharov, 1981):

$$\text{Si—OH} + \text{M(allyl)}_x \longrightarrow \text{Si—O—M(allyl)}_{x-y} + y\text{C}_3\text{H}_6 \quad (3)$$

Mononuclear carbonyl complexes of some of these metals also react with OH groups on oxide surfaces, being oxidized and giving mononuclear cationic subcarbonyls. Complexes of molybdenum (Burwell, 1984; Thomas and Brenner, 1983) with less than fully defined structures have been formed from Mo(CO)$_6$ in this way. The molybdenum carbonyl complexes on alumina are catalytically active for olefin reactions such as metathesis (as are similar complexes of W and Re).

Burwell and coworkers (1984) prepared catalysts with a range of catalytic properties with molybdenum in various oxidation states; they also succeeded in producing zerovalent molybdenum from $Mo(CO)_6$ (present in aggregates of undetermined size) by use of an alumina support dehydroxylated by treatment at high temperatures. The resulting catalyst is much different in character from that containing cationic molybdenum, being highly active for olefin hydrogenation and paraffin hydrogenolysis. This example illustrates a measure of catalyst design for control of the metal oxidation state; it was possible to prepare zerovalent metal, but with loss of control of the nuclearity (aggregate size).

3.4. Oxide-Supported Osmium Clusters for CO Hydrogenation

Control of nuclearity in conventional supported metal catalysts is difficult; an opportunity is provided by molecular metal clusters. Only a few transition metals (e.g., Os, Ru, Rh, Ir) have large families of clusters (Johnson, 1980). The clusters require coordination spheres of stabilizing ligands, and simple geometric considerations require that at least some of these ligands be small; the most common cluster ligands include CO and H.

Therefore, in attempts to design supported metal cluster catalysts that would be stable, the metal was chosen to be osmium (because it forms clusters with strong metal–metal bonds), and the reactants were chosen to be CO and H_2, to provide stabilizing ligands. Attempts to stabilize osmium–carbonyl clusters on γ-Al_2O_3 during CO hydrogenation catalysis were not successful; osmium aggregates of various sizes formed on the support (Odebunmi et al., 1985). When a strongly basic support (magnesia) was used, however, stable anionic clusters formed from the mononuclear complex $HOs(CO)_4^-$, as mentioned above. The magnesia served to deprotonate complexes such as $H_2Os(CO)_4$ and $H_4Os_4(CO)_{12}$, giving stable anions ion-paired to Mg^{2+} on the surface (Lamb and Gates, 1986). The clusters $[H_3Os_4(CO)_{12}]^-$ and $[Os_{10}C(CO)_{24}]^{2-}$ were formed on the surface from the mononuclear complex during catalysis; these are suggested to be the thermodynamically preferred structures; the former has been suggested to be involved in the catalysis, but further work is required to test the suggestion.

The catalyst is low in activity and produces a typical distribution of hydrocarbon products. The important result is the stabilization of a supported molecular metal cluster catalyst at a high temperature for the first time.

In summary, the "design variables" in this example are the metal (Os, which forms highly stable carbonyl clusters with strong Os–Os bonds); the reactants (CO and H_2, which provide small stabilizing ligands); and the support (MgO, which stabilizes anionic clusters).

3.5. Zeolite-Entrapped Metal Clusters

Control of the nuclearity in the foregoing example resulted from thermodynamics, and it was not precise because more than one cluster formed on the surface. There is an alternative for regulation of the size of metal clusters: encage them in a space that will limit their size.

When faujasite zeolite containing rhodium ions was subjected to 80 atm of CO + H_2 at 130°C, rhodium carbonyls were formed, and the IR spectra suggested the presence of $Rh_6(CO)_{16}$, the largest cluster that would fit in the supercage (Mantovani, Palladino, and Zanobi, 1977). This example illustrates the potential value of the pore size of the support in regulating nuclearity; materials such as zeolites with regular, molecular-scale pores are required. There is no evidence that the zeolite-encaged cluster has interesting catalytic properties; nor is it assured that the structure would be stably entrapped.

3.6. Surface Ensembles of Controlled Nuclearity

Metal clusters also provide an opportunity for control of the nuclearity of surface catalytic sites consisting of assemblies of metal complexes. Catalysts consisting of isolated mononuclear complexes $Re(CO)_3\{OMg\}\{HOMg\}_2$ were formed by the reaction of $HRe(CO)_5$ with magnesia; this catalyst was active for hydrogenation of olefins but not for hydrogenolysis of cyclopropane (Kirlin and Gates, 1987). The former reaction is structure-insensitive; it is catalyzed by mononuclear metal complexes, as shown, for example, in Figure 1. The hydrogenolysis reaction is structure-sensitive and known to be catalyzed by metal surfaces, requiring ensembles of more than one metal center for the reaction to occur. A catalyst consisting of ensembles of three of the same rhenium surface complexes and formed by the reaction of the cluster $H_3Re_3(CO)_{12}$ with magnesia was found to be active for cyclopropane hydrogenolysis as well as for olefin hydrogenation. Evidently the trirhenium cluster reacted with the surface to give catalytic sites with neighboring metal centers. These results suggest intriguing possibilities for the design of surface-bound assemblies from organometallic precursors; in prospect, ensembles of a particular nuclearity can be prepared from metal clusters of the same nuclearity, provided that the metals are strongly bonded to the support. The lesson of organometallic chemistry is then that metals to the left of the periodic table are therefore more appropriate than the group VIII metals for formation of such ensembles on metal oxide surfaces.

3.7. Supported Complexes of Zirconium and Titanium for Selective Polymerization of Olefins (Goodall, 1981; Karol, 1985)

As a final example, we address the issues of design of the intricate catalysts used industrially for polymerization of α-olefins, including the stereospecific polymerization of propylene. The choice of metals is evident from the known solution chemistry: Zirconium and titanium complexes, for example, efficiently catalyze the olefin polymerization. These group IVa metals easily form stable cationic complexes with solids (such as oxides) having hard anionic surface ligands. Surface complexes can be formed in reactions similar to that shown in Eq. (3) (Ballard, 1973):

$$\begin{array}{c} \text{Si-OH} \\ \text{Si-OH} \end{array} + \text{Zr(CH}_2\text{Ph})_4 \longrightarrow \begin{array}{c} \text{Si-O} \\ \text{Si-O} \end{array}\text{Zr}\begin{array}{c} \text{CH}_2\text{Ph} \\ \text{CH}_2\text{Ph} \end{array} + 2\ \text{CH}_3\text{Ph} \quad (4)$$

Why should the metal be supported? Because site isolation prevents deactivation by interaction of growing polymer chains bonded to neighboring metal centers and because dispersion of metal centers on a support surface makes them nearly uniformly accessible to reactants, which gives narrow molecular weight distributions in the polymer.

What is the optimum support? This is determined by the inorganic chemistry and was not obvious a priori. The supports that have been found to be most effective for titanium have the same layer lattice structure as δ-TiCl$_3$ and nearly the same cationic radii as Ti^{3+}. Titanium complexes are advantageously supported on finely ground particles of magnesium chloride, among other materials.

Ligands in addition to those provided by the support are chosen to maximize the activity and the stereoselectivity (e.g., of propylene to isotactic polypropylene rather than atactic polypropylene). The "promoter" ligands increasing the activity include aluminum alkyls for polyethylene (which play this role in solution) and electron donors such as ethyl benzoate for polypropylene. Since the stereoselectivity is inferred to be a consequence of the steric constraints offered by the solid surrounding a catalytic site, it is inferred that titanium ions with more than one site of coordinative unsaturation are unselective. Bidentate ligands such as diamines may be added to selectively poison these sites.

Olefin polymerization by these and related catalysts incorporating mononuclear Ti, Zr, and Cr complexes finds large-scale application; these are some of the best examples of high-technology catalysts, having been improved through several generations as a result of research guided to a considerable degree by the ideas of organometallic catalyst design. The details of this research and of the catalysts remain proprietary.

The supported polymerization catalysts are remarkable in being so active that they accumulate thousands of kilograms of polymer per gram of titanium without being significantly deactivated. The modern catalysts are not regenerated; they are retained in the polymer product as impurities present in such low concentrations as to have a negligible effect on the product properties.

4. SUMMARY AND PROGNOSIS

The foregoing examples range from the straightforward to the complex, and the reality of catalyst design decreases accordingly. The catalysts that were straightforwardly designed are not used in technology; the complex technological catalysts were not designed, but instead are the products of arduous research, and the important point is that the research was guided by the principles of organometallic chemistry. Increasingly, the insights provided by this chemistry will influence research, bringing us closer to the goal of rational design of industrial catalysts—which is still many years from reality.

It is emphasized that the supported molecular organometallic catalysts described here are for the most part much simpler in structure than the typical solid catalysts used in industry. This simplicity presents an opportunity; with carefully crafted catalysts of this type, we can derive fundamental understanding of structure and its influence on catalytic performance much more readily than with conventional catalysts. The resulting insights will help to advance catalysis generally, providing, for example, identification of catalytic sites and molecular-scale understanding of metal–support interactions and structure sensitivity in catalysis.

The organometallic precursors offer exciting and scarcely explored opportunities for creation of new surface structures and new catalysts. Some of these are likely to find commercial application, but it is reemphasized that the lack of stability and lack of regenerability of almost all of the known surface structures in this class pose a major challange. It would be wrong to raise hopes too high for immediate application of many supported molecular organometallic catalysts. It would be equally wrong to ignore the opportunities offered by this new class of materials; the challenge to researchers is to find ways to create novel, stable surface-bound organometallic catalysts, and organometallic chemistry will provide a platform for their design.

REFERENCES

J. R. Anderson, P. S. Elmes, R. F. Howe, and D. E. Mainwaring, *J. Catal.* **50,** 508 (1977).

D. G. H. Ballard, *Advan. Catal.* **23,** 269 (1973).

W. D. Bonds, C. H. Brubaker, Jr., E. S. Chandrasekaran, C. Gibbons, R. Grubbs, and L. C. Kroll, *J. Am. Chem. Soc.* **97,** 2128 (1975).

M. Boudart, paper presented at Heinemann Symposium, "Advances in Catalytic Chemistry," Salt Lake City, 1982.

A. Brenner, in *Metal Clusters,* M. Moskovits, Ed., John Wiley & Sons, New York, 1986.

J. R. Budge, B. F. Lücke, J. P. Scott, and B. C. Gates, *Proc. 8th Int. Congr. Catal.* **5,** 89 (1984).

J. R. Budge, B. F. Lücke, B. C. Gates, and J. Toran, *J. Catal.* **91,** 272 (1985).

R. L. Burwell, *J. Catal.* **86,** 301 (1984).

A. Choplin, M. LeConte, J. M. Basset, S. G. Shore, and W.-L. Hsu, *J. Mol. Catal.* **21,** 389 (1983).

G. Collier, D. J. Hunt, S. D. Jackson, R. B. Moyes, J. A. Pickering, and P. B. Wells, *J. Catal.* **80,** 154 (1983).

J. P. Collman and L. S. Hegedus, *Principles and Applications of Organotransition Metal Chemistry,* University Science Books, Mill Valley, California, 1980.

M. Deeba and B. C. Gates, *J. Catal.* **67,** 303 (1981).

M. Deeba, B. J. Streusand, G. L. Schrader, and B. C. Gates, *J. Catal.* **69,** 218 (1981).

F. De Thomas, T. R. Krause, H. S. Gold, and B. C. Gates, to be published.

L. D'Ornelas, A. Choplin, J. M. Basset, L.-Y. Hsu, and S. Shore, *Nouv. J. Chim.* **9,** 155 (1985).

REFERENCES

F. B. M. Duivenvoorden, D. C. Koningsberger, Y. S. Uh, and B. C. Gates, *J. Am. Chem. Soc.* **108,** 6254 (1986).

G. Ertl, in *Metal Clusters in Catalysis,* B. C. Gates, L. Guczi, and H. Knözinger, Eds., Elsevier, Amsterdam, 1986.

H. C. Foley, S. J. DeCanio, K. D. Tau, K. J. Chao, J. H. Onuferko, C. Dybowski, and B. C. Gates, *J. Am. Chem. Soc.* **105,** 3074 (1983).

D. C. Forster, *J. Am. Chem. Soc.* **98,** 846 (1976).

B. C. Gates, J. R. Katzer, and G. C. A. Schuit, *Chemistry of Catalytic Processes,* McGraw-Hill, New York, 1979.

B. L. Goodall, paper presented at international symposium, "Transition-Metal Catalyzed Polymerizations: Unsolved Problems," Michigan, 1981.

L. Guczi, in *Metal Clusters in Catalysis,* B. C. Gates, L. Guczi, and H. Knözinger, Eds., Elsevier, Amsterdam, 1986.

W. O Haag and D. D. Whitehurst, *Proc. 5th Int. Cong. Catal.* **1,** 465 (1973).

B. E. Hanson, G. W. Wagner, R. J. Davis, and E. Motell, *Inorg. Chem.* **23,** 1635 (1984).

T. C. Hayward and J. R. Shapley, *Inorg. Chem.* **21,** 3816 (1982).

L. J. Hilliard and H. S. Gold, *Appl. Spectros.* **39,** 124 (1985).

S. Iijima and M. Ichikawa, *J. Catal.* **94,** 313 (1985).

B. F. G. Johnson, Ed., *Transition Metal Clusters,* Wiley-Interscience, London, 1980.

F. Karol, paper presented at Florida Conference on Catalysis, 1985.

P. S. Kirlin, F. A. DeThomas, W. J. Bailey, H. S. Gold, C. Dybowski, and B. C. Gates, *J. Phys. Chem.,* **90,** 4882 (1986).

P. S. Kirlin and B. C. Gates, *Nature* **325,** 38 (1987).

H. Knözinger and Y. Zhao, *J. Catal.* **71,** 337 (1981).

T. R. Krause, M. E. Davies, J. Lieto, and B. C. Gates, *J. Catal.* **94,** 195 (1985).

H. H. Lamb and B. C. Gates, *J. Am. Chem. Soc.* **108,** 81 (1986).

H. H. Lamb, L. Hasselbring, R. Farlee, C. Dybowski, B. C. Gates, to be published.

W. H. Lang, A. T. Jurewicz, W. O. Haag, D. D. Whitehurst, and L. D. Rollmann, *J. Organomet. Chem.* **134,** 85 (1977).

N. J. Long, B. C. Gates, M. J. Kelley, and H. H. Lamb, in *Physics and Chemistry of Small Clusters,* Reidel, Dordrecht, in press.

E. Mantovani, N. Palladino, and A. Zanobi, *J. Mol. Catal.* **3,** 285 (1977).

L. Markó and A. Vizi-Orosz, in *Metal Clusters in Catalysis,* B. C. Gates, L. Guczi, and H. Knözinger, Eds., Elsevier, Amsterdam, 1986.

E. L. Muetterties and M. J. Krause, *Angew. Chem.* **95,** 135 (1983).

J. N. Nicholls, D. H. Farrar, P. F. Jackson, B. F. G. Johnson, and J. Lewis, *J. Chem. Soc. Dalton Trans.* **1982,** 1395.

E. O. Odebunmi, B. A. Matrana, A. K. Datye, L. F. Allard, Jr., J. Schwank, W. H. Manogue, A. Hayman, J. H. Onuferko, H. Knözinger, and B. C. Gates, *J. Catal.* **95,** 370 (1985).

C. U. Pittman and R. M. Hanes, Jr., *J. Am. Chem. Soc.* **98,** 5402 (1976).

R. L. Pruett, *Advan. Organomet. Chem.* **17,** 1 (1979).

R. Psaro and R. Ugo, in *Metal Clusters in Catalysis,* B. C. Gates,, L. Guczi, and H. Knözinger, Eds., Elsevier, Amsterdam, 1986.

R. Psaro, R. Ugo, G. M. Zanderighi, B. Besson, A. K. Smith, and J. M. Basset, *J. Organomet. Chem.* **213,** 215 (1981).

J. Schwank, L. F. Allard, M. Deeba, and B. C. Gates, *J. Catal.* **84,** 27 (1983).

T. J. Thomas and A. Brenner, *J. Mol. Catal.* **18,** 197 (1983).

K. M. Webber, B. C. Gates, and W. Drenth, *J. Catal.* **47,** 269 (1977).

K. M. Webber, B. C. Gates, and W. Drenth, *J. Mol. Catal.* **3,** 1 (1977/78).

Y. I. Yermakov, B. N. Kuznetsov, and V. A. Zakharov, *Catalysis by Supported Complexes,* Elsevier, Amsterdam, 1981.

4

SUPPORTS AND METAL–SUPPORT INTERACTIONS IN CATALYST DESIGN

A. T. BELL, *Materials and Molecular Research Division, Lawrence Berkeley Laboratory and Department of Chemical Engineering, University of California, Berkeley, California*

1. INTRODUCTION

Catalysts involving group VIII metals are usually prepared by dispersing the metal onto a high surface area support, to assure that a high proportion of the metal atoms is available for interaction with gaseous reactants. For most industrial applications, supports are sought which possess high surface area, high thermal and chemical stability, and high mechanical strength. While it was originally thought that the support was simply an inert carrier of the active component, work conducted in the late 1950s by Schwab (1978) and Solymosi (1967) revealed that significant changes in the catalytic properties of a metal could be achieved by varying support composition. Extensive evidence supporting this view has appeared over the past 7 years, engendered by an interest in the so-called strong-metal–support interactions (SMSI) reported by Tauster and coworkers at Exxon (Tauster, Fung, and Garten, 1978; Tauster and Fung, 1978; Tauster, Murrell, and Fung, 1979).

The purpose of this chapter is to review the chemistry of metal–support interactions and to illustrate the means by which these interactions might be used to design catalysts exhibiting high activity and selectivity. Following a brief review of the classification of metal–support interactions, the chemistry of catalyst preparation will be discussed, with an aim toward identifying how support composition might influence the bonding of small metal particles to the support. Next, the influence of metal–support interactions on the adsorptive and catalytic properties of group VIII metals will be examined. Finally, some thoughts will be presented concerning the opportunities for using metal–support interactions in the design of new catalysts.

2. CLASSIFICATION OF METAL–SUPPORT INTERACTIONS

For a given metal, variations in support composition can influence the size and morphology of the supported metal particles, the surface electronic properties of the particles, and the nature of the sites present at the points of contact between the metal and the support. Only the latter two of these three effects should properly be attributed to metal–support interactions, since changes in particle size and morphology can be achieved in other ways (e.g., sintering, redispersion). Thus, we might refer to changes in particle size and morphology with support composition as a nonspecific effect and refer to all other changes as specific effects of support composition.

The distinctions between specific and nonspecific effects of support composition are particularly important to recognize for structure-sensitive reactions (Bond, 1982). For such reactions, the overall changes in the activity and selectivity of a given metal upon changing support may be due only in part to the alteration of support composition, the balance being due to changes in particle size and/or morphology. To properly assess the influence of metal–support interactions for structure-sensitive reactions, it is, therefore, necessary to know a priori the degree to which specific activity and selectivity are influenced by metal dispersion. Unfortunately, though, much of the recent literature on metal–support interactions has failed to distinguish between the effects of particle size and specific supports. This has made it difficult to compare the results of different authors and to

3. NATURE AND COMPOSITION OF SUPPORTS

properly quantify the magnitude of specific support effects for various metal–support systems.

3. NATURE AND COMPOSITION OF SUPPORTS

To function as a platform for dispersed metal particles, the support must have a surface area in excess of 10 m^2/g, but greater than 100 m^2/g is preferable. While, in principle, virtually any solid material can be used as a support, in practice preference is given to metal oxides, and to a lesser extent carbon. The dominance of metal oxides is a consequence of their generally high thermal and chemical stability and the knowledge of how to prepare these materials with high surface areas. Because most of what is known about the influence of metal–support interactions on the performance of dispersed-metal catalysts comes from studies involving metal oxide supports, the balance of this section will be restricted to a discussion of the properties of these materials.

The surface of metal oxides is comprised predominantly of oxygen atoms, hydroxyl groups, and, to a lesser extent, exposed metal atoms. The chemical properties of these species and the manner in which they interact with metal-bearing precursors is strongly affected by the amount of charge localization. Oxygen anions behave as Lewis bases, metal cations behave as Lewis acids, and hydroxyl groups can act as either acids or bases. The strength and surface concentration of acidic and basic center depends strongly on the nature of the M–O bond: Acidic oxides have mainly a covalent bond, while basic oxides have an ionic bond (Krylov, 1970).

The behavior of hydroxyl groups at an oxide surface is strongly dependent on the composition of the oxide and the local chemical environment. If we describe the surface composition of an oxide as $MO_m(OH)_n$, the following situation occurs: As m increases in value, the O–H bond in the hydroxyl groups weakens and the Brønsted acid strength of these groups increases. The charge on M also influences the acid strength of the hydroxyl groups. As the charge increases, the acidity increases and the basicity falls; for example, Mn_2O_7 is an acidic oxide, but MnO is a basic oxide.

Because of the spatial inhomogeneity of oxide surfaces the strength of acidic and basic sites on such surfaces is strongly dependent on the local environment of the site, and it is not unusual to find acidic and basic sites coexisting. This point is nicely illustrated by considering the changes in the

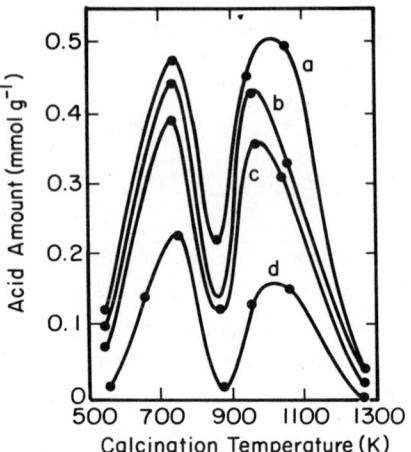

FIGURE 1. Amount of surface acidity of alumina as a function of calcination at varying H_0 values. Curve a, $H_0 < +3.3$; b, $H_0 < +1.5$, c, $H_0 < -3.0$; d, $H_0 < 5.6$. Structures: 720–870 K, η-Al_2O_3; 1070 K, η-Al_2O_3 plus θ-Al_2O_3; 1270 K, α-Al_2O_3. From Tanabe (1970).

acidity of alumina as it undergoes progressive dehydroxylation. Figure 1 shows that for calcination temperatures below 573 K, the acid strength and concentration of Al_2O_3 are low. As the temperature increases up to 773 K, the acidity increases because of the formation of an increasing number of Brønsted acid sites via the following reaction sequence:

$$\begin{array}{cc} OH & OH \\ | & | \\ -O-Al-O-Al- \end{array} \xrightarrow{-H_2O}$$

$$\begin{array}{cc} & O^{\delta-} \\ & \delta^+ \;\; | \\ -O-Al-O-Al- \end{array} \xrightarrow{+H_2O} \begin{array}{cc} H \;\; H \\ \searrow \swarrow \\ O & O^{\delta-} \\ |\delta^+ & | \\ -O-Al-O-Al- \end{array}$$

Further increasing the temperature causes a decrease in the concentration of Brønsted acid sites as a consequence of dehydration. The increase in acidity seen at temperatures above 873 K is a result of the progressive formation of Lewis acid sites created by the exposure of an increasing proportion of Al cations. Finally, the decline in acidity for calcination

3. NATURE AND COMPOSITION OF SUPPORTS

temperatures above 1073 K can be attributed to the collapse in surface area as the alumina is converted to its α form.

Figure 2 illustrates the variety of acidic and basic sites present on a γ-Al_2O_3 surface heat-treated at 1073 K (Peri, 1965). Five types of hydroxyl sites are evident, each differing in the number of nearest-neighbor O^{2-} ions (Lewis basic sites). Type A sites, which are surrounded by four O^{2-} ions, are the most negative and hence most Brønsted basic, while type C sites, which lack nearest-neighbor O^{2-} ions, are the most positive and hence most Brønsted acidic. Exposed Al^{3+} ions are also shown in Figure 2, and these sites behave as Lewis acids.

The Brønsted and Lewis acid–base properties of a support can also be altered by incorporation of a second metal atom into the framework of a host oxide, or by substitution of surface OH groups by more electrophilic groups such as Cl or F. A classic example of the influence of mixed oxide composition on acid–base properties is that of silica-alumina. Figure 3 shows that the Lewis and Brønsted acidity of silica-alumina varies with the proportion of silica. Of particular note is the very high concentration of Brønsted acid sites formed in range of 60–80% SiO_2. Examples of other

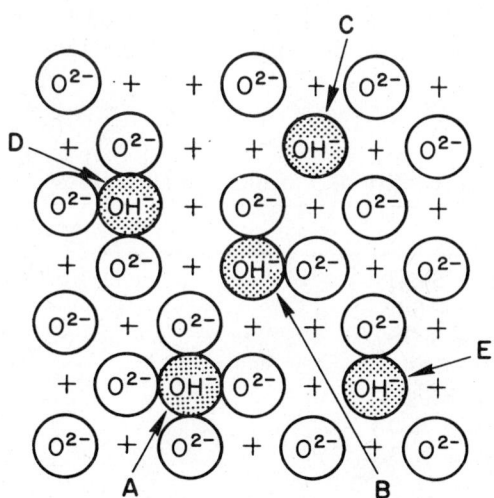

FIGURE 2. Suggested scheme for acidic and basic sites on γ-Al_2O_3. Letters A to E identify the different types of isolated hydroxyl ions, and "+" denotes an Al^{3+} ion in the layer below the surface. From Peri (1965).

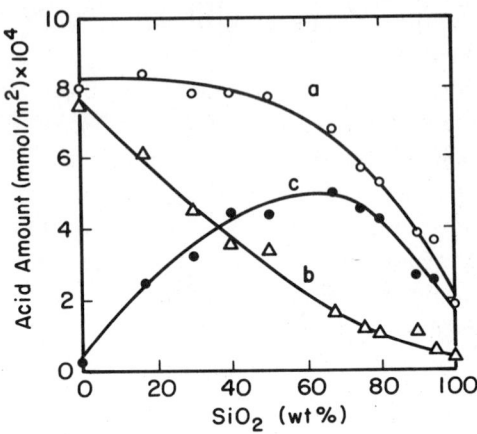

FIGURE 3. Acid amounts for SiO_2-Al_2O_3 versus proportion of SiO_2 (wt%). Curve a (○), total amount of acid $H_0 \leq +1.5$; b (△), Lewis acid amount; c (●), Brønsted acid amount (c = a − b). From Tanabe (1970).

mixed oxides are listed in Table 1, together with an indication of their maximum acid strength. The Brønsted and Lewis acidity of simple oxides can also be altered by replacing surface hydroxyl group by a highly electronegative atom or group. Thus, for example, fluorination has been found to increase the Brønsted acidity of SiO_2 and the Lewis acidity of Al_2O_3 (Boehm, 1966).

TABLE 1. Surface Acidity of Various Mixed Oxides[a]

Material	Maximum Acid Strength
SiO_2/ZrO_2 (12 mole% SiO_2)	$H_0^b \leq -8.2$
SiO_2/Ga_2O_3 (7.5 mole% SiO_2)	≤ -8.2
SiO_2/MgO (30 mole% SiO_2)	≤ -6.4
SiO_2/BeO (15 mole% SiO_2)	≤ -6.4
SiO_2/Y_2O_3 (7.5 mole% SiO_2)	≤ -5.6
SiO_2/La_2O_3 (7.5 mole% SiO_2)	≤ -5.6
Al_2O_3/B_2O_3 (15 mole% B_2O_3)	≤ -8.2
Al_2O_3/Cr_2O_3 (17.5 mole% Cr_2O_3)	≤ -8.2
TiO_2/ZrO_2 (50 mole% TiO_2)	≤ -8.2

[a] From Anderson (1975).
[b] H_0 is the Hammett acidity function.

4. PREPARATION OF DISPERSED METAL CATALYSTS

The manner in which a metal is introduced onto a support will influence its dispersion as well as the nature of the metal–support interaction. The most commonly practiced techniques are incipient-wetness impregnation with a metal salt solution and ion exchange with a metal-containing cation or anion. Considerable interest has also developed recently in the adsorption and reaction or organometallic complexes. Since detailed discussions of the chemistry of catalyst preparation have been given by Anderson (1975) and Yermakov, Kuznetsov, and Zakharov (1981), the present section will (a) focus on a summary of the interactions occurring between the metal precursor and the support and (b) comment on how these interactions affect the final properties of the reduced catalyst.

Incipient-wetness impregnation of a support with a solution of a metal salt is the easiest method of introducing a metal precursor. Drying of the solution results in the precipitation of small particles of the salt onto the support surface. Unless some care has been taken to achieve a chemical reaction between components of the salt and the support, the interaction between the deposited salt and the support will be small. As a consequence, during calcination and/or reduction of the catalyst, significant migration and sintering of the supported metal can occur, making it difficult to achieve high metal dispersion.

A much stronger interaction with the support will occur if conditions are selected to achieve an ion exchange between the metal precursor and the support. The extent to which ion exchange can be achieved is a strong function of support composition and the pH of the solution in which exchange is to be carried. As shown by reaction (1), the hydroxyl groups on the support surface can function either as a Brønsted acid or a Brønsted base, depending on the pH:

$$S^+_{(s)} \underset{-OH^-}{\overset{+OH^-}{\rightleftarrows}} S^+OH^-_{(s)} \underset{+H^+}{\overset{-H^+}{\rightleftarrows}} S^+O^{2-}_{(s)} \tag{1}$$

The propensity of an oxide to become positively or negatively charged is a function of its composition and can be indexed by the pH required to achieve zero net surface charge, as shown in Table 2. This table indicates that at a given pH, SiO_2 and TiO_2 will more readily release protons than will Al_2O_3 or MgO.

TABLE 2. Values for pH of Zero Net Surface Charge for Some Oxides in Aqueous Environment[a]

Material	pH of Zero Net Surface Charge
Silica	~2[b]
Alumina	~9
Chromia	~7
Titania	~5
Zirconia	4–7
Magnesia	~12

[a] From Anderson (1975).
[b] Charge density very low at pH < 6.

The exchange of an aqueous metal ion with a support can be described by reaction (2):

$$n\text{S}^+\text{OH}^-_{(s)} + \text{M}^{n+}_{(aq)} + n\text{OH}^- \longrightarrow [\text{S}^+\text{O}^{2-}]_n\text{M}^{n+}_{(s)} + n\text{H}_2\text{O} \qquad (2)$$

From the nature of this reaction, it is apparent that the extent of reaction will increase with increasing pH. Care must be exerted, though, that the metal concentration or pH not be so high as to precipitate metal hydroxide. With some metals this problem may be overcome by using aqueous ammonia to make the solution alkaline, in which case metal ammine ions are formed and it is these rather than aquo cations which are adsorbed. A second method exchanges the support surface with alkali metal or alkaline earth metal ions at high pH. Then, following washing to remove occluded solution, the alkali or alkaline earth ions are exchanged for another metal at near-neutral pH.

The dependence of the extent of adsorption on pH for cationic exchange of several metal ions with silica is illustrated in Figure 4. It is evident that for a given metal, the percent adsorption increases sharply with increasing pH, as would be expected from reaction (2). Figure 4 also shows that for a given pH, the percent adsorption is a strong function of metal composition. This trend can be interpreted in terms of the ease with which the aquo cation undergoes hydrolysis, reaction (3):

$$[\text{M}(\text{H}_2\text{O})_y]^{n+}_{(aq)} \underset{}{\overset{K^*}{\rightleftharpoons}} [\text{M}(\text{OH})(\text{H}_2\text{O})_{y-1}]^{(n-1)+}_{(aq)} + \text{H}^+_{(aq)} \qquad (3)$$

4. PREPARATION OF DISPERSED METAL CATALYSTS

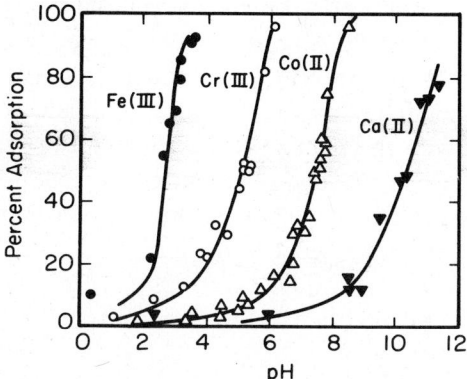

FIGURE 4. Ionic adsorption of metals from aqueous solutions of the metal nitrates onto the surface of silica gels as a function of pH. Temperature 298 K. Aqueous solutions 1–2 × 10^4 mole/dm³. From Anderson (1975).

As shown in Figure 5, the pH for 50% maximum adsorption (pH*) correlates with the pK* for reaction (3). This sort of behavior has been ascribed (Anderson, 1975) to preferential adsorption of the partly hydrolyzed cation, perhaps due to a reduction in the secondary solvation energy which facilitates close approach to the surface.

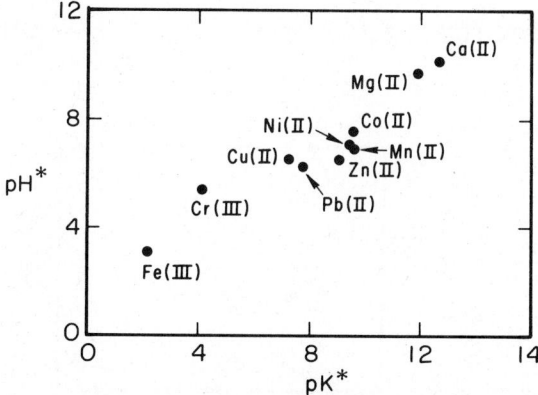

FIGURE 5. Variation of pH required for 50% of maximum adsorption (pH*) of metal ions from aqueous solution onto silica gel, as a function of the ease of hydrolysis of the aquometal cation, the latter being expressed by pK* in terms of the equilibrium constant K* for the first hydrolysis step. Temperature 298 K. Aqueous solution 10^{-3}–10^{-4} mole/dm³. From Anderson (1975).

The introduction of group VIII metals via cationic exchange of ammine complexes is well established. An example of this process is shown in Figure 6 for $[Pt(NH_3)_4]^{2+}$. It is apparent that the degree of exchange is pH dependent and occurs much more readily with SiO_2 than $\gamma\text{-}Al_2O_3$. The latter trend is consistent with the relative Brønsted acidity of the two supports.

Anionic exchange can also be used to introduce group VIII metals. This process is most frequently carried out using metal chloride complexes $[M^{n+}Cl_x^-]^{(x-n)-}$ and can be described by reaction (4):

$$(x-n)S^+OH^-_{(s)} + [M^{n+}Cl_x^-]^{(x-n)-} \longrightarrow \{(x-n)S^+[M^{n+}Cl_x^-]^{(x-n)-}\} + (x-n)OH^- \quad (4)$$

Table 3 lists the uptake of $[PtCl_6]^{2-}$ for several supports. It is significant to observe that the more basic α-alumina exhibits a significantly higher adsorption equilibrium constant than the more acidic silica.

With halide-containing metal precursors, adsorption can be accompanied by ligand displacement. Two illustrations of this phenomenon are given by reactions (5) and (6), the first of which occurs in an aqueous solution and the second occurs in a nonaqueous solution (e.g., acetone, acetonitrile).

$$S^+OH^-_{(s)} + [PtCl_6]^{2-}_{(aq)} \longrightarrow S^+[(HO)PtCl_5]^-_{(s)} + Cl^-_{(aq)} \quad (5)$$

$$2S^+OH^-_{(s)} + FeCl_3 \longrightarrow [S^+O^{2-}]_2FeCl_{(s)} + 2HCl \quad (6)$$

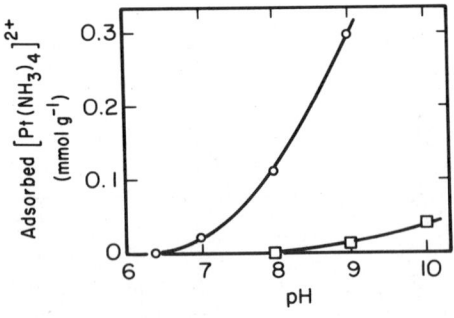

FIGURE 6. Adsorption from solution containing $[Pt(NH_3)_4]^{2+}$ (as chloride) onto SiO_2 (Davison 70, 370 m²/g) and $\gamma\text{-}Al_2O_3$ (Alcoa F-20, 204 m²/g), as a function of pH. Temperature 298 K. ○, SiO_2; □, $\gamma\text{-}Al_2O_3$. From Benesi, Curtis, and Studer (1968).

TABLE 3. Adsorption of Chloroplatinic Acid from Aqueous Solution at Room Temperature[a]

Adsorbent	Surface area (m^2 g^{-1})	Saturation Adsorption (wt% Pt, dry basis)	Adsorption Equilibrium Constant[b] (dm^3 $mole^{-1}$)
α-Alumina[c]	110	2.1	10^4
Silica-alumina[d]	364	0.26	50
Silica gel[e]	491	0	0
Activated carbon[f]	1010	22	7×10^3

[a] From Anderson (1975).
[b] Defined as $C^{-1}(\theta/1 - \theta)$, where C is the solution concentration in mole dm^{-3}, and θ is the fractional surface coverage.
[c] Alcoa, F-10.
[d] 10 mole% Al_2O_3, 90 mole% SiO_2.
[e] Davison.
[f] Columbia.

While dispersed metal catalysts traditionally have been prepared by aqueous impregnation and ion-exchange methods, there has been a growing interest in recent years in the use of organometallic complexes as precursors. This interest is stimulated by the possibilities of introducing homo- and heteropolynuclear metal clusters of fixed nuclearity and composition. In most instances, these precursors are introduced from nonaqueous solution. However, their strength of interaction with a support is governed by the acid–base properties of the support surface. Examples of the types of processes that can occur are given by reactions (7) through (9):

$$SOH + M(CO)_n \longrightarrow SO-M(CO)_n + \tfrac{1}{2}H_2 \qquad (7)$$

$$SOH + MR_n \longrightarrow SO-MR_{n-1} + HR$$
$$(R = C_3H_5, OCH_3, \text{etc.}) \qquad (8)$$

$$S^+ + M(CO)_n \longrightarrow S^+{\leftarrow}OCM(CO)_{n-1} \qquad (9)$$

Further discussions of the chemistry of organometallic complexes with supports may be found in Yermakov, Kuznetsov, and Zakharov (1981) and Chapter 3 in this volume.

The metal–support interactions produced upon contacting of the metal precursor with the support can undergo substantial change during the decomposition of the precursor and subsequent calcination and/or reduction of the catalyst. Several examples will serve to illustrate the types of reactions that can occur.

Hydrolysis of supported metal-containing cations at elevated temperatures will result in the introduction of hydroxyl anions into the coordination sphere. Whether or not this reaction causes a rupture of the S–O–M bond depends on the extent of hydrolysis and the strength of O–M bond. For example, in reaction (10),

$$[Si-O]_3MoCl_2 \xrightarrow[-HCl]{+H_2O} [Si-O]_3Mo(OH)_2 \qquad (10)$$

the M–O bond is retained, whereas in reaction (11),

$$[Si-O]_2Pt(NH_3)_4 \xrightarrow[-NH_3]{H_2O} 2[Si-OH] + Pt(NH_3)_2(OH)_2 \qquad (11)$$

the M–O bond is ruptured. Complete hydrolysis of the M–O bond is usually undesirable since it contributes to the migration of the dispersed metal along the support surface and the formation of large metal particles.

Calcination of a supported metal precursor can result in the substitution of oxygen for other ligands in the coordination sphere of the metal. As in the case of hydrolysis, calcination may lead to rupture of the S–O–M bond established during contacting of the support with the metal precursor and thereby facilitate a loss in ultimate metal dispersion. For example, reactions (12) and (13),

$$[Si-O]_3Mo(OH)_2 \xrightarrow[-H_2O]{180°C} [Si-O]_3Mo=O \qquad (12)$$

$$2[Si-O]_3Mo=O \xrightarrow[+O_2]{>300°C} \begin{array}{c} [Si-O]_3Mo\overset{O}{\underset{O}{\diagup\!\!\!\diagdown}} \\ [Si-O]_3Mo\diagdown_O \end{array} \qquad (13)$$

show the dehydration of a Mo^{5+} hydroxy complex to form the corresponding Mo^{5+} oxy complex and the subsequent transformation of that product to an Mo^{6+} surface oxide. On the other hand, calcination of a Pt–ammine

4. PREPARATION OF DISPERSED METAL CATALYSTS

complex can result in the formation of PtO_2 unattached to the support, via reaction (14):

$$[Si–O]_2Pt(NH_3)_4 \xrightarrow{+O_2} 2[Si–OH] + PtO_2 + NO + H_2O \quad (14)$$

Yet another consequence of calcination can be the dissolution of the dispersed metal ion into the support and/or formation of a mixed metal oxide. The extent to which these processes occur is strongly dependent on the composition of both the dispersed metal and the support. Thus, for Pt there is no evidence for extensive compound formation or solid solution with silica or alumina. There is extensive solid solution formation with titania and some with chromia, but no evidence of compound formation. Magnesia reacts, on the other hand, with PtO_2 to form the spinel Mg_2PtO_4. In contrast to Pt, Ni will readily react with alumina and to a lesser extent with silica to form nickel aluminate and silicate, respectively.

While dissolution of the supported metal into the support is usually undesirable, there are other situations in which this form of interaction is used to stabilize the metal to loss under reaction conditions (Yao, Gandhi, and Shelef, 1982). Thus, in the case of Ru-containing catalysts, the main objective when operating under oxidizing conditions is to prevent volatilization of the metal. For this end, one tries to provide as strong an interaction as possible by forming stable perovskite-type mixed oxides such as $BaRuO_3$ and $LaRuO_3$ (Shelef and Gandhi, 1974). In the case of Rh supported on γ-Al_2O_3, the interaction at high temperatures causes the formation of surface and subsurface spinels. To inhibit the interaction, one can substitute the γ-Al_2O_3 by a more refractory material with lesser tendency to interact, such as ZrO_2 or α-Al_2O_3 (Yao, Stepian, and Gandhi, 1980).

During reduction of the dispersed metal precursor, the bonds in the support may either remain intact or undergo cleavage depending on the composition of the metal support system and the severity of the reduction conditions. Several examples of the types of reactions observed are:

$$[S–O]_n Ti–CH_2C_6H_5 \xrightarrow{H_2} [S–O]_n TiH + CH_3C_6H_5 \quad (15)$$

$$[S–O]Pt–C_3H_5 \xrightarrow{H_2} S–OH + Pt + C_3H_8 \quad (16)$$

$$[S–O]_2Pt(NH_3)_4 \xrightarrow{H_2} 2S–OH + Pt(NH_3)_2(H_2) + 2NH_3 \quad (17)$$

$$[S–O]_2Pt(NH_3)_4 \xrightarrow{H_2} 2S–OH + Pt + 4NH_3 \quad (18)$$

From the preceding discussion it is evident that metal–support interactions can occur at various stages of catalyst preparation. The extent to which these interactions remain in the final catalyst and influence its performance is a strong function of the composition of both the metal and the support, the method of introducing the metal precursor, and the calcination and reduction conditions used to prepare the final form of the catalyst. While it is difficult to generalize, it seems reasonable to anticipate that metal–support interactions will be strongest when the metal is introduced by ion exchange or reaction of an organometallic precursor with the support surface and when the conditions of calcination and reduction are relatively mild.

Several illustrations of the types of metal–support interactions one might expect to find in the final catalyst are presented in Figure 7. In Figure 7a,

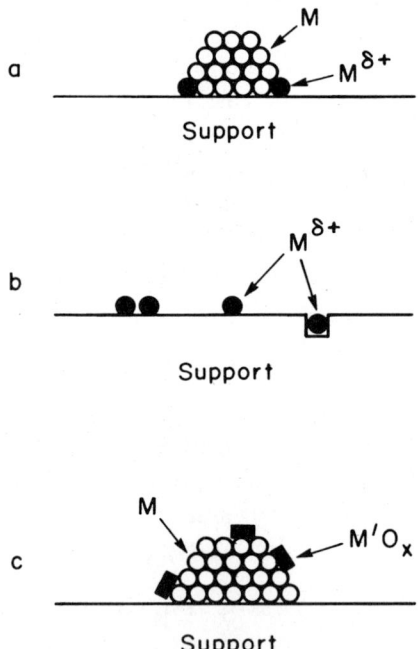

FIGURE 7. Illustration of three types of metal–support interactions. (*a*) Cationic sites of $M^{\delta+}$, located along the adlineation of a metal particle and the support. (*b*) Cationic sites formed from isolated metal atoms and atom clusters. (*c*) Decoration of metal particles by metal oxide moieties, $M'O_x$.

the interactions are limited to the points of contact between the metal particle and the support, and it is indicated that the dispersed-metal atoms at such interfaces may retain cationic character. The influence of these metal cations on the electronic properties of the metal surface atoms (and hence their adsorptive properties) would strongly depend on the size of the metal particles. For metal particles with diameters below ~15 Å the effects might be significant, whereas for larger particles the effects will be smaller because of screening of the cations by the bulk metal atoms. Figure 7b shows what could occur if the dispersed metal dissolves into the lattice or forms a mixed metal oxide. Here again, depending on the metal particle size, either all or only a small part of the particle might be affected. Yao, Gandhi, and Shelef (1982) have shown that the extent to which a given metal forms a highly dispersed phase, which interacts with the support as shown in Figure 7b, is highly dependent on the composition of the support. These authors also demonstrate that the maximum level of such a highly dispersed phase can be enhanced by incorporation of a surface modifier such as CeO_2, MoO_3, or WO_3.

Figure 7c illustrates yet a third type of metal–support interaction that may coexist with either of the preceding two types. The figure shows the decoration of the metal particle surface with oxidic moieties derived from the support. These moieties may have the stoichiometry of the support or may be partially reduced. While the process by which decoration occurs is not well understood, several mechanisms can be proposed. The first is that portions of the support dissolve during introduction of the metal precursor and are redeposited as amorphous material during solvent removal. Migration of the amorphous material and its contacting with newly formed metal or metal oxide particles might occur during subsequent steps in the catalyst preparation (i.e., calcination and reduction). A second possibility is that during catalyst reduction, a portion of the support undergoes partial reduction and consequently migrates to contact the growing metal particles. Examples of such a process are well known for supports such as titania, zirconia, and neodymia (Resasco and Haller, 1983; Santos, Phillips, and Dumesic, 1983; Ko, Hupp, and Wagner, 1984). Decoration of the supported metal particles by metal oxide moieties could lead to a modification of the electronic properties of surface metal atoms near the points of metal–metal oxide contact. Likewise, new catalytic sites might be created along the adlineation of the metal oxide patch and the metal particle surface. Evidence for both of these effects will be presented below.

5. INFLUENCE OF METAL–SUPPORT INTERACTIONS ON ADSORPTIVE PROPERTIES

The influence of metal–support interactions on the chemisorption of H_2 and CO has been investigated for virtually all of the group VIII metals. It is generally observed that an increasing degree of metal–support interaction is manifested by a decrease in H_2 and CO chemisorption capacity with increasing temperature of catalyst reduction. An illustration of this effect is shown in Figure 8 for H_2 chemisorption on Ir. It is significant to note that the strongest suppressions of H_2 chemisorption is observed for the oxides of Nb, Mn, V, Ta, and Ti, all of which can be reduced more readily than the remaining oxides listed in Figure 8 (Tauster et al., 1981). This correlation suggests that the creation of anionic vacancies in an oxide facilitates the migration of oxidic moieties onto the surface of the supported metal particles. The validity of this interpretation is supported by the work of Resasco and Haller (1983), which shows that the chemisorption capacity of Rh/TiO_2 decreases proportionally with the square root of the time of reduction, a time dependence characteristic of the surface diffusion of TiO_x species. Further substantiation of the decoration model of chemisorption suppression comes from the observation that similar effects are observed

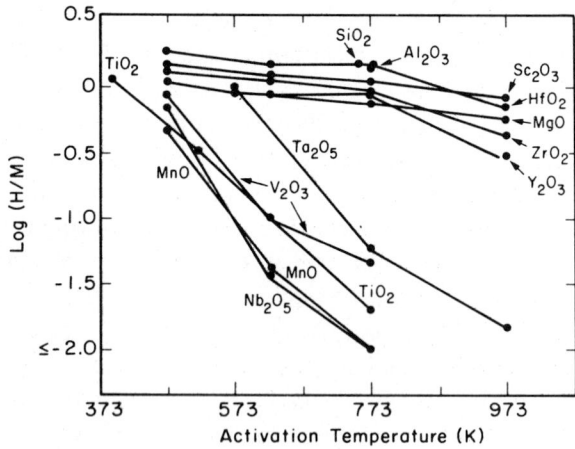

FIGURE 8. Hydrogen chemisorption on Ir supported on various oxides as a function of activation in hydrogen for 1 hr at various temperatures. H/M = atomic ratio of hydrogen adsorbed to Ir in catalyst. From Tauster et al. (1981).

5. METAL–SUPPORT INTERACTIONS ON ADSORPTIVE PROPERTIES

with small as well as large metal particles. If, for example, the modification in chemisorption behavior were due solely to interactions occurring at the point of metal–support contact, then the influence of such interactions on large particles would be expected to be smaller than that observed for small particles.

The means by which metal oxide moieties present on the surface of metal particles alter the chemisorptive properties of the the metal particles is not yet fully understood. It is obvious that the primary effect is simple site blockage, but there is now growing evidence that metal sites adjacent to metal oxide moieties might also be affected. An illustration of this is shown in Figure 9 for the case of TiO_x deposited on the surface of a Rh foil. With increasing TiO_x coverage, the CO chemisorption capacity falls off more rapidly than would be predicted by simple site blockage. Levin et al. (1986a) have demonstrated that the data can be explained by assuming that CO chemisorption is suppressed on those sites that are immediately adjacent to the TiO_x islands covering the surface of the Rh foil, in addition to those Rh sites covered by the TiO_x islands. The authors postulate that the Rh sites along the perimeter of the TiO_x islands are affected by a perturbation in the density of states at the Fermi level, along the lines described by Feibelmann and Hamann (1984,1985).

In addition to altering the chemisorptive capacity of a metal, metal oxide moieties may also alter the distribution of adsorption states. Two cases are

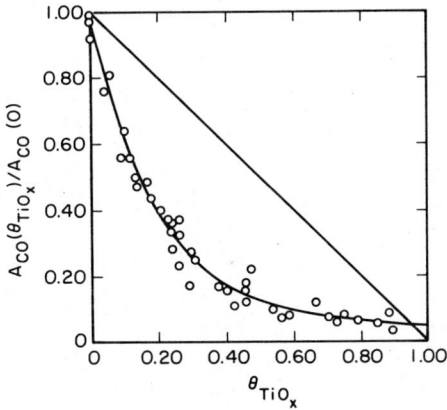

FIGURE 9. Normalized CO adsorption capacity of a Rh foil decorated with TiO_x, as a function of TiO_x coverage. Adsorption temperature 298 K. From Levin et al. (1985a).

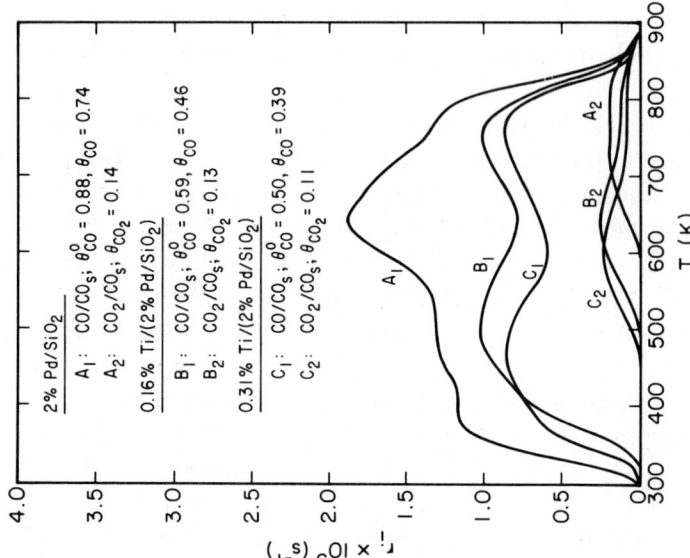

FIGURE 11. Temperature-programmed desorption spectra for CO adsorbed on Pd/SiO$_2$ and TiO$_2$-promoted Pd/SiO$_2$. From Rieck and Bell (1985c).

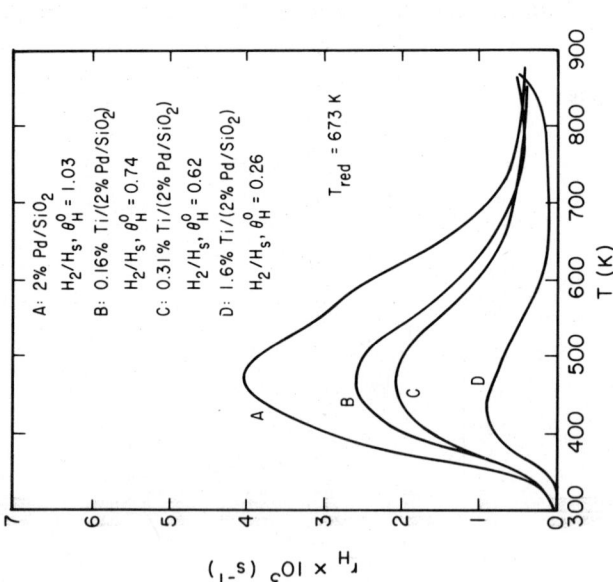

FIGURE 10. Temperature-programmed desorption spectra for H$_2$ adsorbed on Pd/SiO$_2$ and TiO$_2$-promoted Pd/SiO$_2$. From Rieck and Bell (1985c).

illustrated in Figures 10 and 11. In both instances the chemisorptive properties of Pd/SiO_2 have been altered by the promotion of the catalyst with TiO_2. It is evident from both figures that with increasing addition of TiO_2, the initial level of H_2 or CO chemisorption, given by θ_H^0 or θ_{CO}^0, decreases. While TiO_2 promotion appears to have no effect on the distribution of H_2 adsorption states, the distribution of CO adsorption states is affected significantly. The most pronounced effect is the preferential suppression of CO chemisorption on Pd(100) surfaces, characterized by the peak at 630 K. A further effect of TiO_2 promotion is the shift in the onset of CO_2 formation to lower temperatures. Since CO_2 is formed via CO dissociation and subsequent reaction of the released oxygen atom, the downscale shift in the CO_2 peak is evidence of an enhancement in the ease of CO dissociation.

6. INFLUENCE OF METAL–SUPPORT INTERACTIONS ON CATALYTIC PROPERTIES

As noted in Section 1, observation of the effects of metal–support interactions on the performance of dispersed-metal catalysts were first reported by G. M. Schwab and F. Solymosi in the late 1950s. Since much of the work done by these investigators and others in the following decade or so has been summarized (Solymosi, 1967; Schwab, 1978), only a few illustrations will be given here of this early work. The balance of this section will be devoted to an overview of more recent studies and the interpretations of the effects observed.

The first studies of metal–support effects on catalytic properties dealt with formic acid decomposition. Figure 12, taken from the work of Schwab and coworkers (Schwab et al., 1967; Schwab, Block, and Schultze, 1968) shows how the activation energy for formic acid dehydrogenation varies with the electrical conductivity and the doping of the aluminum oxide support. Similar effects reported by Solymosi and Szabó (1960) and Szabó and Solymosi (1961) are presented in Table 4. In each case, the variation in the activation energy can be attributed to changes in the conductivity of the support brought about by adding dopants containing cations of either lower or higher valence than that of the cation in the host support.

The influence of support composition on catalyst selectivity is nicely illustrated by the data for cyclohexane dehydrogenation given in Table 5. As can be seen, the benzene selectivity decreases in the order TiO_2, Al_2O_3,

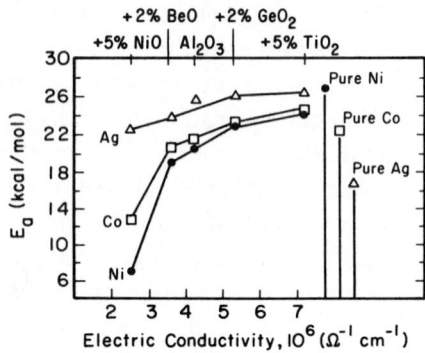

FIGURE 12. Activation energy for formic acid decomposition over Ni, Co, and Ag supported on doped Al_2O_3 carriers. From Solymosi (1967).

TABLE 4. Activation Energy of Formic Acid Decomposition on Ni/TiO$_2$ and Ni/Cr$_2$O$_3$ Contacts[a]

Catalyst	Activation Energy (kcal/mole)
Ni/TiO$_2$	22.3
Ni/TiO$_2$ + 2% Sb$_2$O$_5$	25.0
Ni/TiO$_2$ + 1% BeO	20.5
Ni/TiO$_2$ + 0.5% Al$_2$O$_3$	19.4
Ni/TiO$_2$ + 2% Cr$_2$O$_3$	17.0
Ni/TiO$_2$ + 1% NiO	15.1
Ni/TiO$_2$ + 1% NiO	17.0
Ni/TiO$_2$ + 5% NiO	13.0
Ni/Cr$_2$O$_3$	19.8
Ni/Cr$_2$O$_3$ + 0.5% TiO$_2$	19.0
Ni/Cr$_2$O$_3$ + 1% TiO$_2$	21.5
Ni/Cr$_2$O$_3$ + 2.5% TiO$_2$	23.5
Ni/Cr$_2$O$_3$ + 5% TiO$_2$	23.3
Ni/Cr$_2$O$_3$ + 0.3% NiO	19.5
Ni/Cr$_2$O$_3$ + 1% NiO	18.0
Ni/Cr$_2$O$_3$ + 3% NiO	17.5

[a] From Solymosi (1967).

TABLE 5. Selective Dehydrogenation of Cyclohexane over Pt at 773 K[a]

Catalyst	Benzene (%)	Noncondensed Products (%)	Cyclohexane (%)
Pt/ZnO	—	—	100
Pt/TiO$_2$	76.1	3.2	20.7
Pt/Al$_2$O$_3$			
pH 6	63.1	2.5	34.4
pH 8	59.8	2.8	37.4
Pt/MgO	32.3	22.0	45.7
Pt/SiO$_2$	23.1	20.1	56.8
Pt/C	55.0	14.0	31.0

[a]From Nehring and Dryer (1960).

MgO, and SiO$_2$, which corresponds to the decrease of the n-type character of the supports. It is interesting to note, however, that in the case of ZnO, the electron concentration of which is greater by several orders of magnitude than that of TiO$_2$, not only the hydrogenolysis but also the dehydrogenation reaction is suppressed. This indicates that in the case of platinum catalysts the weak n-conductor oxides are the most effective supports.

Virtually all of the more recent work on metal–support effects and their influence on catalytic properties has appeared in the past 6 years. Most of this effort has been focused on CO hydrogenation, with a smaller number of studies being devoted to NO reduction and hydrocarbon reactions. For the sake of convenience these studies will be discussed according to reaction class.

6.1. CO Hydrogenation

A summary of the studies devoted to the influence of metal–support interactions on the hydrogenation of CO over group VIII metals is presented in Table 6. The general conclusion of these investigations is that for a given metal the turnover frequency for CO conversion can be varied by more than two orders of magnitude depending on the support composition. The distribution of reaction products is also affected by support composition, and it is possible to achieve high selectivities for either hydrocarbons or oxygenated compounds on a given metal. To provide an indication of the scope and range of metal–support effects, illustrations will be presented for Pd and Rh.

The influence of support composition on the hydrogenation of CO over Pd was first reported by Ichikawa (1979) and Ryndin et al. (1981). As may be seen from the data reported in Table 7, the high activity and selectivity to methanol is achieved when Pd is supported on oxides such as La_2O_3, MgO, and ZnO, whereas high activity and selectivity to methane is achieved with supports such as TiO_2 or ZrO_2. It should be noted that the effects of support composition reported by these authors are distorted by the fact that no account was taken of the effects of metal dispersion. Recent studies by Hicks and Bell (1985), Ichikawa, Poppa, and Boudart (1985), Rieck and Bell (1985), and Kelley et al. (1985) have demonstrated that, in fact,

TABLE 6. Studies of the Effects of Metal–Support Interactions and CO Hydrogenation

Metal	Supports	References
Pt	TiO_2, ZrO_2, La_2O_3, CeO_2, TiO_2	Ichikawa (1978a,c)
	SiO_2, Al_2O_3, BeO, CaO, ZnO, MgO	Ichikawa (1978b)
	SiO_2, TiO_2, ZrO_2, WO_3, Li_2O, CaO, MgO, ZnO, CeO_2, La_2O_3, Nd_2O_3, Y_2O_3	Ichikawa and Shikakura (1980)
	SiO_2, Al_2O_3, TiO_2	Vannice, Moon, and Twu (1980)
	Al_2O_3, TiO_2	Vannice (1982)
	TiO_2	Vannice, Twu, and Moon (1983a,b)
	SiO_2, TiO_2	Vannice and Chou (1984)
Rh	SiO_2, ZrO_2, MgO, La_2O_3	Ichikawa and Shikakura (1980)
	SiO_2, Al_2O_3, TiO_2, MgO, CeO_2	Katzer et al. (1981)
	SiO_2, Al_2O_3, ZrO_2, MgO, Nb_2O_5	Iizuka, Tanaka, and Tanabe (1982)
	SiO_2, Al_2O_3, TiO_2, MgO	Solymosi, Tombacz, and Kocsis (1982)
	Al_2O_3, TiO_2, MgO	Orita, Naito, and Tamaru (1983)
	TiO_2, Nb_2O_5	Kunimori, Abe, and Uchijima (1983) Kunimori, et al. (1984)
	SiO_2, TiO_2	Haller et al. (1984)
	MgO, $Mg(OH)_2$	Poels et al. (1984)
	SiO_2, MnO, TiO_2, ZrO_2	Ichikawa, Fukushima, and Shikakura (1984)
	SiO_2, La_2O_3, Nd_2O_3, Sm_2O_3	Underwood and Bell (1986)

TABLE 6. (*Continued*)

Metal	Supports	References
Pd	SiO_2, Al_2O_3	Poutsma et al. (1978)
	SiO_2, TiO_2, ZrO_2, WO_3, Li_2O, CaO, MgO, ZnO, CeO_2, La_2O_3, Nd_2O_3, Y_2O_3	Ichikawa and Shikakura (1980)
	SiO_2, Al_2O_3, TiO_2, ZrO_2, MgO, ZnO, La_2O_3	Ryndin et al. (1981)
	SiO_2, V_2O_5, MgO, Nd_2O_3, SiO_2, MgO, La_2O_3	Poels, von Broekhaven, and van Barneveld (1981)
		Driessen et al. (1983)
	La_2O_3, CeO_2, Pr_6O_{11}, Nd_2O_3, Sm_2O_3, Eu_2O_3	Mitchell and Vannice (1984)
	SiO_2, La_2O_3	Hicks and Bell (1984, 1985)
	La_2O_3, CeO_2, Pr_6O_{11}, Nd_2O_3, Sm_2O_3, Eu_2O_3, Gd_2O_3	Sudhakar and Vannice (1985)
	La_2O_3, CeO_2, Pr_6O_{11}, Nd_2O_3, Sm_2O_3, Eu_2O_3	Rieck and Bell (1985b)
	SiO_2, TiO_2	Rieck and Bell (1985c)
Ir	SiO_2, Al_2O_3, BeO, CaO, ZnO, MgO	Ichikawa (1978b)
	Al_2O_3, TiO_2	Vannice (1982)
	SiO_2, TiO_2	Anderson et al. (1984)
Ru	Al_2O_3, Ta_2O_5, Nb_2O_5, V_2O_3	Vannice and Tauster (1979)
	Al_2O_3, TiO_2	Vannice and Garten (1980)
	Al_2O_3, MnO	Kugler, Tauster, and Fung (1980)
	Al_2O_3, TiO_2, Nb_2O_5, Ta_2O_5, V_2O_3	Kikuchi et al. (1983)
Ni	SiO_2, Al_2O_3, TiO_2	Vannice and Garten (1979)
	Al_2O_3, Ta_2O_5, Nb_2O_5	Vannice, Moon, and Twu (1980)
	Al_2O_3, Ta_2O_5, Nb_2O_5	Kugler, Tauster, and Fung (1980)
	SiO_2, Al_2O_3, TiO_2	Bartholomew, Pannell, and Butler (1980)
	SiO_2, SiO_2-Al_2O_3, TiO_2	Burch and Flambard (1981a,b; 1982a,b; 1984)
	SiO_2, Nb_2O_5	Ko, Hupp, and Wagner (1983)
Fe	Al_2O_3, TiO_2	Vannice (1982)
Co	SiO_2, Al_2O_3, TiO_2	Zowtiak and Bartholomew (1983)
	SiO_2, Al_2O_3, TiO_2	Reuel and Bartholomew (1984a,b)

TABLE 7. Activity and Selectivity of Supported and Unsupported Pd[a,b]

Catalyst	Turnover Frequency ($\times 10^3$ sec^{-1})				S(%)			
	CH_3O	CH_3OCH_3	CH_4	C_2^+	CH_3OH	CH_3OCH_3	CH_4	C_2^+
Pd black	0.60	0	0.07	0.13	75.0	0	8.8	16.2
Pd/MgO	7.70	0.097	0.02	0.01	98.4	1.2	0.3	0.2
Pd/ZnO	8.40	0	0.01	0.01	99.8	0	0.1	0.1
Pd/Al$_2$O$_3$	2.61	4.920	0.26	0.07	33.2	62.7	3.3	0.8
Pd/La$_2$O$_3$	99.10	0	0.50	0.53	99.0	0	0.5	0.5
Pd/SiO$_2$	2.33	0	0.08	0.13	91.6	0	1.5	0.2
Pd/TiO$_2$	4.20	0.819	4.00	0.50	44.1	8.6	42.1	5.2
Pd/ZrO$_2$	11.40	0.070	3.40	0.37	74.7	0.5	22.3	2.5

[a] From Ryndin et al. (1981).
[b] Reaction conditions: T : 523 K; P = 10 atm; H$_2$/CO = 3; Q = 200 cm^3 (STP)/min.

the turnover frequencies for both methane and methanol formation can be affected by metal dispersion. Hicks and Bell (1985), Rieck and Bell (1985,1986b,c), and Kelley et al. (1985) report that the specific activity for methanation increases with decreasing Pd dispersion. On the other hand, Ichikawa, Poppa, and Boudart (1985) suggest that the opposite is true. In the case of methanol synthesis, Hicks and Bell (1985) found no effect of dispersion on specific activity for dispersions between 10 and 30% but did note that the specific activity was twofold higher for Pd(100) surfaces than Pd(111) surfaces. Working over a broader dispersion range, Kelley et al. (1986) observed that the specific activity for methanol synthesis increased with decreasing dispersion, but did not comment on the influence of the morphology of the supported Pd crystallites.

In an attempt to isolate the influence of Pd dispersion, Rieck and Bell (1986a,b,c) have examined the influence of promoting a fixed Pd/SiO$_2$ catalyst with TiO$_2$ or rare earth oxides. The idea behind this work was that the specific effects of metal–support interactions on CO hydrogenation could be attributed to decoration (or promotion) of the metal particles by metal oxide moieties derived from the support during catalyst preparation and/or pretreatment. The results of their studies on methane synthesis are presented in Figures 13 and 14. It is evident that the methanation activity is substantially greater when a promoter is used. The trend in the effects of different rare earth oxide promoters observed in Figure 14 is similar to

6. METAL–SUPPORT INTERACTIONS ON CATALYTIC PROPERTIES

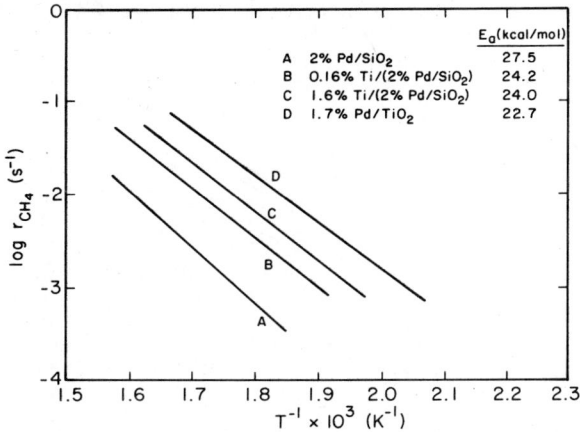

FIGURE 13. Comparison of the specific activities of Pd/SiO$_2$, TiO$_2$-promoted Pd/SiO$_2$, and Pd/TiO$_2$ for methane formation from CO and H$_2$, as functions of temperature. H$_2$/CO = 3, 1 atm. From Rieck and Bell (1985c).

that reported by Mitchell and Vannice (1984) for methanation over Pd supported on rare earth oxides.

Table 8 clearly demonstrates the importance of support composition on the specific activity of Rh for CO hydrogenation. In considering these data, it is important to recognize that no account was made in the studies cited

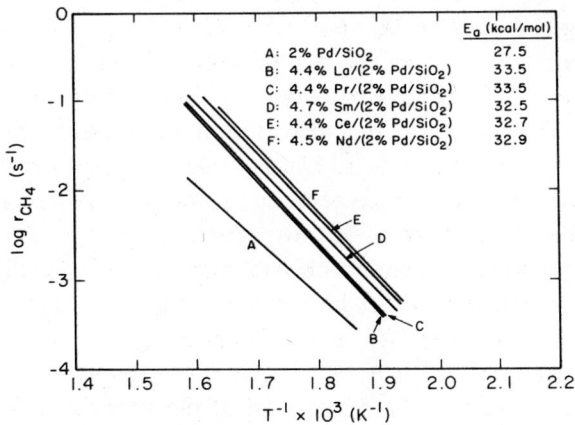

FIGURE 14. Comparison of the specific activities of Pd/SiO$_2$ and rare earth oxide-promoted Pd/SiO$_2$ for methane formation from CO and H$_2$, as functions of temperature. H$_2$/CO = 3, 1 atm. From Rieck and Bell (1985b).

TABLE 8. Activity of Rh on Various Supports for CO Hydrogenation

Support	Relative Activity[a]	References
TiO_2	100	Katzer et al. (1981)
MgO	10	
Al_2O_3	5	
CeO_2	3	
SiO_2	1	
TiO_2	100	Orita, Naito, and Tamaru (1983)
MgO	9	
Al_2O_3	5	
TiO_2	100	Haller et al. (1984)
SiO_2	6	
Nb_2O_5	100	Iizuka, Tanaka, and Tanabe (1982)
ZrO_2	2	
SiO_2	0.3	
Al_2O_3	0.2	
MgO	< 0.1	

[a]Based on surface Rh sites as determined by H_2 chemisorption.

for the effects of Rh dispersion. Recent studies by Arakawa et al. (1984) on Rh supported on SiO_2 have shown that the specific activity of Rh depends inversely on Rh dispersion. Since the investigations of support composition listed in Table 8 were not carried out with a constant Rh dispersion, only a part of the observed effect of support composition can be properly attributed to metal–support interactions.

In addition to affecting the activity of Rh, metal dispersion and support composition also influence the distribution of products formed by CO hydrogenation (Ichikawa, 1982; Underwood and Bell, 1986). Figure 15 shows that the formation of oxygenated products is favored relative to the formation of hydrocarbons as Rh dispersion decreases. The influence of support composition on product distribution is shown in Figure 16. It is apparent from this figure that high selectivities for oxygenated products can be obtained by using TiO_2 as the support.

As in the case of Pd, it appears that the influence of metal–support interactions is caused by the decoration of the supported Rh particles by metal oxide moieties. Pande, Rieck, and Bell (1986) have shown that the CO hydrogenation activity of Rh/SiO_2 can be enhanced tenfold by pro-

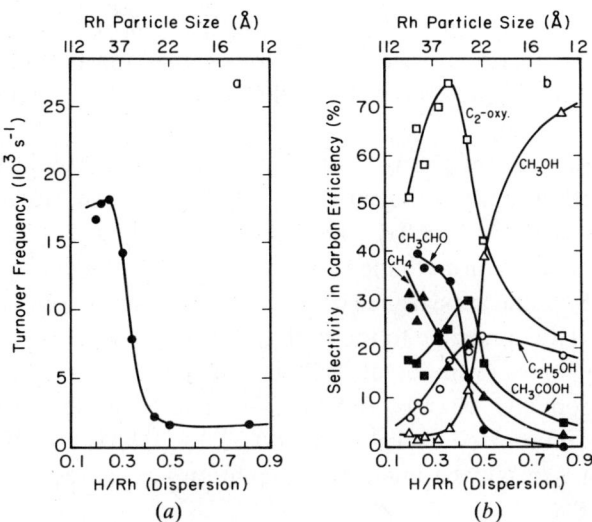

FIGURE 15. Effects of Rh dispersion on (*a*) the specific activity for CO hydrogenation and (*b*) the product selectivity. From Arakawa et al. (1984).

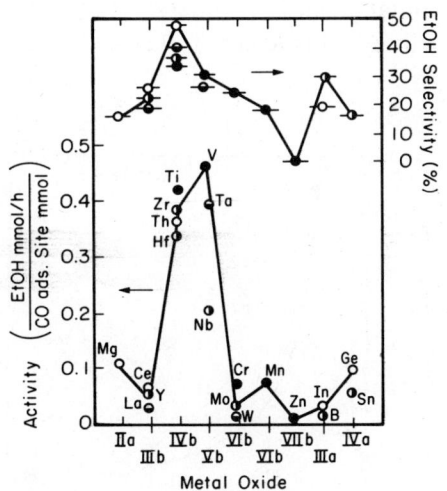

FIGURE 16. Specific activities and selectivities of ethanol over Rh-M_xO_y/SiO_2 catalysts. H_2/CO/He = 2/1/1, 1 atm, SV = 190–200 hr^{-1}, at 473 K. From Ichikawa, Tanaka, and Tanabe (1982).

motion of the catalyst with TiO_2. Electron microscopy studies by Singh, Pande, and Bell (1985) have demonstrated that for both Rh/TiO_2 and TiO_2-promoted Rh/SiO_2, a portion of the Rh particles is decorated by TiO_x species.

Further evidence of the role of TiO_x moieties on the surface Rh has been reported recently by Levin et al. (1986b). As shown in Figure 17, decoration of a Rh foil by TiO_x results in a nearly tenfold increase in the turnover frequency for methane formation at a TiO_x coverage of 15%. The authors attribute the appearance of a maximum in the curve of methanation activity with TiO_x coverage to the superposition of the activities of Rh sites located along the perimeter of TiO_x islands and Rh sites located in the spaces between TiO_x islands. Results similar to those shown in Figure 17 have also been reported by Takatani and Chung (1984) for TiO_x-promoted Ni. In the case of Ni, the highest methanation activity occurred at a TiO_x coverage of 8%.

The results presented above indicate that the specific activity and selectivity of group VIII metals for CO hydrogenation can be adjusted by careful control of metal dispersion and support composition. Low dispersion favors hydrocarbon formation, while high dispersion favors oxygenate formation. The influence of support composition is expressed through the decoration of the metal particles by metal oxide moieties, and it appears that the greatest activity enhancement may be achieved at modest levels of decoration.

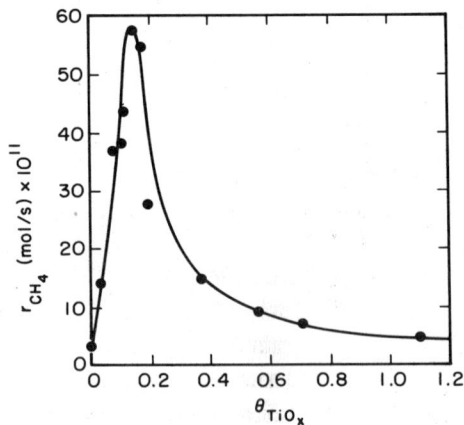

FIGURE 17. Methanation activity of Rh foil decorated with TiO_x, as a function of TiO_x coverage. $H_2/CO = 3$, 1 atm, 523 K. From Levin et al. (1986b).

6.2. NO Reduction

Only a small number of studies have been devoted to the effects of metal–support interactions on the reduction of NO over group VIII metals. Solymosi and coworkers (Solymosi and Sárkány, 1979; Solymosi, Volgyesi, and Raskó, 1980), Rives-Arnau and Munuera (1980) and Nakamura et al. (1981a,b), have shown that Rh/TiO_2 is significantly more active for NO reduction by CO than Rh/SiO_2 or Rh/Al_2O_3. An illustration of the extent to which the specific activity of Rh/TiO_2 exceeds that of Rh/SiO_2, taken from the recent work of Pande and Bell (1986a), is given in Figure 18a. This figure also shows that even higher specific activities can be achieved if the Rh/SiO_2 catalyst is promoted with TiO_2. When H_2, rather than CO, is used as the reducing agent, the effects of using TiO_2-containing catalysts is even more dramatic, as can be seen from Figure 18b. Pande and Bell (1986b) have proposed that under reaction conditions a portion of the TiO_2 in contact with the Rh particles is reduced and becomes active in the reduction of NO.

6.3. Hydrocarbon Reactions

The influence of support composition on the reactions of hydrocarbons has been investigated for three principal reactions: paraffin hydrogenolysis, olefin and aromatic hydrogenation, and paraffin dehydrogenation. Of these reactions only the first is strongly structure sensitive (Boudart and Djéga-Mariadassou, 1984). A summary of the metal–support systems investigated for specific reactions is given in Table 9.

For paraffin hydrogenolysis it is generally observed that where oxide migration from the support to the surface of the metal particles can occur, the rate of hydrogenolysis is suppressed by orders of magnitude. Figure 19 illustrates this point very dramatically for ethane hydrogenolysis over Rh/TiO_2 (Resasco and Haller, 1983). As the catalyst reduction temperature increases from 250 to 500°C, the rate of ethane hydrogenolysis falls by nearly five orders of magnitude. This pattern is very similar to the effect of adding Cu to a Ni catalyst (Sinfelt, 1983) and suggests that the decline in hydrogenolysis activity over Rh may be caused by the breakup of ensembles of neighboring Rh sites required to break C–C bonds by TiO_x species that have migrated onto the metal particle surface. Consistent with the interpretation, the activation energy for hydrogenolysis does not change with reduction temperature.

In contrast to ethane hydrogenolysis, the cyclohexane dehydrogenation

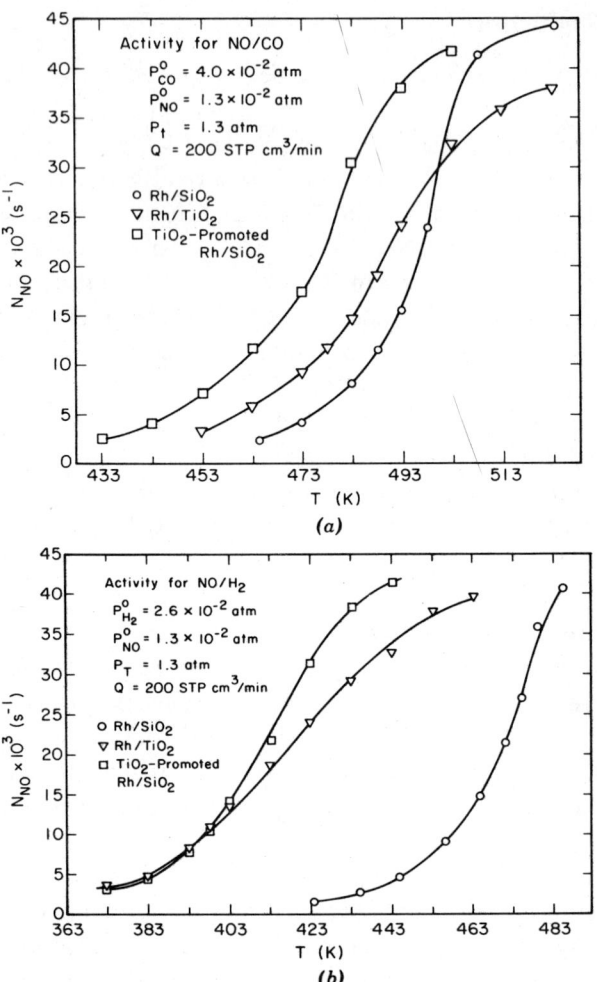

FIGURE 18. Comparison of the specific activities of Rh/SiO$_2$, TiO$_2$-promoted Rh/SiO$_2$, and Rh/TiO$_2$ for NO reduction by (a) CO and (b) H$_2$. From Pande and Bell (1986a).

over Rh/TiO$_2$ is affected relatively little by reduction temperature (Resasco and Haller, 1983). Here again, Figure 19 shows that the pattern for Rh/TiO$_2$ with reduction temperature is analogous to that for alloying Ni with Cu. The insensitivity of cyclohexane dehydrogenation to the breakup of ensembles of metal atoms is consistent with the absence of a strong structural sensitivity for this reaction. Thus, while the absolute catalyst activity

TABLE 9. Studies of the Effects of Metal–Support Interactions on Hydrocarbon Reactions

Metal	Supports	Reaction	References
Pt	SiO_2, Al_2O_3, TiO_2, CeO_2	Hydrogenolysis of paraffins	Ko and Garten (1981) Anderson (1981) Dauscher et al. (1982) Meriaudeau et al. (1982a,b)
	Al_2O_3, TiO_2	Isomerization of paraffins	Dauscher et al. (1982)
	SiO_2, Al_2O_3, TiO_2	Hydrogenation of olefins and aromatics	Meriaudeau, Pommier, and Teichner (1979) Merlaudeau, Ellestad, and Naccache (1981) Meriaudeau et al. (1982a,b)
	SiO_2, Al_2O_3, TiO_2	Hydrogenation of olefins	Briggs et al. (1980)
Rh	SiO_2, Al_2O_3, TiO_2	Hydrogenolysis of paraffins	Tauster, Murrell, and Fung (1979) Ko and Garten (1981) Haller, Resasco, and Ronco (1981) Resasco and Haller (1982, 1983)
	SiO_2, TiO_2	Hydrogenolysis of paraffins, hydrogenation of aromatics, dehydrogenation of paraffins	Meriaudeau et al. (1982b)
Pd	SiO_2, TiO_2	Hydrogenolysis of paraffins	Ko and Garten (1981) Tauster, Murrell, and Fung (1979)
	SiO_2, Al_2O_3, TiO_2	Hydrogenation of aromatics	Vannice and Chou (1984).
Ir	SiO_2, Al_2O_3, TiO_2	Hydrogenolysis of paraffins	Anderson et al. (1981) Foger (1982) Meriaudeau et al. (1982b) Anderson et al. (1984)
Ru	SiO_2, TiO_2	Hydrogenolysis of paraffins	Tauster, Murrell, and Fung (1979)
Os	SiO_2, TiO_2	Hydrogenolysis of paraffins	Tauster, Murrell, and Fung (1979) Ko and Garten (1981)

TABLE 9. (*Continued*)

Metal	Supports	Reaction	References
Ni	SiO_2, TiO_2, Nb_2O_5, CeO_2	Hydrogenolysis of paraffins	Ko and Garten (1981) Ko, Hupp, and Wagner (1981) Burch and Flambard (1981a,b; 1982a,b) Maubert et al. (1984)
Fe	SiO_2, TiO_2	Hydrogenolysis of paraffins	Ko and Garten (1981)
Co	SiO_2, TiO_2	Hydrogenolysis of paraffins	Ko and Garten (1981)

decreases with increasing reduction temperature because of the loss of Rh sites, the turnnover frequency on the remaining sites remains nearly constant.

The hydrogenation of unsaturated hydrocarbons is also structure insensitive and hence the effects of support composition on reaction rates are similar to those observed for cyclohexane dehydrogenation. Thus, for example, high-temperature reduction of Pt/TiO_2 results in a 5–20-fold reduction in the catalytic activity for hydrogenation of benzene, ethylene, and styrene, relative to that observed for low-temperature reduction (Mer-

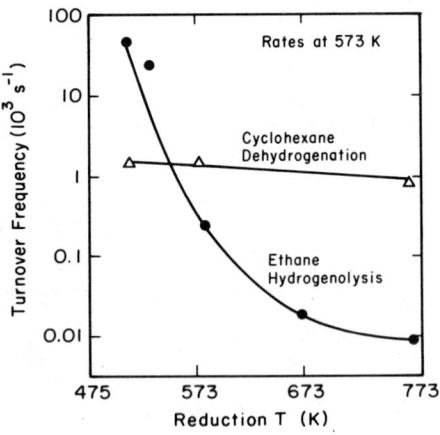

FIGURE 19. Influence of reduction temperature on the activity of Rh/TiO_2 for ethane hydrogenolysis. From Resasco and Haller (1983).

iaudeau, Pommier, and Teichner, 1979; Meriaudeau, Ellstad, and Naccache, 1981; Meriaudeau et al., 1982a,b).

7. IMPLICATIONS FOR CATALYST DESIGN

The preceding discussions amply demonstrate that metal–support interactions do influence the adsorptive and catalytic properties of group VIII metals. This is achieved through the effects of the support on metal dispersion, morphology, and oxidation state, and through the creation of unique catalytic sites at the interface between metal particles and the support. Modification of the properties of metal atoms at the surface of metal particles can also be induced by decoration of the particle with metal oxide moieties derived from the support; this effect is most pronounced for reducible metal oxides.

In designing a catalyst for a specific application consideration must be given to the minimum activity and selectivity requirements and the range of conditions under which the catalyst must operate. Since it is essential that the support possesses long-term thermal, chemical, and mechanical stability, consideration must be given to these properties in selecting a support. Once these criteria are satisfied, further selection should be based on achievement of high activity and selectivity for the desired reaction. Unfortunately no general guidelines can be given for this step, since present understanding of how and why metal–support interactions influence catalytic properties is still too limited. In the event that good support stability and desired catalyst performance cannot be achieved with a single support, consideration should be given to dispersing the catalytically active metal on a stable support and subsequently promoting the catalyst with an oxide that enhances the desired catalytic properties. An example of this idea is the automobile exhaust control catalyst, which consists of Pt, Pd, and Rh dispersed on θ-Al_2O_3 promoted with La_2O_3 and CeO_2. The θ-Al_2O_3 provides a high-surface-area support that is reasonably stable thermally. La_2O_3 is added to the θ-Al_2O_3 to delay its transformation to low-surface-area α-Al_2O_3. Promotion with CeO_2, on the other hand, enhances the water–gas shift activity of the catalyst and the ability of the catalyst to oxidize CO when the exhaust gas composition is low in O_2. The H_2 produced by the water–gas shift reaction facilitates the reduction of NO since it is a better reducing agent than CO.

In summary, exploitation of metal–support interactions holds considerable promise for the development of improved and novel supported metal catalysts. Recent research has revealed much about how such interactions work and has demonstrated their influence on a small number of reactions. It is anticipated that as further information becomes available, it will be used to design supported metal catalysts with performance characteristics superior to those available today.

REFERENCES

J. B. F. Anderson, J. D. Bracey, R. Burch, and A. R. Flambard, *Proc. 8th Int. Congr. Catal. Berlin,* **V,** 111 (1984).

J. R. Anderson, *Structure of Metallic Catalysts,* Academic Press, New York, 1975

J. R. Anderson, *Am. Chem. Soc. Div. Pet. Chem. Prepr.* **26,** 361 (1981).

H. Arakawa, K. Takeuchi, T. Matsuzaki, and Y. Sugi, *Chem. Lett.,* 1607 (1984).

C. H. Bartholomew, R. B. Pannell, and J. L. Butler, *J. Catal.* **65,** 33 (1980).

H. A. Benesi, R. M. Curtis, and H. P. Studer, *J. Catal.* **10,** 328 (1968).

H. P. Boehm, *Advan. Catal.* **16,** 179 (1966).

G. C. Bond, *Stud. Surf. Sci. Catal.* **11,** 1 (1982).

M. Boudart and G. Djéga-Mariadassou, *Kinetics of Heterogeneous Catalytic Reactions,* Princeton University Press, Princeton, N.J., 1984.

D. Briggs, J. Dewing, A. G. Burden, R. B. Moyes and P. B. Wells, *J. Catal.* **65,** 31 (1980).

R. Burch and A. R. Flambard, *J. Chem. Soc. Chem. Commun.,* 123 (1981a).

R. Burch and A. R. Flambard, *React. Kinet. Catal. Lett.* **17,** 23 (1981b).

R. Burch and A. R. Flambard, *J. Catal.* **78,** 389 (1982a).

R. Burch and A. R. Flambard, *Stud. Surf. Sci. Catal.* **11,** 193 (1982b).

R. Burch and A. R. Flambard, *J. Catal.* **85,** 16 (1984).

A. Dauscher, F. Garin, F. Luck, and G. Moire, *Stud. Surf. Sci. Catal.* **11,** 113 (1982).

J. M. Driessen, E. K. Poels, J. P. Hindermann, and V. Ponec, *J. Catal.* **82,** 26 (1983).

P. J. Feibelmann and D. R. Hamann, *Phys. Rev. Lett.* **52,** 61 (1984).

P. J. Feibelmann and D. R. Hamann, *Surface Sci.* **149,** 48 (1985).

K. Foger, *J. Catal.* **78,** 406 (1982).

G. L. Haller, D. E. Resasco, and A. J. Ronco, *Faraday Discuss.* **72,** 109 (1981).

G. L. Haller, V. E. Henrich, M. McMilland, D. E. Resasco, H. R. Sadeghi, and S. Sakellson, *Proc. 8th Int. Congr. Catal. Berlin,* **V,** 135 (1984).

R. F. Hicks and A. T. Bell, *J. Catal.* **90,** 20 (1984).

R. F. Hicks and A. T. Bell, *J. Catal.* **91,** 104 (1985).

M. Ichikawa, *J. Chem. Soc. Chem. Commun.,* 566 (1978a).

M. Ichikawa, *Bull. Chem. Soc. Jpn.* **51,** 2268 (1978b).

M. Ichikawa, *Bull. Chem. Soc. Jpn.* **51,** 2273 (1978c).

M. Ichikawa, *Shokubai* **21,** 253 (1979).

REFERENCES

M. Ichikawa, *Proceedings China–Japan–U.S. Symposium on Heterogeneous Catalysis*, Dalian, China, A08 (1982).

M. Ichikawa and K. Shikakura, *Proc. 7th Int. Cong. Catal. Tokyo* **B,** 925 (1980).

M. Ichikawa, T. Fukishima, and K. Shikakura, *Proc. 8th Int. Congr. Catal. Berlin* **II,** 69 (1984).

S. Ichikawa, H. Poppa, and M. Boudart, *J. Catal.* **91,** 1 (1985).

T. Iizuka, Y. Tanaka, and K. Tanabe, *J. Molec. Catal.* **17,** 381 (1982).

J. R. Katzer, A. W. Sleight, P. Gajardo, J. B. Michel, E. F. Gleason, and S. McMillan, *Faraday Discuss.* **72,** 121 (1981).

K. P. Kelley, T. Tatsumi, T. Uematsu, and J. H. Lunsford, *J. Catal.*, in press (1986).

E. Kikuchi, H. Nomura, M. Matsmoto, and Y. Morita, *Appl. Catal.* **7,** 1 (1983).

E. I. Ko and R. L. Garten, *J. Catal.* **68,** 233 (1981).

E. I. Ko, J. M. Hupp and N. J. Wagner, *J. Chem. Soc. Chem. Commun.,* 94 (1983).

E. I. Ko, J. M. Hupp and N. J. Wagner, *J. Catal.* **86,** 315 (1984).

O. V. Krylov, *Catalysis by Nonmetals. Rules for Catalyst Selection,* Academic Press, New York, 1970.

E. L. Kugler, S. T. Tauster, and S. C. Fung, U.S. Patents No. 4,206,135 and No. 4,206,136 assigned to Exxon Research & Engineering, 1980.

K. Kunimori, H. Abe, E. Yamaguchi, S. Matsui, and T. Uchijima, *Proc. 8th Int. Congr. Catal. Berlin* **V,** 251 (1984).

K. Kunimori, H. Abe, and T. Uchijima, *Chem. Lett.,* 1619 (1983).

M. Levin, M. Salmeron, A. T. Bell, and G. A. Somorjai, *Surface Sci.* **169,** 123 (1986a)

M. Levin, M. Salmeron, A. T. Bell, and G. A. Somorjai, *Faraday Symp. Chem. Soc.,* in press (1986b).

A. Maubert, G. A. Martin, H. Praliaud, and P. Turlier, *React. Kinet. Catal. Lett.* **22,** 183 (1984).

P. Meriadeau, O. H. Ellstad, and C. Naccache, *Stud. Surf. Sci. Catal.* **7,** 1464 (1981).

P. Meriadeau, B. Pommier, and S. J. Teichner, *C. R. Hebd. Seances Acad. Sci. Sec. C.* **298,** 395 (1979).

P. Meriadeau, J. F. Dutel, M. Dufaux, and C. Naccache, *Stud. Surf. Sci. Catal.* **11,** 95 (1982a).

P. Meriadeau, O. H. Ellestad, M. Dufaux, and C. Naccache, *J. Catal.* **75,** 243 (1982b).

M. D. Mitchell and M. A. Vannice, *Ind. Eng. Chem. Fundam.* **23,** 88 (1984).

R. Nakamura, S. Nakai, K. Sugiyama, and E. Echigoya, *Bull. Chem. Soc. Jpn.* **54,** 1950 (1981a).

R. Nakamura, K. Yamagami, S. Nishiyama, H. Niiyama, and E. Echigoya, *Chem. Lett.* (2), 275 (1981b).

D. Nehring and H. Dreyer, *Chem. Tech. (Berlin)* **12,** 343 (1960).

H. Orita, S. Naito, and K. Tamaru, *J. Chem. Soc. Chem. Commun.* **18,** 993 (1983).

N. K. Pande and A. T. Bell, *Appl. Catal.* **20,** 109 (1986a).

N. K. Pande and A. T. Bell, *J. Catal.* **97,** 137 (1986).

N. K. Pande, J. S. Rieck, A. T. Bell, in preparation (1986).

J. B. Peri, *J. Phys. Chem.* **69,** 220 (1965).

E. K. Poels, E. H. van Broekhaven, and W. A. A. van Barneveld, *React. Kinet. Catal. Lett.* **18,** 223 (1981).

E. K. Poels, P. J. Mangnus, J. V. Welzen, and V. Ponec, *Proc. 8th Int. Congr. Catal. Berlin* **II,** 59 (1984).

M. L. Poutsma, L. F. Elek, P. A. Ibarbia, A. P. Risch, and J. A. Rabo, *J. Catal.* **52,** 157 (1978).

D. E. Resasco and G. L. Haller, *Stud. Surf. Sci. Catal.* **11,** 77 (1982).

D. E. Resasco and G. L. Haller, *J. Catal.* **82,** 279 (1983).

R. C. Reuel and C. H. Bartholomew, *J. Catal.* **85,** 63 (1984a).

R. C. Reuel and C. H. Bartholomew, *J. Catal.* **85,** 78 (1984b).

J. S. Rieck and A. T. Bell, *J. Catal.* **96,** 88 (1985).

J. S. Rieck and A. T. Bell, *J. Catal.* **99,** 262 (1986a).

J. S. Rieck and A. T. Bell, *J. Catal.* **99,** 278 (1986b).

V. Rives-Arnau and G. Munuera, *Appl. Surf. Sci.* **6,** 122 (1980).

Yu. A. Ryndin, R. F. Hicks, A. T. Bell, and Yu. I. Yermakov, *J. Catal.* **70,** 287 (1981).

J. Santos, J. Phillips, and J. A. Dumesic, *J. Catal.* **81,** 147 (1983).

G. M. Schwab, *Advan. Catal.* **27,** 1 (1978).

G. M. Schwab, J. Block, and D. Schultze, *Angew. Chemie* **71,** 101 (1968).

G. M. Schwab, J. Block, W. Muller, and D. Schultze, *Naturwissenschaften* **44,** 582 (1967).

M. Shelef and H. S. Gandhi, *Platinum Met. Rev.* **18,** 2 (1974).

J. H. Sinfelt, *Bimetallic Catalysts,* John Wiley & Sons, New York (1983).

A. K. Singh, N. K. Pande, and A. T. Bell, *J. Catal.* **94,** 422 (1985).

F. Solymosi, *Catal. Rev.* **1,** 233 (1967).

F. Solymosi and J. Sárkány, *Appl. Surf. Sci.* **3,** 68 (1979).

F. Solymosi and Z. G. Szabó, *Magy. Kem. Foly.* **66,** 289 (1960).

F. Solymosi, I. Tombacz, and M. Kocsis, *J. Catal.* **75,** 78 (1982).

F. Solymosi, L. Volgyesi, and J. Raskó, *Z. Phys. Chem. N.F.* **120,** 79 (1980).

C. Sudhakar and M. A. Vannice, *J. Catal.* **95,** 227 (1985).

Z. G. Szabó and F. Solymosi, *Actes Congr. Intern. Catalyses, 2^e, Paris, 1960,* 1627 (1961).

S. Takatani and Y. W. Chung, *J. Catal.* **94,** 75 (1984).

K. Tanabe, *Solid Acids and Bases,* Kodansha, Tokyo, and Academic Press, New York (1970).

S. J. Tauster and S. C. Fung, *J. Catal.* **55,** 29 (1978).

S. J. Tauster, S. C. Fung, and R. L. Garten, *J. Am. Chem. Soc.* **100,** 170 (1978).

S. J. Tauster, L. L. Murrell, and S. C. Fung, U.S. Patent No. 4,149,998, assigned to Exxon Research and Engineering Co. (1979).

S. J. Tauster, S. C. Fung, R. T. K. Baker, and J. A. Horsley, *Science* **221,** 1121 (1981).

R. Underwood and A. T. Bell, *Appl. Catal.* **21,** 157 (1986).

M. A. Vannice, *J. Catal.* **74,** 199 (1982).

M. A. Vannice and P. Chou, *Proc. 8th Int. Congr. Catal. Berlin,* **V,** 99 (1984).

M. A. Vannice and R. L. Garten, *J. Catal.* **56,** 236 (1979).

M. A. Vannice and R. L. Garten, *J. Catal.* **63,** 255 (1980).

M. A. Vannice and C. Sudhakar, *J. Phys. Chem.* **88,** 2429 (1984).

M. A. Vannice, S. H. Moon, and C. C. Twu, *Am. Chem. Soc. Div. Pet. Chem. Prepr.* **25,** 303 (1980).

REFERENCES

M. A. Vannice and S. S. Tauster, U. S. Patent No. 4,171,320, assigned to Exxon Research and Engineering Co., 1979.

M. A. Vannice, C. C. Twu, and S. H. Moon, *J. Catal.* **79,** 70 (1983a).

M. A. Vannice, C. C. Twu, and S. H. Moon, *J. Catal.* **82,** 213 (1983b).

H. C. Yao, H. S. Gandhi, and M. Shelef, *Stud. Surf. Sci. Catal.* **11,** 159 (1982).

H. C. Yao, H. K. Stepien, and H. S. Gandhi, *J. Catal.* **61,** 547 (1980).

Yu. I. Yermakov, B. N. Kuznetsov, and V. A. Zakharov, *Catalysis by Supported Metal Complexes,* Elsevier, New York, 1981.

J. M. Zowtiak and C. H. Bartholomew, *J. Catal.* **83,** 107 (1983).

5

KINETICS-ASSISTED DESIGN OF CATALYTIC CYCLES

M. BOUDART, *Department of Chemical Engineering, Stanford University, Stanford, California*

1. INTRODUCTION: CHEMICAL KINETICS AND MOLECULAR ENGINEERING

When we think of a catalyst, some of us think of a chunk of metal; others think of a solvated proton or of a soluble organometallic complex. For the sake of simplicity in understanding how a simple catalytic cycle can be designed by kinetics and tuned by molecular engineering, let us choose the simplest case where the catalyst is an atom X, say chlorine, and the catalyzed reaction is one that takes place in the stratosphere, resulting in the destruction of ozone O_3 by reactive collision between O_3 and oxygen atoms O to yield two molecules of dioxygen O_2:*

$$O + O_3 \longrightarrow O_2 + O_2 \qquad [0]$$

Reaction [0] proceeds as an elementary step, as written. But it can proceed faster as a result of a catalytic cycle consisting of two elementary

---

*Two types of reaction arrows are used in this chapter: → is used to denote elementary steps; ⇒ is used to denote irreversible overall reaction.

irreversible steps using up and regenerating, respectively, a catalytic atom X such as chlorine:

$$X + O_3 \longrightarrow O_2 + XO \qquad [1]$$

$$XO + O \longrightarrow O_2 + X \qquad [2]$$

When we sum up steps [1] and [2] side by side, we obtain the overall reaction [0]. The role of chlorine and fluorine atoms in accelerating the destruction of stratospheric ozone by the above catalytic cycle was first pointed out by Molina and Rowland (1974). Such atoms are coming from aerosol propellants and refrigerator cycle fluids.

A good way to depict any catalytic cycle is through a diagram (Figure 1) that shows clearly the catalytic entities X and XO at the nodes of the diagram as well as reactants O_3 and O and products O_2 and O_2, entering and leaving the cycle, respectively. These diagrams were first introduced into the catalytic literature by Tolman (1972).

The first question to ask about any catalytic cycle such as the one depicted in Figure 1 is: How fast does it turnover? The turnover rate or frequency v_t can be obtained readily by making two observations. First, at the kinetic steady state, the rates v_1 and v_2 of steps [1] and [2] must be equal and are also equal to the rate v_0 of the overall reaction [0]. With appropriate rate constants k_1 and k_2 and concentrations of species denoted by square brackets, we get:

$$v_0 = v_1 = k_1[X][O_3] = k_2[XO][O] = v_2 \qquad (1)$$

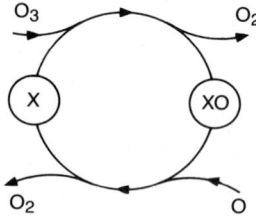

FIGURE 1. A prototype of a catalytic cycle. X is the catalyst for the reaction:

$$O + O_3 \longrightarrow O_2 + O_2$$

1. INTRODUCTION

Second, let us note that the catalytic entity alternates between its two forms X and XO but that, if the catalyst stays alive, its total concentration remains a constant denoted by [L]:

$$[X] + [XO] = [L] \tag{2}$$

From (1) and (2) we get:

$$v_0 = [L] \frac{k_1 k_2 [O_3][O]}{k_1 [O_3] + k_2 [O]} \tag{3}$$

But the turnover rate v_t, that is, the number of times that the cycle turns over per second, is given by

$$v_t = \frac{v_0}{[L]} \tag{4}$$

since this frequency is the number of overall acts of reaction taking place per second per catalytic entity. Hence from (3) and (4):

$$v_t = \frac{k_1 k_2 [O_3][O]}{k_1 [O_3] + k_2 [O]} \tag{5}$$

The second question to ask about a catalytic cycle is: What is, at a given temperature, its maximum turnover rate? This is a much more difficult question than the first one, which can be answered by direct measurement of the rate, without an a priori knowledge of the mechanism of the cycle or of the rate constants k_1 and k_2. Moreover, there exists no general answer to the question of maximum rate. The type of information that is needed to answer the question can be appreciated in the case of the cycle shown in Figure 1. A number of nonessential assumptions will be made to illustrate the main concepts in the most simple way.

Clearly, what is needed to proceed is information on rate constants k_1 and k_2. Generally, they should be of the Arrhenius form:

$$k_1 = A \exp\left(-\frac{E_1}{RT}\right), \quad k_2 = A \exp\left(-\frac{E_2}{RT}\right) \tag{6}$$

where the preexponential factors A have been taken to be the same in good first approximation of theoretical expectations (Boudart, 1968) while the activation energies E_1 and E_2 are normally expected to be different. What can we say about E_1 and E_2? For the sake of discussion, let us assume that they can be expressed in terms of a relation first proposed by Semenov (1954) for reactions between free radicals and molecules, such as steps [1] and [2]:

$$E = \text{const} - \frac{q}{4} \qquad (7)$$

where q, a positive quantity, is the heat of the elementary reaction in the exothermic direction. Note that if the elementary reaction were endothermic, its activation energy E' would be given by E of (7) *plus* the value of q. Thus $E' > E$ and we conclude already that to maximize k_1 and k_2, and hence the turnover rate, the two steps of the cycle should be exothermic, if that can be done, so that activation energies E_1 and E_2 are smaller than they would be otherwise. It is assumed that the overall reaction [0] is exothermic.

To find out whether this is possible, let us note that we can obtain the values of q from values of bond dissociation energy D which are frequently accessible in the case of simple molecules.

The dissociation energy of bond A–B is denoted by $D(A–B)$. By inspection of steps [1] and [2], we see that

$$q_1 = D(X–O) - D(O–O_2) \qquad (8)$$
$$q_2 = D(O–O) - D(X–O) \qquad (9)$$

Thus, in our kinetic design of the catalytic cycle, since we should first look for a catalyst X giving positive values of q_1 and q_2, the value of $D(X–O)$ should be such that

$$D(O–O_2) < D(X–O) < D(O–O)$$

Fortunately, this is possible, as thermochemical tables (Prigogine and Defay, 1954) give

$$D(O–O_2) = 25 \text{ kcal/mol}$$
$$D(O–O) = 118 \text{ kcal/mol}$$

1. INTRODUCTION

We can go one step further in the kinetic design and find the optimum value of $D(X–O)$ that corresponds to maximum rate of turnover. As an unessential simplification, assume that $[O] = [O_3]$. The (5) and (6) yield

$$v_t = A[O_3] \frac{\exp(-E_1/RT) \times \exp(-E_2/RT)}{\exp(-E_1/RT) + \exp(-E_2/RT)} \quad (10)$$

If we now substitute in (10) the values of E_1 and E_2 given by Semenov (7) with values of q_1 and q_2 of (8) and (9), we obtain an expression for v_t as a function of the variable $D(X–O)$:

$$v_t = v_t[D(X–O)] \quad (11)$$

Next the derivative of v_t is set equal to zero and the result is a very simple one: For maximizing the turnover rate, $D(X–O)$ must be exactly intermediate in value between $D(O–O_2)$ and $D(O–O)$:

$$\text{For } (v_t)_{\max}: D(X–O) = \tfrac{1}{2}[D(O–O_2) + D(O–O)] \quad (12)$$

It follows from (12) to (7) that for the optimum catalyst, $E_1 = E_2$ and $q_1 = q_2$. Also, from (1), with $k_1 = k_2$ since $E_1 = E_2$, and $[O_3] = [O]$ by assumption, we see that $[X] = [XO]$. In words, the best catalytic cycle is one consisting of two easy steps with the same activation energy and the same heat of reaction, the catalyst being half free and half occupied. The catalyst binds the reactive intermediate strongly, but not too strongly, as Sabatier expressed it in the early days of industrial catalysis, in which he took an active part and for which he received the Nobel Prize (Boudart, 1968). By the way, the reader should refrain from using (7) with the same constants to calculate the activation energy for the uncatalyzed reaction. Indeed the Semenov equation does not apply to it, as can be inferred from the observation that the reverse of that reaction is one between two molecules while the Semenov equation applies to reactions between free radicals and molecules in both directions.

There is a third question to ask about a catalytic cycle once there is evidence for the existence of a cycle with maximum turnover rate: How can it be improved? It applies whether or not one knows that an optimum catalyst exists and what the requirements are for an optimum catalyst. The strategy for improved design is clear from the illustration just discussed concerning the optimum catalyst. One should change the catalyst in a systematic and rational manner in order to change the binding energy of the catalytic intermediates. In the above discussion, the question is: How can we change $D(X-O)$, not in an Edisonian manner, but by stepwise strategy with checks at every step? This strategy can be called the molecular engineering of catalytic materials.

The time-honored method of changing $D(X-O)$ is to change X by going through the periodic table across in a series or down in a group. This is by far the most powerful way to change bond energies in a substantial way.

The second, more refined method consists of modifying the catalyst by surrounding it with ligands or with different ligands. Thus, in the case of the ozone destruction, instead of a halogen atom, one could take a nitrogen atom. This would be a disaster from a catalytic standpoint because the value of $D(N-O)$ is far too great in violation of the principle of Sabatier. But by adding a ligand to N, for example an oxygen atom, one could obtain a good catalyst, namely a molecule of nitric oxide NO. Indeed, the catalytic role of NO from supersonic aircraft was first suggested by Johnston (1971) as a cause of the destruction of stratospheric ozone, before the role of halogens became an added concern. In both cases, possible adverse health effects are expected because of an increase in ultraviolet (UV) radiation at the Earth's surface due to the depletion of the ozone layer, which is an effective UV screen.

The third method of modifying the catalyst is the most subtle one for the ultimate fine tuning of the catalytic cycle. It consists of changing the structure of the catalytic entity. For instance, if X is a transition metal, it would be stabilized by ligands in a square planar or an octahedral structure. If X were part of a solid surface, the change in structure would correspond to a change in crystalline plane exposed or in crystal size in the case of clusters between 1 and 5 nm in diameter. Details on the molecular engineering of solid metallic catalysts can be found elsewhere (Boudart and Djéga-Mariadassou, 1984).

These questions will be dealt with again following this section, which merely introduced the notion of catalytic cycles, their rate of turnover, their maximum rate of turnover, and the molecular engineering that can be done to improve the catalyst by purposeful changes in the nature of the catalytic entity, its coordination to ligands, and its structure. In real, complicated situations where experimental values of $D(X-O)$ are not available, theoretical estimates are just becoming available. Even if they are only approximate, they help in the kinetic design of certain catalytic cycles

2. KINETIC COUPLING

(Goddard, 1985) together with the stepwise molecular engineering of the catalyst.

The purpose of the present chapter is not to present a retrospective of kinetics-assisted design of catalytic cycles. Rather, its aim is to sketch a perspective of the subject in order to understand what is expected when molecular engineering will use theoretical and experimental chemical kinetics to design catalytic cycles of reasonable complexity.

2. KINETIC COUPLING

Since there is much more kinetic information available on gas-phase elementary reactions involving free radicals than on any type of catalytic reaction, it is interesting to start discussing the turnover of catalytic cycles consisting of the propagation steps of chain reactions. The first condition for turnover of the cycle is that *all* elementary steps in it be driven by a positive affinity $A \equiv -\Delta G$ where G is the Gibbs free energy. This condition, which can be proved readily (Boudart, 1983), may seem evident, but it has been argued lately that it was too stringent on the grounds that a thermodynamically unfavored step ($A < 0$) might be driven by *thermodynamic coupling* with another one with a sufficiently large positive value of A so that overall affinity of the cycle is positive. In fact, an unfavorable step is one with a negative value for its *standard* affinity A^0. An effective device to drive an unfavorable step ($A^0 < 0$) is not thermodynamic coupling but *kinetic coupling*. The latter method operates through the interplay of rate constants of successive steps in such a way that the active reactant of the unfavorable step can accumulate above its equilibrium concentration or the product of the unfavorable step can be removed below its equilibrium concentration. The latter case is illustrated by the catalytic cycle in the H_2–Br_2 reaction. Let us examine this case briefly (Boudart, 1983).

The rate of that reaction is inhibited by its product despite the fact that the overall reaction is not limited by equilibrium but proceeds to completion. The inhibition by HBr is due to the fact that HBr is produced in an unfavorable step. Thanks to the beneficial effect of the kinetic coupling, that step does not proceed near equilibrium. In fact, even at 50% conversion for the overall reaction, the unfavorable step proceeds 11 times faster in the forward direction than in the reverse direction. This indicates a sizable positive affinity which is the result of the kinetic coupling.

Yet the *standard* affinity of the step which is reversible with a rate inhibited by product HBr is clearly negative, as shown by thermodynamic data:

$$Br + H_2 \underset{k_{-1}}{\overset{k_1}{\rightleftarrows}} HBr + H$$

$$A_0 = -15.7 \text{ kcal/mol at 298 K}$$

However, the next step in the closed chain (catalytic sequence),

$$H + Br_2 \overset{k_2}{\longrightarrow} HBr + Br$$

has a rate constant k_2 that is 10 times larger than k_{-1} over a certain temperature range. It can be shown (Boudart, 1983) that the measure of irreversibility of the above reversible step, namely the ratio of forward and reverse rates, is given by

$$\frac{\vec{v}_1}{\overleftarrow{v}_1} = 1 + \frac{k_2}{k_{-1}} = \frac{[H]_e}{[H]_{ss}}$$

for an originally equimolar H_2–Br_2 mixture at half conversion to HBr. The concentration of H at the kinetic steady state is denoted as $[H]_{ss}$. The concentration of H in equilibrium with the molecular species in the system, namely H_2, Br_2, and HBr, is denoted as $[H]_e$. Thus it is seen that the second propagation step pumps down the steady-state concentration of H 11 times below its equilibrium value. As a result, from the De Donder relation (Boudart, 1986a),

$$\frac{\vec{v}_1}{\overleftarrow{v}_1} = \exp\left(\frac{A_1}{RT}\right)$$

we obtain a value for the steady-state affinity (at 298 K) of the unfavorable step equal to $+1.43$ kcal/mole while the standard affinity is -15.7 kcal/mol.

Generally, a first consequence of kinetic coupling is that if a reaction product inhibits the rate of turnover of a catalytic cycle, which is run far away from equilibrium, the inference is that this reaction product appears as a product of a thermodynamically unfavorable step. A second consequence is that the suspected presence in a catalytic cycle of an unfavorable

2. KINETIC COUPLING

step does not preclude that the cycle will turnover at an acceptable rate, provided that the unfavorable situation is corrected by kinetic coupling of steps in the cycle.

A second example of kinetic coupling of catalytic cycles deals with the decomposition of ammonia at low pressure and high temperature on a molybdenum foil (Boudart et al., 1981). In this case, the rate of reaction is not inhibited by either product, dinitrogen or dihydrogen. In the case of dihydrogen, this is not surprising, because surface coverage by hydrogen at reaction conditions is very low, so that even if surface hydrogen is in equilibrium with dihydrogen in the gas phase, the competition by hydrogen for sites, and therefore inhibition of the rate, is negligible. But in the case of nitrogen, the situation is quite different. Monitoring of the surface by Auger electron spectroscopy (AES) during the reaction at the steady state shows high concentrations of surface nitrogen $[*N]_{ss}$. The reason why the desorption of nitrogen is not proceeding near equilibrium with affinity almost equal to zero, but far away from equilibrium with a positive affinity, is that $[*N]_{ss}$ is considerably higher than $[*N]_e$, the concentration of surface nitrogen as measured separately by AES in equilibrium with dinitrogen at the pressure corresponding to that in the steady-state reaction. The reason why $[*N]_{ss}$ is substantially larger than $[*N]_e$ is kinetic coupling between the adsorption step of ammonia and the desorption step of nitrogen. As a result, desorption is essentially irreversible and proceeds with a high positive affinity.

Another way to describe the effect of kinetic coupling is that it provides a *virtual* pressure of dinitrogen $[N_2]_e$ which is that equilibrium pressure that would be required to achieve the value of $[*N]_{ss}$ during the steady-state reaction. The value of $[N_2]_v$ is substantially higher than $[N_2]_{ss}$, the actual pressure of dinitrogen during the reaction. Why this is so will be discussed again in Section 3.

Application of the De Donder relation to the nitrogen desorption step yields a result resembling that reported above for the H_2–Br_2 reaction:

$$\frac{\vec{v}}{\overleftarrow{v}} = \exp\left(\frac{A}{RT}\right) = \frac{[N_2]_v}{[N_2]_{ss}}$$

At 1000 K and a pressure of ammonia of 1.1×10^{-4} Pa, the ratio of the measured values of $[N_2]_v$ and $[N_2]_{ss}$ is 1.4×10^2, corresponding to almost completely irreversible desorption with a large positive steady-state affinity equal to 10 kcal/mol.

The situation of catalytic ammonia decomposition at high temperature and low pressure can be generalized. Whenever a reaction product does not inhibit the rate, this product does not compete for catalytic sites or it is taking place far away from equilibrium as an irreversible step with positive affinity and the corresponding surface species is a major surface intermediate.

By contrast, if ammonia decomposition is carried out at lower temperatures and higher pressures than in the previous case, the rate of synthesis even far away from overall equilibrium is strongly inhibited by dihydrogen, one of the reaction products. This is due to a preequilibrium between surface nitrogen on the one hand, and gas-phase ammonia and dihydrogen on the other hand. But again dinitrogen is not an inhibitor of the rate, because its desorption step once more takes place far from equilibrium, as ammonia provides a very large virtual pressure of dinitrogen $[N_2]_v$ with the consequence that $[*N]_{ss} \gg [*N]_e$. As before, $[N_2]_v$ is much larger than $[N_2]_{ss}$. Details are found elsewhere (Boudart, 1986a).

There are reasons to believe that these examples of kinetic coupling are far from isolated. Indeed, a general feature of catalytic cycles is that the last step is the desorption of a product molecule from the catalyst surface or the reductive elimination of a product from an organometallic complex. Except for reactions that are running very far away from equilibrium, the normal situation is that the desorption–elimination step should be endothermic just as it is normal to expect that the adsorption–addition step should be exothermic. Since the entry step is generally associated with a loss of entropy while the exit step proceeds with a gain of entropy, the balance between enthalpy and entropy factors suggests that entry and exit steps may frequently run near or very close to equilibrium. Indeed, this is the familiar picture of catalytic kinetics following Langmuir–Hinshelwood in heterogeneous catalysis, Michaelis–Menten in enzyme catalysis, and Hougen–Watson in chemical engineering kinetics. According to that picture, there is a rate-determining step involving catalyst-bound species, sandwiched in between various quasiequilibrated steps involving reactants, products, or other competitors on the one hand and catalytic entities on the other hand.

The Gibbs free energy drop for the overall reaction resides at the rate-determining step. This conventional picture is quite different from that discussed above in connection with kinetic coupling. There are many examples (Boudart and Djéga-Mariadassou, 1984) where both entry and exit steps are essentially irreversible and kinetically coupled in such a way that

3. COOPERATIVITY 151

the catalyst-bound species that decomposes in the exit step exists at steady-state concentrations far exceeding equilibrium concentrations. Thus, the corresponding reaction product does not inhibit the rate of the catalytic cycle. Product inhibition can be so severe as to preclude the use of a catalytic material. Thus, ammonia inhibition in ammonia synthesis on rhenium (Asscher and Somorjai, 1984) and molybdenum (Volpe and Boudart, 1986) prevents these active metals from being considered as practical industrial catalysts.

3. COOPERATIVITY

Catalytic cycles can turn over at acceptable rates even if they contain steps that would be ruled out because of a negative affinity in the absence of kinetic coupling. Kinetic coupling is even more effective when assisted by an effect which we shall call cooperativity. This effect is particularly well documented in the case of heterogeneous catalysis, although there is no reason to believe that it is limited to that type of catalysis. Let us return to the case of low-pressure and high-temperature decomposition of ammonia on a molybdenum foil. Adsorption isotherms of nitrogen obtained by AES yield isosteric heats of adsorption, q, which decrease enormously and quasilinearly with increasing $[*N]_e$. It follows that the catalytic cycle of ammonia decomposition on a clean surface of molybdenum cannot turn over, since the nitrogen desorption step is prohibitively endothermic. This is borne out by non-steady-state experiments. These reveal that the *logarithm* of the rate of adsorption of ammonia decreases linearly as $[*N]$ increases, as the *logarithm* of the rate of desorption of nitrogen increases linearly with $[*N]$, until finally both rates become equal at the steady state of the overall reaction. Thus a small change in $[*N]$ corresponds to a large change in rates. This is all because of the large variation of q with respect to $[*N]$, which is now well understood and is found to be a result of two factors: structure sensitivity and lateral repulsive interactions between adspecies at nearest-neighbor distances. Both effects lead to cooperativity in the sense that a catalyst without is would perform at a miserable turnover rate while a catalyst with it adjusts itself to a steady-state value of surface coverage corresponding to acceptable turnover rates. Because of the empirical observation that rates of adsorption and desorption steps frequently depend exponentially on surface coverage, the leverage of kinetic coupling is much more effective than it would be without cooperativity. Another way to describe the situation is to state that on a surface where q decreases linearly with fractional coverage, the equilibrium value of the latter is proportional to the *logarithm* of the pressure of the gas in equilibrium with the surface. Hence, even a relatively small variation in surface concentration brought about by kinetic coupling can be considered as substantial, because of its logarithmic effect on rates. In this way, cooperativity enhances the effect of kinetic coupling.

As an illustration, consider a two-step catalytic cycle, say the water–gas shift reaction taking place on catalyst sites S in two irreversible steps:

$$S + H_2O \xrightarrow{k_{in}} SO + H_2$$
$$SO + CO \xrightarrow{k_{out}} S + CO_2$$

Assume that the concentrations of H_2O and CO are unity, $k_{in} = 10^4 \text{ s}^{-1}$ and $k_{out} = 10^{-4} \text{ s}^{-1}$. Then practically all of the sites are in the SO form and the turnover rate v_t of the cycle, equal to the rate per site of the second step, is simply equal to 10^{-4} s^{-1}, a very low value (Burwell and Boudart, 1974). Kinetic coupling between steps drives the exit step but unfortunately not well enough. Suppose now that the catalyst exhibits cooperativity in a manner commonly encountered with solid surfaces, namely with rates of entry and exit depending exponentially on fractional surface coverage θ:

$$k_{in} = k_{in}^0 \exp(-\alpha\theta), \qquad k_{out} = k_{out}^0 \exp(\sigma\theta)$$

The first expression is often attributed to Elovich, while the second was first reported by Langmuir (Boudart and Djéga-Mariadassou, 1984). For k_{in}^0 and k_{out}^0, the rate constants at zero coverage, let us assume the same values as above, namely 10^4 and 10^{-4} s^{-1}, respectively. As the surface fills up, the rate of entry decreases exponentially while the rate of exit increases exponentially. Both become equal at the steady state of the catalytic reaction, as already discussed in connection with the decomposition of ammonia. The steady-state value of coverage θ depends on the values of α and σ. Suppose they are both equal to $8 \ln 10$. The θ at the steady state is equal to $\frac{1}{2}$, a reasonable value (Boudart and Djéga-Mariadassou, 1984), and the steady-state turnover rate is 1 s^{-1}, a commonly found value in

3. COOPERATIVITY

practical catalysis (Burwell and Boudart, 1974). Thus, with cooperativity, the turnover rate is 10^4 times larger than it would be with kinetic coupling alone.

The exponential dependence of rates of adsorption and desorption is well documented in the surface chemistry literature. We chose to refer to this phenomenon as cooperativity because it is a result of a variety of causes that have to do with complexity of the system exhibiting cooperativity. A much-studied example of cooperativity is the increase in binding energy of dioxygen to hemoglobin as the number of bound molecules increases: The more oxygen that is bound, the higher the oxygen affinity of the residual free sites (Friedman, 1985). This cooperative behavior is attributed to a transition between two structures of hemoglobin with a low and a high affinity for oxygen (Perutz, 1982). Moreover, changes in protein structure may also affect the oxygen binding sites.

Cooperativity is clearly associated with complex structures. The convenient distinction between homogeneous or heterogeneous catalysts is not a fundamental one. What matters is complexity. A solvated proton is a simple catalyst. So is a transition metal coordination complex in solution. A solid surface is always a complex catalyst, in the sense that it exhibits cooperativity. The same is true for an enzyme. While cooperativity clearly can help in the design of complex catalysts, the design itself provides a formidable challenge to our present molecular engineering capabilities.

As a result of kinetic coupling and cooperativity, the design of catalytic cycles with good turnover rates appears to be less formidable a problem than it might appear at first glance. In a way, we might say that the cycle designs itself and finds an acceptable steady-state turnover rate as a compromise between values of rate constants (kinetic coupling) and changing values of rate constants with surface coverage (cooperativity). Besides, although a catalytic cycle such as that for disproportionation of alkenes consists of six steps, only three of these are chemically distinct, the three others being of the same type, respectively, as the three former ones (Figure 2). Finally, many steps in a catalytic sequence are frequently devoid of kinetic significance in the sense that neither their rate constants nor their equilibrium constants appear in the kinetic expression for the turnover rate. This is due to the common occurrence of a rate-determining step and/ or of a most abundant reactive intermediate. When such simplifications do occur, a multistep catalytic cycle can frequently be simplified to a two-step cycle (Boudart, 1972).

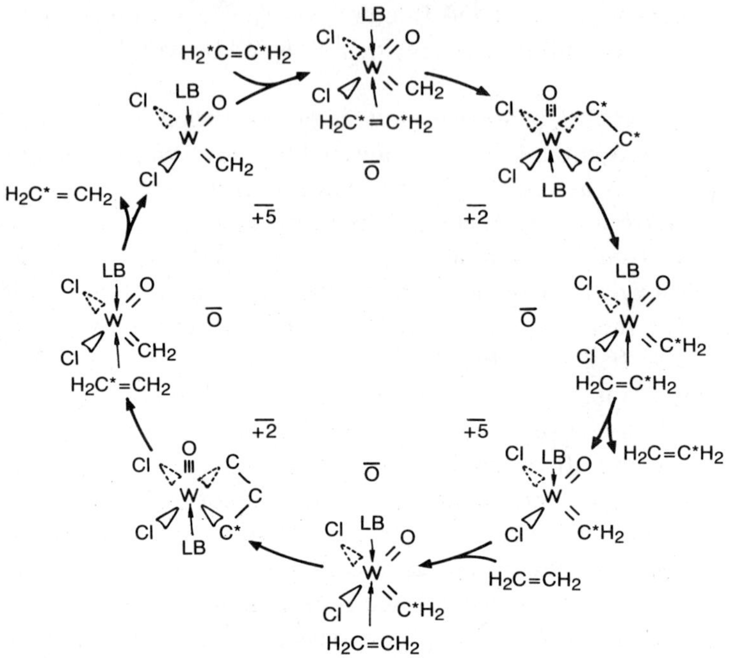

FIGURE 2. Catalytic cycle for dismutation of alkenes:

$$^*CH_2 = {}^*CH_2 + CH_2 = CH_2 \Rightarrow 2\ ^*CH_2 = CH_2$$

The catalyst is a complex of tungsten including a Lewis base LB. The figures 0, +2, and +5 within the cycle are calculated relative values of the standard Gibbs free energy of the corresponding intermediates. Adapted from A. K. Rappé and W. A. Goddard, 1981, in *Potential Energy Surfaces and Dynamics Calculations*, D. J. Trular, Ed., p. 661, Plenum, New York.

4. COUPLED CYCLES

In any situation where consecutive or parallel catalytic reactions occur in a network of reactions on the *same* catalyst, the catalytic cycles of each reaction are coupled. What we want to discuss in this final section is the coupling of cycles taking place with *different* catalysts but in the same reaction space. The complex catalyst is then called a multifunctional catalyst, and the two or more catalytic functions are separated by distances which may be of molecular size but should not be too large lest diffusion

4. COUPLED CYCLES

of intermediates from one function to the next becomes a rate-limiting factor.

Multifunctional catalysis is encountered in the living cell where complex reactions are split into a number of successive coupled simple reactions, each one being catalyzed by its own enzyme (Termonia and Ross, 1981). Multifunctional catalysis has been exploited in free-radical chain reactions, as well as both homogeneous and heterogeneous catalysis. The design of such multiply coupled catalytic cycles is of course much more difficult than the design of single ones. In both cases, the obstacle toward a priori design is the lack of thermochemical and kinetic information on potential elementary steps. Only in the case of free-radical chain reactions does there exist enough data and semiempirical estimation methods for design of chain propagation cycles to be feasible.

Let us give an example of two coupled cycles involving free-radical chain reactions. It deals with a chain reaction used on a very large scale in industry, namely steam cracking. Gas-phase pyrolysis of hydrocarbons is by now so well understood, and the rate constants of so many relevant elementary steps are known or can be estimated, that a priori design of steam crackers with the help of large computers is now an attractive possibility (Buekens and Froment, 1971).

The simplest chain cycle leading to the pyrolysis of an alkane μH to a smaller alkane βH and an alkene m,

$$\mu H \Longrightarrow m + \beta H$$

is represented by the cycle in Figure 3a. As an example, μH is $C(CH_3)_4$, m is $CH_2{=}C(CH_3)_2$, and βH is CH_4. The two irreversible steps in the chain cycle are

$$\mu \longrightarrow m + \beta$$
$$\beta + \mu H \longrightarrow \beta H + \mu$$

where μ is a free radical that decomposes unimolecularly (μ stands for monomolecular) and β is a free radical that reacts bimolecularly. Let us now see what happens when we add to pure μH a substance YH that can compete with μH for reacting with β:

$$\beta + YH \longrightarrow \beta H + Y$$

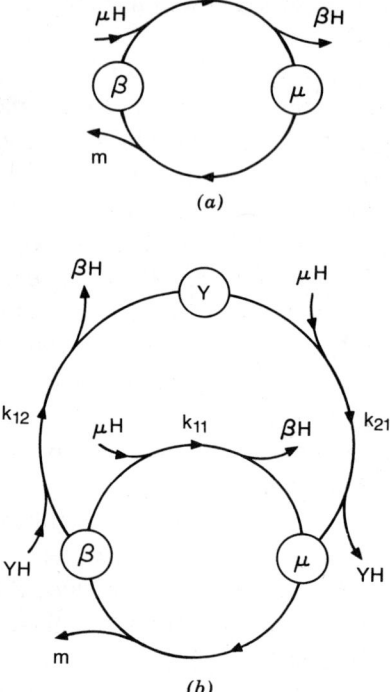

FIGURE 3. (a) Chain propagation cycle for the gas-phase pyrolysis reaction:

$$\mu H \longrightarrow \beta H + m$$

where μH is an alkane, and βH is an alkane of lower molecular weight, and m is an alkene. (b) Same cycle as above coupled with a catalytic cycle involving the catalyst YH.

This step is followed by another one, which regenerates the additive YH:

$$Y + \mu H \longrightarrow YH + \mu$$

It is seen in Figure 3b that the addition of YH leads to the grafting of a catalytic bypass cycle onto the original chain cycle. Indeed, YH can be truly called a catalyst. Depending on whether its presence accelerates or retards the turnover rate of the original cycle, YH can be called a positive or a negative catalyst, respectively. Let us find out how the coupled cycles operate and compare theory with experiment (Scacchi et al., 1968).

4. COUPLED CYCLES

Let us consider the case for which β is the most abundant radical, so that $[\beta] \gg [\mu]$. Then, the only pertinent homogeneous termination steps are

$$\beta + \beta \xrightarrow{k_{\beta\beta}}$$
$$\beta + Y \xrightarrow{k_{\beta Y}}$$
$$Y + Y \xrightarrow{k_{YY}}$$

where products are inactive. For simplicity, let us write

$$k_{\beta\beta} = k_{YY} = \tfrac{1}{2} k_{\beta Y} = \tfrac{1}{2} k_t$$

A material balance around the node Y in Figure 3b yields

$$\frac{[Y]}{[\beta]} = \frac{k_{12}}{k_{21}} \frac{[YH]}{[\mu H]} = bx$$

with $b = k_{12}/k_{21}$ and $[YH]/[\mu H] = x$. Hence

$$\frac{[Y] + [\beta]}{[\beta]} = 1 + bx \tag{13}$$

Both in the presence and in the absence of the catalyst YH, initiation of the chain proceeds with the identical rate v_i. At the steady state, v_i is equal to the termination rate v_t in the presence of YH or v'_t in the absence of YH. Thus,

With YH: $v_i = v_t = k_t\{[\beta]^2 + 2[\beta][Y] + [Y]^2\} = k_t\{[\beta] + [Y]\}^2$

Without YH: $v_i = v'_t = k_t[\beta']^2$

Hence

$$[\beta'] = [\beta] + [Y] \tag{14}$$

On the other hand, the rates of reaction in the presence, v, and in the absence, v', of YH are

With YH: $v = \{k_{11}[\mu H] + k_{12}[YH]\}[\beta] = k_{11}[\mu H][\beta]\{1 + ax\}$

where $a = k_{12}/k_{11}$.

Without YH: $v' = k_{11}[\mu H][\beta']$

Finally, with the use of (13) and (14), the ratio of rates with and without the catalyst is

$$\boxed{\frac{v}{v'} = \frac{1 + ax}{1 + bx}} \tag{15}$$

We conclude that YH is a positive catalyst ($v/v' > 1$) if $a > b$ or $k_{21} > k_{11}$ and a negative catalyst ($v/v' < 1$) if $a < b$ or $k_{21} < k_{11}$.

Physically, the situation is very simple. Upon introduction of the catalyst, some of the most abundant β radicals are transformed into Y radicals. If the latter react faster with the pyrolized molecule μH, YH is a positive catalyst. Otherwise it is a negative catalyst.

With neopentane $C(CH_3)_4$ as μH, H_2S is a positive catalyst. But with ethane, H_2S is a negative catalyst. This is seen in Figure 4, where the data points are experimental and the curves correspond to (15) with the values of k_{12} and k_{11} determined separately. The agreement between experimental and theoretical values justifies the earlier statement that chain pyrolysis of hydrocarbons can now be submitted to kinetics-assisted design.

For heterogeneous catalytic-coupled cycles, the classical example is the naphtha-reforming catalyst, platinum supported on acidic alumina, the typical reaction being isomerization of n-alkanes to i-alkanes (Boudart and Djéga-Mariadassou, 1984). The kinetics of the reaction on the bifunctional catalyst, Pt/Al_2O_3, as well as on alumina and on mixtures of Pt/SiO_2 and Al_2O_3 particles, supports the mechanism illustrated in Figure 5. The alkane is dehydrogenated in a quasiequilibrated reaction on the platinum function. The resulting alkene is then isomerized on the acidic alumina function. Finally the isomerized alkene is rehydrogenated to the product on the platinum function of the catalyst. The two catalytic cycles on the platinum and on the alumina function are coupled very efficiently because of their close proximity. The kinetic coupling between the two cycles has been demonstrated quantitatively (Sinfelt, Hurwitz, and Rohrer, 1960).

In homogeneous catalysis by organometallic complexes, there are several examples of coupled cycles, but the kinetic basis underlying this system

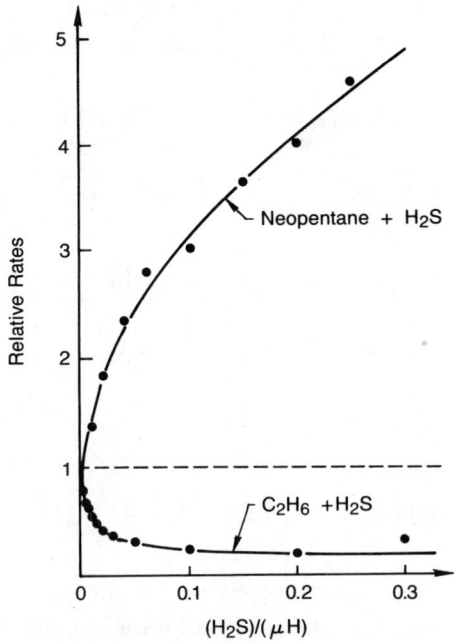

FIGURE 4. The additive H_2S as a positive catalyst in the pyrolysis of neopentane and a negative catalyst in the pyrolysis of ethane. Experimental data are shown together with full lines calculated as indicated in the text. Adapted from G. Scacchi, F. Baronnet, R. Martin, and M. Niclause, *J. Chim. Phys.* **65,** 1671 (1968); and G. Scacchi, M. Dzierzynski, R. Martin, and M. Niclause, *Int. J. Chem. Kin.* **2,** 115 (1970).

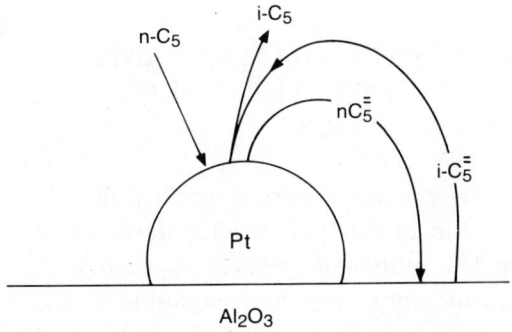

FIGURE 5. Schematic representation of two coupled catalytic cycles in the isomerization of *n*-pentane to *i*-pentane on a bifunctional catalyst consisting of platinum supported on acidic alumina.

is not as developed as it is for the above examples (Masters, 1981). The carbonylation of methanol (Forster, 1979) can be used as an illustration:

$$CH_3OH + CO \Longrightarrow CH_3OOH$$

In fact, the rhodium-catalyzed cycle converts CH_3I into CH_3COI:

$$CH_3I + CO \Longrightarrow CH_3COI$$

Grafted on this catalytic cycle is another closed cycle consisting of two reactions:

$$CH_3COI + H_2O \Longrightarrow CH_3OOH + HI$$
$$CH_3OH + HI \Longrightarrow CH_3I + H_2O$$

Kinetic details on these reactions, as is also the case for many reactions catalyzed by organometallic species in solution, are not plentiful, and speculations abound concerning the nature of intermediates in the catalytic cycles. But kinetics-assisted design of new such cycles in the future is a real possibility because of the nature of the intermediates, whether they are in solution or attached to solid surfaces (Gates, 1987). These intermediates are molecular in the traditional sense, and rules pertaining to their formation and reactivity are well developed.

5. PREDICTION, EXPLANATION, AND DESIGN IN CATALYSIS: THE ROLE OF CHEMICAL KINETICS

We are making progress with regard to prediction of the rate of chemical elementary steps. But in catalysis, success depends on the rapid and repeated turnover of a multistep cycle. Even though they may not be all kinetically significant, they must be compatible in a kinetic sense. Even though turnover may be helped by kinetic coupling, the molecular engineering of this kinetic coupling is not ripe as a design tool. Cooperativity may help as well as coupling of cycles. But a multifunctional catalyst is a very complex system, especially as used in the fully developed industrial

5. ROLE OF CHEMICAL KINETICS

process. Physical design of solid catalysts has made substantial progress (Aris, 1987; Wei, 1987). What about chemical design?

The information surveyed in this chapter shows that chemical design of catalytic cycles can be helped by kinetics. This is a reality in certain free-radical chain reactions. In fact, copolymers are designed by kinetics (Boudart, 1968). Kinetics-assisted design of catalytic cycles involving soluble or attached organometallic complexes is probably carried out semi-intuitively by the inventors of such processes. In heterogeneous catalysis, there is great hope of future progress because kinetics has finally made heterogeneous catalysis a quantitative and reproducible science. This has come about because of two parallel developments. The first one (Boudart, 1985) is our ability to measure reproducible rates of catalytic cycles on supported metal catalysts in the absence of disguise by heat and mass transfer for sufficiently well-characterized reproducible samples. The second one is the availability of high-pressure rates on well-defined large single crystals of metals (Somorjai, 1987). In fact, in many cases in recent years, kinetic measurements on these model catalysts have provided the standards by which to assess the kinetic measurements made on industrial-type catalysts (Boudart, 1986b). A particularly interesting result is the comparison (Löffler and Boudart, 1984) between the turnover rate measured for ammonia synthesis on the (111) face of an iron single crystal and the turnover rate on a commercial Topsøe ammonia synthesis catalyst. Under identical conditions, both values differ by only a factor of 2. Chemical kinetics remains largely an experimental science. It provides data. These kinetic data can be used and will be used to *help* in the design of catalytic cycles that function according to the principles presented in the chapter.

Fortunately, in both homogeneous and heterogeneous catalysis, recent progress in quantum chemistry has been such that thermochemical values can now be estimated for hypothetical intermediates in potential catalytic cycles. These values can be used for rejecting or accepting conceptual cycles, especially if they are supplemented by prior art and chemical intuition. Also, in the case of heterogeneous catalysts, values of rate constants for elementary steps are being measured and reported at an accelerating pace as a result of spectroscopic work on well-defined surfaces, together with temperature-programmed reaction and pulsed molecular beam relaxation techniques. The new theoretical and experimental information, together with the concepts of kinetic coupling, cooperativity, and coupled catalytic cycles, are giving us hope for kinetics-assisted design in the not too distant future.

REFERENCES

R. Aris, Chapter 7 in this volume, 1987.
M Asscher and G. Somorjai, *Surf. Sci.* **143**, L389 (1984).
M. Boudart, *Kinetics of Chemical Processes,* Prentice-Hall, Englewood Cliffs, N.J., 1968.
M. Boudart, *AIChE J.* **18**, 465 (1972).
M. Boudart, *J. Phys. Chem.* **87**, 2786 (1983).
M. Boudart, *J. Mol. Catal.* **30**, 27 (1985).
M. Boudart, *I&EC Fundam.* **25**, 70 (1986a).
M. Boudart, *Chemtech,* 688 (August 1986b).
M. Boudart and G. Djéga-Mariadassou, *Kinetics of Heterogeneous Catalytic Reactions,* Princeton University Press, Princeton, N. J., 1984.
M. Boudart, C. Egawa, S. T. Oyama, and K. Tamaru, *J. Chim. Phys.* **78**, 987 (1981).
A. G. Buekens and G. Froment, *Chemical Reaction Engineering, Proceeding of the 4th European Symposium,* Pergamon Press, Oxford, 1971.
R. L. Burwell, Jr., and M. Boudart, in *Techniques of Chemistry,* A. Weissberger, Ed., Vol. VI, Chapter XII, John Wiley & Sons, New York, 1974.
D. Forster, *Adv. Organometal. Chem.* **17**, 255 (1979).
J. F. Friedman, *Science* **228**, 1273 (1985).
B. Gates, Chapter 3 in this volume, 1987.
W. A. Goddard III, *Science* **227**, 917 (1985).
H. S. Johnston, *Science* **173**, 517 (1971).
D. G. Löffler and M. Boudart, *J. Phys. Chem.* **88**, 5763 (1984).
C. Masters, *Homogeneous Transition-Metal Catalysis,* Chapman and Hall, London, 1981.
M. J. Molina and F. S. Rowland, *Nature* **249**, 810 (1974).
M. F. Perutz, *Hemoglobin and Oxygen Binding,* Chien Ho, Ed., p. 113, Elsevier/North Holland, New York 1982.
I. Prigogine and R. Defay, *Chemical Thermodynamics,* Longmans, London, 1954.
G. Scacchi, F. Baronnet, R. Martin, and M. Niclause, *J. Chim. Phys.* **65**, 1671 (1968).
N. N. Semenov, *Problems in Chemical Kinetics and Reactivity, Volume I,* Princeton University Press, Princeton N.J., 1954.
J. H. Sinfelt, H. Hurwitz, and J. C. Rohrer, *J. Phys. Chem.* **64**, 892 (1960).
G. A. Somorjai, Chapter 2 in this volume, 1987.
Y. Termonia and J. Ross, *Proc. Nat. Acad. Sci. USA* **78**, 2952 (1981).
C. A. Tolman, *Chem. Soc. Rev.* **1**, 337 (1972).
L. Volpe and M. Boudart, *J. Phys. Chem.,* in press (1986).
J. Wei, Chapter 8 in this volume, 1987.

6

CATALYST DESIGN WITH ZEOLITES

W. O. Haag and N. Y. Chen, *Mobil Research and Development Corporation, Central Research Division, Princeton, New Jersey*

1. INTRODUCTION

Zeolites can be viewed as solid analogues of such classical acids as sulfuric acid or aluminum chloride. These all have in common the ability to promote a large number of acid-catalyzed reactions, including polymerization; cracking; isomerization of olefins, paraffins, and aromatics; alkylation of aromatics and paraffins; transalkylation; and many others. Some reactions, such as gas oil cracking, are among the largest-scale catalytic processes.

The first solid acid catalysts were activated natural clays and synthetic binary oxides, such as silica-alumina. Their acidic properties have been well described (Tanabe, 1970). They have remained a very important class of catalysts both as acidic catalysts and as acidic components in dual functional catalysts by the addition of a second function for hydrogenation, oxidation, and so on. The evolution and optimization of these catalysts has remained a highly empirical art, partly because these amorphous catalysts are difficult to characterize. By contrast, zeolites lend themselves very well to catalyst design for a variety of reasons. Zeolites have a well-defined pore system and large intracrystalline surface area (Barrer, 1982; Breck, 1974). It was discovered in the late 1950s at the Mobil laboratories that catalytic reactions can take place *inside* these structures. This discovery marked the real beginning of zeolite catalysis.

Two aspects make zeolites unique. The intracrystalline surface is an inherent part of the crystal structure and hence topologically well defined, in sharp distinction to amorphous and even most crystalline solids whose outer surface may be considered a crystal defect with atoms whose coordination number differs from atoms in the crystal. The second distinctive feature of zeolites is that the diameter of their pores is uniform and of similar magnitude as that of many organic molecules of interest. Molecular sieving and shape selectivity previously unknown with synthetic catalysts became possible.

Each zeolite has a unique pore system and crystal structure. Nature has provided us with 34 different zeolites (Barrer, 1968; Breck, 1974; Meier and Olson, 1978; Meier, 1979). But among those of interest to catalysis, only a few are found in abundance. The ability to synthesize zeolites of known and new structures in the laboratory made new discoveries in zeolite catalysis possible.

So far nearly 100 synthetic zeolites have been identified covering a large variety of different framework structures, with pore openings ranging from less than 5 Å to larger than 10 Å.

In heterogeneous catalytic processes one usually considers three major performance characteristics: *activity, selectivity* toward one or several products, and *stability* of operation, that is, low catalyst aging. These performance features are complex, interacting functions of several basic catalyst properties: sorption energies for feeds and products, rates of transport of molecules to and from the active sites, mostly by diffusion, and intrinsic activities for various reactions. With zeolites, it has now become possible to vary these properties in a systematic way and one at a time.

Thus, the molecular engineering design of zeolite catalysts may be stated as the science of coupling of chemical reactions with sorption and diffusion characteristics of the zeolite, including molecular sieving effects, to alter the reaction pathway and the product selectivity of known chemical reactions.

The major arsenal of tools that can be used in the *design* of zeolite catalysts includes (a) the manipulation of crystal shape and size, (b) the nature of the channel system, (c) the shape and size of the pores, and (d) the nature and number of active sites.

This chapter will therefore describe structural and compositional features of zeolites (Section 2), sorption and diffusional characteristics (Section 3), intrinsic catalytic activity (Section 4), and practical design principles (Section 5) and their application in commercial processes (Section 6 and 7). Special emphasis is given to methods by which these features can be controlled.

2. CLASSIFICATION OF ZEOLITES

Zeolites are porous tectosilicates (Barrer, 1982), that is, three-dimensional networks built up of TO_4-tetrahedra (T = Si or heteroatom) such that each of the four oxygen atoms is shared with another tetrahedron. The most common forms are aluminosilicates, although structures containing boron, gallium, or iron in place of aluminum and germanium in place of silicon have been reported (Barrer, 1982, and references therein). The tetrahedron can link up to form a variety of secondary building units (Meier, 1968; Gramlich-Meier and Meier, 1982) from which zeolites of various framework topologies can be derived. Meier and Olson (1978) have summarized the structures of 38 zeolites; in addition, the framework topologies of ZSM-12 (LaPierre et al., 1985), ZSM-22 (Kokotailo et al., 1985), Theta-1 (Barri et al., 1984), ZSM-23 (Rohrman et al., 1985), ZSM-48 (Schlenker et al., 1985), and others have been described. Altogether, the crystal structures of about 50 zeolites are now known (von Ballmoos, 1984).

The structural characteristic of greatest interest for catalysis is the channel system, which is described for some of the more important zeolites in Table 1. Depending on the largest channel, zeolites are characterized as small, medium, or large pore zeolites if they contain apertures made by rings of 8, 10, or 12 linked tetrahedra. Framework structures and pore openings of representative zeolites from each group are depicted in Figure 1. Within each group, there is considerable variation in the aperture, both in size and ellipticity. For example, the opening in Linde A is circular (4.1 Å), but in erionite it is elliptical (3.6 × 5.2 Å).

The projections of a series of similar zeolites (Figure 2) show the subtle variations of channel dimensions and shapes available. The channel system may be one-dimensional, for example ZSM-22 and ZSM-48, two-dimensional such as in ferrierite, or three-dimensional as in ZSM-5 (Figure 3). Multidimensional channels often intersect each other, but this is not always the case. The interconnecting channels can be of the same size (e.g., faujasite) or smaller (ferrierite); they may be straight (ZSM-11) or tortuous (ZSM-5). The connectivity of the channel system has major consequences

TABLE 1. Channel System of Representative Zeolites

Structure Type	Ring Size of Channels[a]	Largest Channel, Å[b]
Linde type A	8-8-8	4.1
Chabazite	8-8-8	3.6 × 3.7
Erionite	8-8	3.6 × 5.2
ZSM-22	10	4.5 × 5.5
ZSM-23	10	4.5 × 5.6
ZSM-48	10	5.3 × 5.6
Ferrierite	10-8	4.3 × 5.5
ZSM-5	10-10	5.4 × 5.6
ZSM-11	10-10	5.1 × 5.5
ZSM-12	12	5.7 × 6.1
Linde type L	12	7.1
Mazzite	12	7.4
Mordenite	12-8	6.7 × 7.0
Offretite	12-8-8	6.4
Faujasite	12-12-12	7.4

[a]Number or either T or O atoms forming the smallest rings of the channels.
[b]Crystallographic free diameter, based on an oxygen radius of 1.35 Å, of the smallest ring or window in the channel.

for diffusion and aging characteristics. For example, zeolites with one-dimensional channels are more subject to deactivation than those with a three-dimensional pore system because of their susceptibility to pore mouth plugging.

The pores in some zeolites are relatively uniform "tubes," for example in ZSM-48 (Figure 3); in others they contain larger cavities (e.g., erionite, Linde A) or "supercages" (e.g., zeolite Y, Figure 1), which are connected by "windows." Zeolites containing large cavities are more prone to deactivation since molecules such as condensed aromatics can form but cannot escape through the smaller port holes and are trapped (Venuto, Hamilton, and Landis, 1966; Venuto and Landis, 1968). Most medium-pore zeolites are noteworthy for their unusually low coke-forming propensity in acid-catalyzed reactions. Their resistance to aging is one of the major contributing factors to the successful industrial application of these zeolites.

The pore dimensions for zeolites given in the literature are based on crystallographic data and are nominal values only. Usually, molecules about

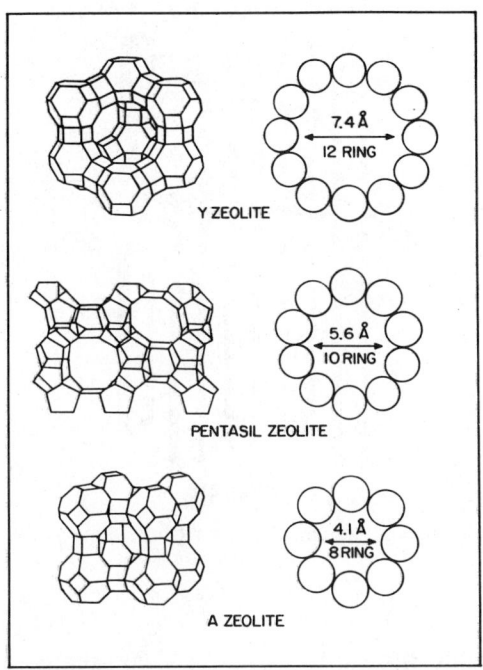

FIGURE 1. Framework structures and projections of representative large-, medium-, and small-pore zeolites.

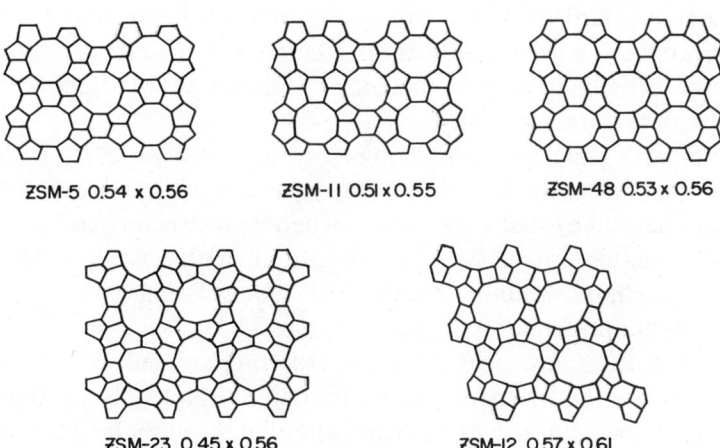

FIGURE 2. Projections of ZSM-5, -11, -12, -23, and -48 structures. The axes of the channels (in nm) are indicated.

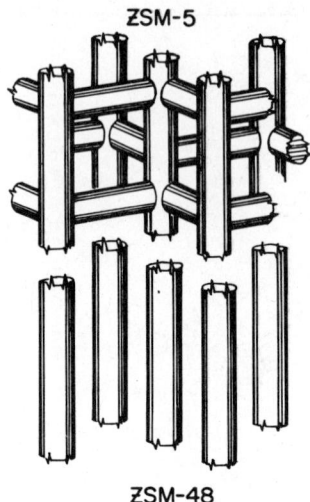

FIGURE 3. Schematic channel system of ZSM-5 (three-dimensional) and ZSM-48 (one-dimensional).

1 Å larger can readily diffuse, especially at higher temperature. The reason is that the pore size is derived—by crystallographic convention—using the large oxygen radius (1.35 Å) of an oxygen anion, although the oxygen in silicates has nearly covalent character and is actually considerably smaller. Also, the atoms of the zeolite and the diffusing molecules are not hard spheres. Zeolite lattice vibration and guest molecule bending and vibration also can reduce repulsive interactions. Gauging of the effective pore size by sorbing or reacting organic molecules is therefore an effective way to determine their catalytic utility.

Zeolites with eight-membered rings sorb only straight-chain molecules such as n-paraffins, n-olefins, and primary alcohols. Some larger-pore zeolites also behave like small-pore zeolites when their structure contains many stacking faults that reduce the effective pore diameter. Some examples of these are gmelinite, stilbite, dachiardite, Linde T, ZSM-34, epistilbite, heulandite/clinoptilolite, ferrierite, and so on.

Medium-pore zeolites with 10-membered-ring systems are of most interest to the design of shape-selective catalysts. Not only do they sorb straight-chain molecules, they also can discriminate, either by size exclusion or by rate of diffusion, among a variety of branched and cyclic molecules.

2. CLASSIFICATION OF ZEOLITES

Included in this group are ZSM-5, ZSM-11, ZSM-22 (Theta-1), ZSM-23, and ZSM-48.

The 12-membered-ring zeolites include all the large-pore zeolites. The synthetic faujasites, zeolite X and Y, and mordenite are the best-known and most-studied zeolites in this group.

A useful diagnostic test for the effective pore size of zeolites has been reported (Frilette, Haag, and Lago, 1981). The relative rates of cracking of n-hexane and 3-methylpentane define a "constraint index," which is less than about 1 to 2 for zeolites that can be considered large-pore, between 1 or 2 and 12 for medium-pore, and >12 for small-pore zeolites (Table 2). Catalytic diagnostic tests such as the constraint index test, using model compounds of different size and shape, have become a useful tool complementary to other physicochemical characterization methods in determining the structural features of these zeolites.

The silica-to-alumina ratio of different zeolite structures is another distinguishing feature. Within each group, the ratio can be varied, especially with medium-pore silica-rich zeolites. Figure 4 shows the compositional ranges for selected zeolite structures.

In summary, the selection of a catalyst can take advantage of a great variety of available zeolite structures. In addition, the crystal size of many zeolites can be varied by 2–3 orders of magnitude via synthesis, as can the crystal habit and the chemical composition. A detailed review of synthesis

TABLE 2. Constraint Index (C.I.) Values for Some Typical Catalysts Determined at 316°C

CAS	C.I.
ZSM-5	8.3
ZSM-11	8.7
ZSM-12	2.3
TMA offretite	3.7
Beta	0.6
ZSM-4 (mazzite)	0.5
H-Zeolon (mordenite)	0.5
REY	0.4
Amorphous silica-alumina	0.6[a]
Erionite	38

[a] At 510°C.

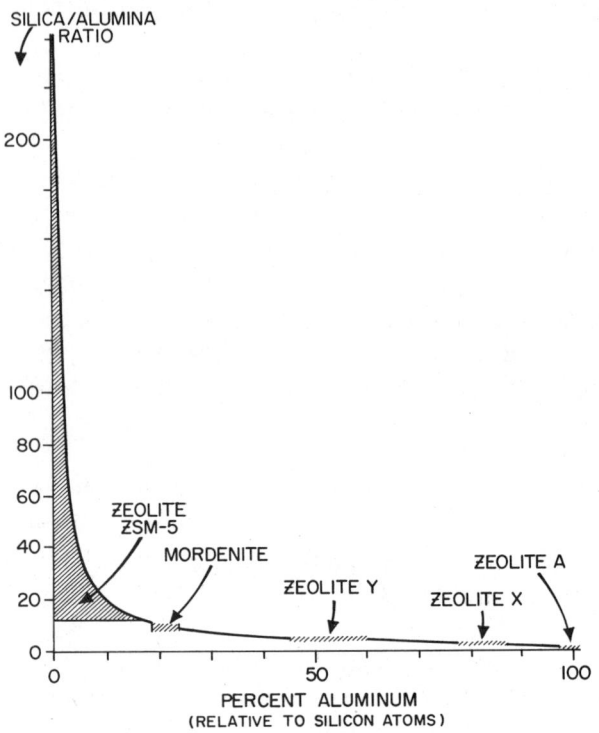

FIGURE 4. Aluminum concentration and silica-to-alumina ratio for various zeolites. Reprinted by permission from W. O. Haag, R. M. Lago, and P. B. Weisz, *Nature* **309,** 589. Copyright © 1984, Macmillan Journals Ltd.

is beyond the scope of this chapter. An excellent description of the methods available to direct the synthesis toward a particular structure, crystal size, and compositional zoning has been presented (Barrer, 1982).

3. SORPTION AND DIFFUSION PROPERTIES

Optimum utilization and rational design of zeolite catalysts requires an awareness and preferably a quantitative knowledge of the influence of sorption and diffusional phenomena on observed activity and selectivity. While sorption and diffusion need to be considered in all heterogeneous catalytic processes, they are especially important in zeolite catalysis, be-

3. SORPTION AND DIFFUSION PROPERTIES

cause they can be systematically varied and produce rather large effects. The term "sorption" is used here to represent the thermodynamic equilibrium aspects of the interaction of sorbate and solid while "diffusion" deals with the dynamic aspect of sorption and desorption.

3.1. Sorption

In view of the great importance of zeolites for separations and as a tool to examine zeolite structure, the literature on sorption in zeolite is voluminous. The books by Breck (1974) and Barrer (1978) provide valuable reviews of this topic. Much of the collected information describes aluminosilicates of relatively low silica-to-alumina ratios and mostly in their cationic forms. However, the catalytically important zeolites are the cation-free protonic forms of high-framework SiO_2–Al_2O_3 ratios, such as hydrogen mordenite, ultrastable Y, and HZSM-5. Their sorption properties differ significantly from those of the former group.

High-alumina, cationic zeolites possess a strong electrostatic field and are hydrophilic; they sorb polar molecules such as water or alcohols in preference to nonpolar hydrocarbons. High-silica zeolites do not have such preference; they are hydrophobic, as was shown first for a series of dealuminized mordenites (Chen, 1976). The sorbate–zeolite interaction in these silica-rich hydrogen zeolites is essentially due to dispersion forces, with little contribution from polarization forces. This is illustrated by the data of Table 3. In the sodium forms, benzene and 1-butene are more strongly adsorbed than cyclohexane and butane, respectively, because of specific interaction of their π-electrons with the large electrostatic field of the sodium ions. Replacing sodium by hydrogen generates undissociated hydroxyl groups and produces zeolites that have a very low, if any, electrostatic field. Since the sorbate–zeolite interaction in these samples is essentially caused by dispersion forces, cyclohexane is more strongly sorbed than benzene, and, surprisingly, butane is more strongly sorbed than butene, because of their larger number of hydrogen atoms. The zeolite geometry now has a dominant influence. Butane and hexane are much more strongly sorbed in HZSM-5 than in USY-Ex, because of the former's smaller pore diameter.

The lack of preferential sorption of aromatics and olefins in high-silica hydrogen zeolites is probably contributing to the great stability of HZSM-5 in all its applications and to the lower coke selectivity of ultrastable Y in gas oil cracking.

TABLE 3. Differential Heats of Sorption at Low Loading (kJ/mole)

	I[a] Mordenite		II[b] ZSM-5		III[d] Y		ZSM-5
	Na+	H+	Na+	H+	Na+	US-Ex[e]	H+
Si–Al ratio:	5	5	24	>100	2.4	40–190	>100
Benzene	105	52	91	50	80	45	
Cyclohexane	75	56			56	47	
1-Butene					58	—	49
n-Butane			53	49	41	34	52
n-Hexane	88	67		80[c]	57	46	70

[a]Eberle (1963).
[b]Lechert and Schweitzer (1984).
[c]Olson, Haag, and Lago (198).
[d]Schirmer et al. (1980); Thamm et al. (1982); Thamm, Stach, and Fiebig (1983); Stach et al. (1984).
[e]US-Ex = ultrastable Y, extracted.

The strong sorption of paraffins in zeolites relative to amorphous catalysts leads to higher concentrations of sorbed molecules in the pore volume. This factor contributes to the much higher rate observed with zeolites, especially for bimolecular reactions such as paraffin cracking and aromatics formation via hydrogen-transfer reactions.

3.2. Diffusion

The diffusion properties in zeolites are of greatest importance in determining the unique selectivity obtainable with zeolite catalysts. In their simplest and earliest version, small-pore zeolites were used as *molecular sieves*. Some molecules, such as linear paraffins, can enter the zeolite channels and react, whereas branched paraffins are excluded and remain unconverted. The advent of HZSM-5 and other medium-pore zeolites brought out a more subtle distinction based not on yes-or-no accessibility but on a difference in the rate of diffusion. The quantitative treatment of this case is the same as for diffusion in larger pores, and its principles, which have been described elsewhere (Satterfield, 1970), will be put forward here in simple form.

Diffusion effects on rate and selectivity result when the rate of catalytic transformation is larger than the rate at which feed molecules can reach

3. SORPTION AND DIFFUSION PROPERTIES

or product molecules escape from the catalytic sites via a diffusive process. The observed apparent rate constant, k_{obs}, then becomes smaller than the intrinsic rate constant, k, which is found only when the transport rate is sufficiently fast. For a simple first-order reaction,

$$A \xrightarrow{k} B$$

we note that

$$k_{obs} = k\eta \tag{1}$$

where η = effectiveness factor with values between 0 and 1. The factor η is a function of the Thiele modulus ϕ, which is defined for first-order reactions as

$$\phi = R\sqrt{k/D} \tag{2}$$

where R = diffusion path (cm), k = intrinsic rate constant (sec^{-1}), and D = diffusivity (cm^2 sec^{-1}).

The relationship between η and ϕ for different particle geometries is given in Figure 5. For flat plate geometry, $\eta = \tanh \phi/\phi$. If $\phi > 1$, η is less than 1, and the intrinsic rate is reduced by diffusion effects. For high values of ϕ (>3), $\eta = 1/\phi$.

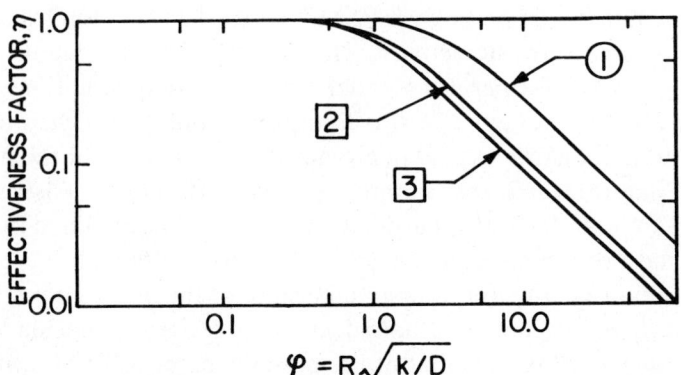

FIGURE 5. Relationship between the effectiveness factor η and the Thiele modulus ϕ. Curve 1, first-order kinetics in spherical particle geometry; curves 2 and 3, first- and second-order kinetics in flat-plate geometry.

Diffusion in zeolites represents a special diffusion regime, termed "configurational diffusion" (Weisz, 1973); no general method exists to estimate these diffusion constants in contrast to diffusion in the gas phase or in larger pores (Knudsen diffusion).

3.3. Determination of Diffusion Coefficients in Zeolites

In the absence of a predictive theoretical framework, diffusivities need to be experimentally determined for each system of interest. Several techniques can be used: a gravimetric adsorption method, a chromatrographic method, measuring the rate of transport across a true zeolite membrane (Paravar and Hayhurst, 1984), or determining molecular displacements by means of the NMR pulsed-field gradient technique (Kärger, 1985). While in many cases useful, all these methods have considerable limitations that have been critically discussed in detail (Ruthven and Lee, 1981; Haag, Lago, and Weisz, 1982). In contrast, the quantitative determination of the degree of diffusion inhibition during catalytic reactions determines directly the steady-state diffusion coefficients under catalytic conditions, regardless of the complexity or diversity of the states of sorption and mobility of the species. In this "triangle method," pioneered by Weisz and Prater (1954), the diffusion modulus ϕ, and hence the diffusivity D, is obtained from the variation of the observed reaction rate for a known change in particle size. The method has been expanded by varying also the intrinsic activity in a precisely known way (Haag, Lago, and Weisz, 1982). It has been applied to the measurement of the rate of cracking of hexane and hexene isomers and of 2,2-dimethylheptane over ZSM-5, using catalyst samples of varying crystal size and intrinsic activity. The effective intracrystalline diffusion coefficient of several paraffins under actual reaction conditions thus obtained are shown in Figure 6. It is found that doubly branched molecules diffuse 10^4 times slower than linear ones, but the lengths of the molecules or their nature—olefin vs. paraffin— have little effect. Also shown in Figure 6 are data for several aromatic hydrocarbons, obtained by the gravimetric adsorption method (Olson et al., 1981). Although these two sets of data cannot be directly compared because of the different methods and temperatures used, there is a sharp decrease in D as the molecular diameter of the sorbate approaches that of the pore diameter of ZSM-5, which has a nominal value of 0.54×0.56 nm (Meier and Olson, 1978).

The temperature dependence of the diffusion constants is often quite high. An example is shown in Figure 7 for 2,2-dimethylbutane (Post, van

3. SORPTION AND DIFFUSION PROPERTIES

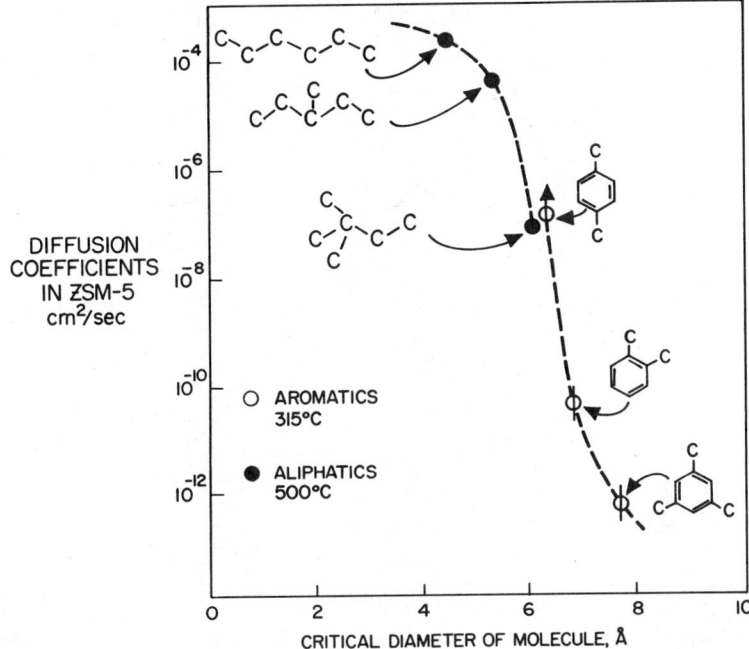

FIGURE 6. Diffusion coefficients in HZSM-5 for hexane isomers (at 500°C) and aromatics (at 315°C).

Amstel, Kowenhoven, 1984). The activation energies obtained in this work and by several other authors for a variety of hydrocarbons in ZSM-5 are listed in Table 4 together with diffusivity data calculated to a standard temperature of 373 K. It becomes apparent that temperature has a strong effect on the diffusion coefficient. For example, the diffusivity of 2,2-dimethylbutane is 10^{-18} m^2 sec^{-1} at 373 K. Extrapolation of the data of Figure 7 yields a value of 10^{-13} m^2 sec^{-1} at 811 K, that is, diffusion is five orders of magnitude faster at the higher temperature. Temperature, therefore, is a major variable in affecting the diffusivity of a given system.

Diffusivity can be moderated by controlled narrowing or blocking of diffusion paths with strongly or irreversibly sorbed molecules (Barrer, 1978). While this always *reduces* diffusivity, it can result in increased selectivity, as will be shown in Section 6. Diffusively modified HZSM-5 has been made by treatment with phosphorus compounds and with magnesium, boron, silicon, and antimony oxides or by controlled coking (Olson and Haag, 1984, and references therein). As discussed by Barrer (1978), diffusion in

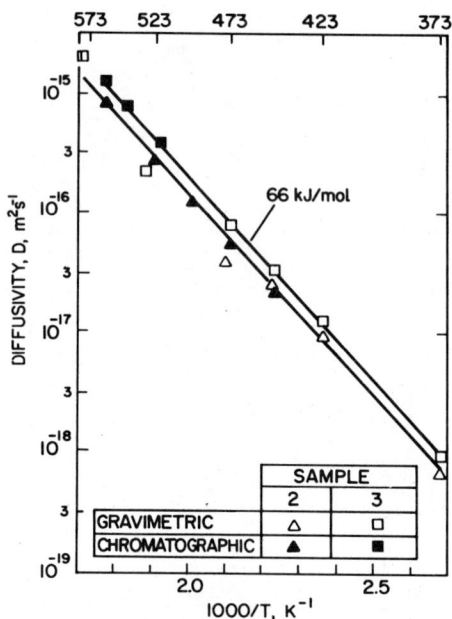

FIGURE 7. Effect of temperature on the diffusivity of 2,2-DMB in high-silica ZSM-5. From M. F. M. Post, J. van Amstel, and H. W. Kouwenhoven, *Proceedings of the 6th International Zeolite Conference,* D. H. Olson and A. Bisio, Eds., p. 517, by permission of the publishers, Copyright © 1984, Butterworth and Co. Ltd.

TABLE 4. Activation Energy for Diffusion Coefficients in ZSM-5

Molecule	Temperature (K)	E_a (kJ/Mole)	$D(m^2 sec^{-1})$ 373 K
n-Butane[a]	297–334	14	1.8×10^{-11}
Isobutane[a]	297–334	23	1.3×10^{-11}
2,2-dimethylbutane[b]	373–573	66	10^{-18}
o-Xylene[c]	523–623	38	10^{-18}
m-Xylene[c]	523–623	59	10^{-19}
t-Butylbenzene[c]	523–623	50	10^{-19}
1,3,5-Trimethylbenzene[c,d]	523–623	80	10^{-17}

[a]Paravar and Hayhurst (1984).
[b]Post et al. (1984).
[c]Weisz (1980).
[d]At 600 K.

zeolites with one-dimensional channels is very sensitively affected by these obstructions, but only moderately so in two- or three-dimensional channel systems. A model for the one-dimensional case has been presented with the simplifying assumption of uniformly spaced moderating species (Barrer and Rees, 1954; Barrer, 1978). Modeling of the two- and three-dimensional pore systems is presently under investigation (Sundaresan and Hall, 1986; Theodorou and Wei, 1983). The presented models predict that the effectiveness of the blocking agent is much greater if it is placed close to the outer surface instead of in random locations throughout the crystal. Experimental data show that a small amount of coke deposited on or near the outside of ZSM-5 crystals reduces the diffusion rate of o-xylene at 393 K from 1.8×10^{-18} to 2.4×10^{-20} m^2 sec^{-1}, that is, by two orders of magnitude (Olson and Haag, 1984).

In summary, the diffusion rate in zeolites can be varied over many orders of magnitude by choice of proper zeolite structure, additional modification for fine trimming, and by choice of temperature.

4. INTRINSIC CATALYTIC PROPERTIES OF ZEOLITES

The catalytic activity of zeolites derives from specific acid sites that are located within the intracrystalline cavities. It is an outstanding feature of zeolite catalysts that their activity can be varied over many orders of magnitude in a predictive manner, independent of other catalyst properties such as catalyst size, surface area, and pore volume and diameter. Not only has this fact contributed greatly to a better understanding of fundamental catalytic phenomena, but it also provides great flexibility in the design of commercial catalysts. Different reactions catalyzable by the same zeolite structure occur at greatly different rates, and activity manipulation is required to adapt the catalyst to each process, especially to optimize selectivity and catalyst life.

The acid activity of zeolites depends on the presence of trivalent ions such as Al^{3+} in the zeolite framework of shared SiO_4 tetrahedra. Without their presence, crystalline—or for that matter amorphous—silicas have no acid activity. It is possible to control the activity in several ways: (1) by incorporating the desired amount of aluminum during synthesis; (2) by removing or incorporating framework aluminum after synthesis; or (3) by introducing metal cations. This section describes these methods, the nature of the catalytic sites, and their catalytic consequences.

4.1. Preparation Methods

While many zeolites such as Y or mordenite can only be synthesized with nearly fixed Si–Al ratio, many medium-pore zeolites such as ZSM-5 can be obtained in any desired Si–Al ratio directly by synthesis (Argauer and Landolt, 1972; Dwyer and Jenkins, 1976; Olson, Haag, and Lago, 1980; Gabelica, Derouane, and Blom, 1984). Samples with an aluminum content ranging from 3 ppm to over 2 wt% have been synthesized, providing a range of four orders of magnitude.

Several methods exist to *reduce* the framework Al content of zeolites. Among these are steaming (McDaniel and Maher, 1968; Maher and McDaniel, 1968; Scherzer and Bass, 1973), treatment with chelating agents such as EDTA (Kerr, 1967; Beaumont and Barthomeuf, 1972), extraction with mineral acids, and treatment with silicon tetrachloride (Beyer and Belenykaja, 1980). A thorough review of the preparation and characterization of aluminum-deficient zeolites has been presented by Scherzer (1984).

Only very recently has it been recognized that it is also possible to *introduce* Al into the siliceous framework of a crystalline zeolite. Treatment of high-silica ZSM-5 with aluminum chloride (Dessau and Kerr, 1984; Anderson, Klinowski, and Liu, 1984) and with other aluminum compounds (Chang et al., 1984) including alumina binder (Shihabi et al., 1985; Chang et al., 1985) produces zeolites of increased Al content and higher activity. Incorporation of aluminum as tetracoordinated species into the zeolite framework was confirmed by ^{27}Al MAS-NMR (magic angle spinning nuclear magnetic resonance), NH_3-TPD (temperature-programmed desorption), and FTIR (Fourier transform IR).

4.2. Active Sites

The extensive studies concerning the nature of the acid sites in amorphous silica-aluminas and low Si–Al ratio zeolites, such as zeolite Y, have been reviewed in detail (Benesi and Winquist, 1978; Barrer, 1978; Beagley et al., 1984). The earlier work leaves little doubt about the importance of Brønsted acid sites, whose structure is depicted in Figure 8. (Haag, Lago and Weisz, 1984).

The detailed understanding of the relationship between sites and catalytic activity in zeolites such as HY has remained controversial, however, because of the instability of HY and the complexity of the system. Mikovsky and Marshall (1976) have shown that HY contains four types of acid sites

4. INTRINSIC CATALYTIC PROPERTIES OF ZEOLITES

FIGURE 8. The structure of the Brønsted acid site in zeolites.

of seemingly different strength, as shown by the TPD spectrum of NH_4Y (Figure 9). Such distribution of types of Al has recently been confirmed experimentally by MAS-NMR spectroscopy (Engelhard et al., 1981; Melchior, Vaughan, and Jacobson, 1982; Klinowski et al., 1982). In addition, incompletely coordinated aluminum species, that is, Lewis sites, can be present in HY (Barrer, 1978; Sohn et al., 1986), whose catalytic significance, if any, has remained poorly understood.

Zeolites with a Si–Al ratio >5 are much more stable and can provide precise knowledge of the number and nature of active sites. When the hydrogen form of ZSM-5 is prepared in the absence of steam (Olson, Haag, and Lago, 1980), it contains only one type of catalytic site as shown by

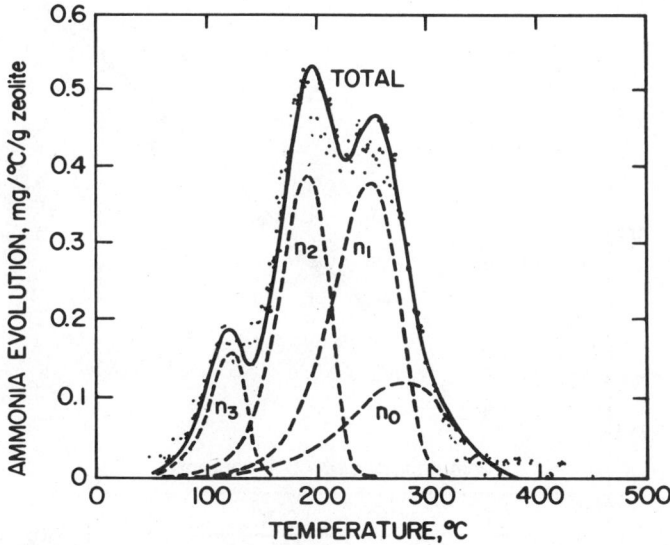

FIGURE 9. Temperature-programmed NH_3 desorption from NH_4Y. The symbols n_0, n_1, n_2, and n_3 refer to the number of Al atoms in next-nearest-neighbor positions. From R. J. Mikovsky and J. F. Marshall, *Journal of Catalysis* **44**, 170. Copyright © 1976, Academic Press.

various physical measurements (Jacobs, 1982; Jacobs and von Ballmoos, 1982; Haag, 1984a; Haag, Lago, and Weisz, 1984). They include chemical analysis, quantitative ^{27}Al MAS-NMR, NH_3-TPD, and ion exchange capacity measurements; a single IR hydroxyl band at about 3603 cm^{-1} with an intensity proportional to Al content is found (Jacobs and von Ballmoos, 1982). Each framework Al atom gives rise to one Brønsted acid site which has the structure shown in Figure 8.

The catalytic activity in these HZSM-5 preparations with a Si–Al ratio from about 17 to 180,000 has been found to be strictly proportional to their aluminum concentration and hence to their content of active sites, which range from 5.6×10^{16} to 5.6×10^{20} sites/g. The correlation line goes through the origin. If a zeolite could be made with no aluminum, it would have no acid activity. The linear correlation was found for the cracking of hexane (Figure 10, log–log plot with slope = 1), toluene disproportionation, hexene cracking (Haag, 1984a), methanol conversion (Chen and Reagan, 1979), ethylbenzene dealkylation and cyclopropane isomerization (Chu and Chang, 1985), and propene polymerization. The proportionality between activity and number of Al atoms indicates that every acid site

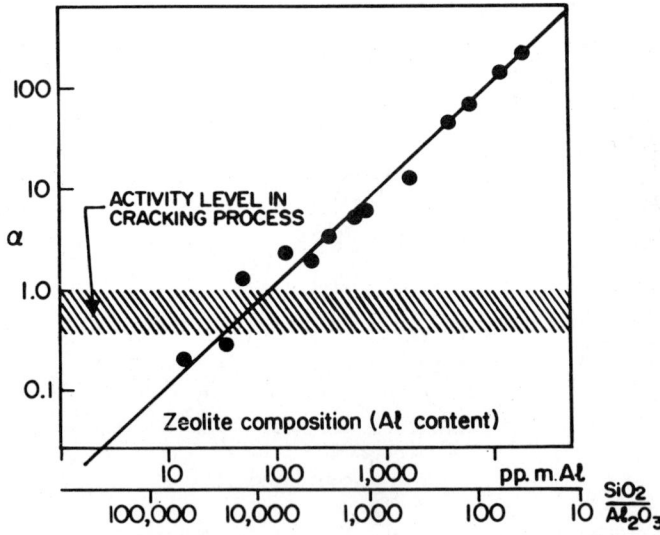

FIGURE 10. The hexane cracking activity, α, plotted against the content in HZSM-5. Shaded band indicates activities near $\alpha \simeq 1$. Reprinted by permission from W. O. Haag, R. M. Lago, and P. B. Weisz, *Nature,* **309,** 590. Copyright © 1984, Macmillan Journals Ltd.

contributes equally to the observed activity, independent of the total site concentration. The sites must therefore be homogeneous. This fact allows calculations of turnover frequencies (TOF), which are given for various reactions in Table 5. The range of TOF's covers six orders of magnitude.

Recently, similar linear relationships between activity and *framework* Al content were reported for dealuminated zeolite Y. Bremer et al. (1983) investigated hexadecane cracking, whereas the laboratory of J. H. Lunsford (DeCanio et al., 1986; Sohn et al., 1986) studied the cracking of hexane, cumene dealkylation, and conversion of methanol to dimethyl ether with a series of dealuminated Y's with Si–Al ratios from 4.7 to 255. Again, the constant TOF observed indicates homogeneity of sites. The active site in these high Si–Al ratio zeolites thus is free from chemical interactions with neighboring framework aluminum atoms and is in a state of an ideally dilute solution.

Furthermore, the specific activities for hexane cracking at 350°C of dealuminated Y and HZSM-5, both with Si–Al ratio of 26, were 11.4 and 8.5 μmole/g·min (Sohn et al., 1986), that is, they are surprisingly similar. It appears that the specific activity of a Brønsted site is relatively insensitive to the different structural and topological environment of different zeolites, provided the overall Si–Al ratio is about 5 or greater.

For these high Si–Al ratio zeolites, the intrinsic activity for any given reaction can be predicted from the framework Al content in the absence of deactivating cations, once the proportionality factor for the specific reaction has been established.

In summary, the catalytic activity of high Si–Al ratio zeolites can be

TABLE 5. Turnover Frequencies (TOF) of Various Reactions[a]

Reaction	Feed	TOF (Molecule/site·sec)	Relative Rate Constants
Cracking	Hexane	0.0469	1
	Nonane	0.150	3.2
	Dodecane	1.07	23
Isomerization	m-Xylene	10.6	226
Polymerization	Propylene	40.7	868
Cracking	1-Hexene	36.7	782
	1-Heptene	56.7	1209
Double-bond shift	1-Hexene	4.7×10^4	1×10^6

[a] With ZSM-5 catalyst, 1 atm pressure, 450°C.

varied over wide ranges by changing the number of active sites (tetrahedral framework Al) by synthesis, or by subsequent Al removal or by Al incorporation. In many cases, although not all, zeolites with a homogeneous set of acid sites can be obtained whose TOF is relatively insensitive to their concentration, method of catalyst synthesis or modification, and crystal structure.

4.3. Steam Enhancement

While *severe* steaming of ZSM-5 and HY has been found to yield lower activity catalysts with a reduced number of active sites of unchanged specific activity, *mild* steaming surprisingly gives catalysts of enhanced activity. The degree of activity enhancement depends on the original Al content. Detailed investigation has shown that mild steaming *reduces* framework Al and the total number of active sites, but produces some sites of *higher* specific activity, at least 40 times more active for cracking hexane at 538°C (Lago et al., 1986).

While the detailed structural nature of the enhanced activity site is not yet completely elucidated, it consists probably of a conventional zeolite Brønsted acid site (Figure 8) whose acidity is inductively increased by a neighboring hydrolyzed cationic Al atom that has Lewis acid character. A similar possibility has been invoked by Lunsford (1968) to account for activity maxima in thermally treated, partially dehydroxylated HY.

4.4. Effect of Cations

Cations play a large role in controlling the catalytic activity in zeolites. Their specific effect depends on the cation nature and on the water content. While the effect is complex and a detailed description is outside the scope of this chapter, a few simplified generalities can be stated.

Monovalent cations such as sodium neutralize the Brønsted acid sites and cause deactivation. Di- and trivalent cations usually give active catalysts. It has been suggested that in the presence of water, they can undergo hydrolysis to form Brønsted acids, for example,

$$M^{n+} + H_2O = [MOH]^{(n-1)+} + H^+$$

The extent of this reaction depends on the electrical field, that is, cation charge and radius. The importance of water as cocatalyst has been well recognized, but the possible role of the unhydrated cations as Lewis acids has remained in dispute, as well as the specific acid activity of cation-derived acid sites (Sohn et al., 1986).

In addition to providing active sites, polyvalent cations, especially rare earth ions, function to stabilize the zeolite structure of zeolite Y and are used in commercial cracking catalysts. Similarly, the nonframework aluminum species created by steaming of NH_4Y to form "ultrastable Y" (USY-A) stabilizes the zeolite (Scherzer, 1978), although in highly dealuminated Y (USY-B, Si/Al > 50) the nonframework aluminum plays no role in stability enhancement (Scherzer, 1984).

4.5. Compositional Gradients

The use of the diffusion equations and Thiele modulus to assess activity and selectivity effects (Section 3) assumes equal activity in any volume element of the zeolite crystal, where the element size is small relative to the size of the crystal. Local concentration fluctuations of active sites within each element are to be expected when the Al distribution is random, but these fluctuations average out and can be ignored. However, macroscopic Al gradients or compositional zoning, if present, should be taken into consideration. For example, if enrichment of Al—and hence of active sites—toward the crystal surface occurs, it would reduce the intracrystalline diffusion distance; its effect is equivalent to a reduction in crystal size.

Compositional zoning has been observed with as-synthesized crystals of zeolites A, X, and Y and mordenite (Tempere, Delafosse, and Contour, 1977; Dwyer et al., 1982) and of ZSM-5 (von Ballmoos and Meier, 1981; Derouane et al., 1981; Dwyer et al., 1982). The presence or absence and the direction of a compositional gradient throughout a crystal seems to depend on the synthesis method, source of reactants, and crystal size (Gabelica, Derouane, and Blom, 1984; Debras et al., 1985). Small crystallites tend to have uniform Al distributions, whereas large crystals can have an increasingly high Al concentration near the surface relative to the bulk (von Ballmoos and Meier, 1981; Gabelica, Derouane, and Blom, 1984).

Modification of synthesized zeolites can also lead to crystals with nonuniform Al distribution. Steaming and treatment with $SiCl_4$ produces an Al-rich surface, whereas leaching with acids or EDTA leads to Al depletion of the crystal surface (Dwyer et al., 1981, 1982; Scherzer, 1984).

Although nonuniform distribution of Al in zeolite crystals has thus been observed in several instances, its catalytic significance has yet to be demonstrated. The methods used so far determine the total Al gradients, not those of the catalytically active sites associated with tetrahedral, framework Al. Yet especially the steam-treated samples contain large quantities of catalytically inactive nonframework Al species (Sohn et al., 1986) that appear to migrate toward the crystal surface. Further work is required to define the effect of zoning on catalytic performance.

4.6. Isomorphous Substitution

While classical zeolites contain Si and Al, the possible substitution of these elements by Ga, Ge, Be, B, Fe, Cr, P, and Mg in framework positions by various researchers has been discussed by Barrer (1982). Replacement in a tectosilicate of Si by a trivalent element could lead to the generation of a Brønsted acid site, analogous to Al substitution. Measurements of the acidities of B-, Fe-, Ga- and Al-ZSM-5 have been carried out by Chu and Chang (1985) using infrared measurements and NH_3-TPD. The results indicate the following order of Brønsted acidity:

$$SiOH < B(OH)Si \ll Fe(OH)Si < Ga(OH)Si < Al(OH)Si$$

Most of the catalytic data have been reported for B-ZSM-5 by Tamarasso, Perego, and Notari (1980), Klotz (1981), Ione et al. (1981), Hölderich et al. (1984), Chu et al. (1985), and others. As reviewed by Chu and Chang (1985), B-ZSM-5 was investigated for cracking and aromatization of olefins and paraffins, xylene isomerization, ethylbenzene synthesis and dealkylation, and methanol conversion to hydrocarbons. While the physical measurements (Chu and Chang, 1985; Scholle et al., 1984) indicate that the B(OH)Si framework structure exhibits Brønsted acidic properties, albeit weak ones, the studies of Chu and Chang (1985) showed that the acid activity of the B(OH)Si sites is extremely low and that the catalytic activity observed with B-ZSM-5 was caused mainly, if not entirely, by trace amounts of framework Al atoms; these can be introduced—often inadvertently—as an impurity in the reagents used for zeolite synthesis and/or through subsequent Al transfer from alumina binder as discussed above.

4.7. Activity of ZSM-5 Containing Trace Aluminum

ZSM-5 prepared from commercial grade silica without separate addition of an alumina source can contain as much as 0.3 wt.% Al (3000 ppm). In view of the high activity per site in ZSM-5, as apparent from the TOF's

TABLE 6. Reactions with Low-Aluminum ZSM-5[a]

	SiO_2–Al_2O_3 Ratio	Aluminum (ppm)	WHSV
m-Xylene isomerization	7000	120	90
Propylene polymerization	5600	160	178

[a] With 20% conversion, 1 bar pressure, 450°C.

listed in Table 5, these trace impurities can produce significant activity levels. The data of Table 6 show that the commercially useful reactions of xylene isomerization and propene polymerization (to C_4–C_7 olefins) can be carried out at unusually high weight hourly space velocities on the order of 100, with HZSM-5 containing even less than 200 ppm Al (Haag, 1984b). With a HZSM-5 containing about 2.5 ppm Al (Si–Al ratio ~ 180,000), these reactions have been accomplished at WHSV of about 1.

5. GENERAL PRINCIPLES IN CATALYST DESIGN

In the previous sections it was shown that zeolites offer an unusually large choice of catalyst parameters. For given reactants and products, the proper selection from the wide variety of available zeolite structures together with subsequent modifications determines the diffusivity D. While the precise value of D cannot be predicted and must be determined for each case, a rough first guess can be made by considering that D is a strong function of the ratio of critical molecular diameter and zeolite pore diameter. Together with the selected crystal size, R, an estimate can be made of the characteristic diffusion time $\tau_D = R^2/D$. The characteristic reaction time $\tau_R = 1/k$ for a given reaction can be varied over many orders of magnitude by varying the number of the active sites and the reaction temperature. The appropriate selection of the τ_D and τ_R values, as well as their ratio, $\tau_D/\tau_R = R^2k/D = \phi^2$, depends on the desired objective, as will be discussed below.

5.1. Catalytic Activity

In the absence of other constraints, the activity of the zeolite catalyst can be adjusted over a wide range, either by choosing zeolites of the proper acidity or by varying reaction temperature.

Commercial practice, however, imposes limitation on the choice because of considerations of economy, heat transfer, pressure drop, and catalyst deactivation. For example, in fixed-bed reactors, economy requires a minimum WHSV of 0.2–0.5. When particle diffusion effects are to be avoided, a WHSV of about 20 should not be exceeded for $\frac{1}{16}$-in. particles used in a large-scale gas-phase process; a reduction in particle size reduces diffusion limitations and allows higher space velocities, but the pressure drop in the reactor increases. To reduce this problem, special catalyst shapes have been designed that combine a low-pressure drop with an effective small-particle diffusion path (Pereira, Kim, and Hegedus, 1985).

5.2. Catalyst Deactivation

Catalyst aging through coke formation is of great importance for the design of any commercial catalyst. Extensive reviews are available for catalysts in general (Butt, 1972; 1978; Delmon and Froment, 1980; Figueiredo, 1982; Hegedus and McCabe, 1984; Oudar and Wise, 1985). With zeolite catalysts it is found that the pore structure exerts a very large effect on the rate of coke formation and deactivation (Walsh and Rollmann, 1977; Rollmann and Walsh, 1979) as illustrated in Figure 11. HZSM-5 and similar medium-pore zeolites have remarkably low coking tendencies; this can be ascribed to the geometrical constraint imposed by the pores which makes it sterically difficult to form the large polynuclear hydrocarbons responsible for irreversible deactivation. Thus coking, if any, occurs predominantly on the outer surface of the crystals (Olson and Haag, 1984). In view of the larger intracrystalline cavities, large-pore zeolites such as zeolite Y and mordenite can form such large cokelike molecules within their cavities and deactivate much more rapidly.

In addition to the pore structure, the activity of the catalyst also affects aging. A reduction in activity generally has been found to reduce coke formation and the rate of deactivation. For example, in gas oil cracking with Y-type catalysts, the coke selectivity is reduced when their activity is reduced by steaming. A similar beneficial effect of dealumination was found with mordenite in the conversion of methanol to hydrocarbons (Bandiera, Hamon, and Naccache, 1984) and with Y-type zeolites during hexadecane cracking (Bremer et al., 1983).

Catalyst deactivation has also been said to be a function of the acid site distribution and arrangement (Ducarme and Vedrine, 1985; Bandiera, Harmon, and Naccache, 1984) and to be increased by an increase in the acid

5. GENERAL PRINCIPLES IN CATALYST DESIGN

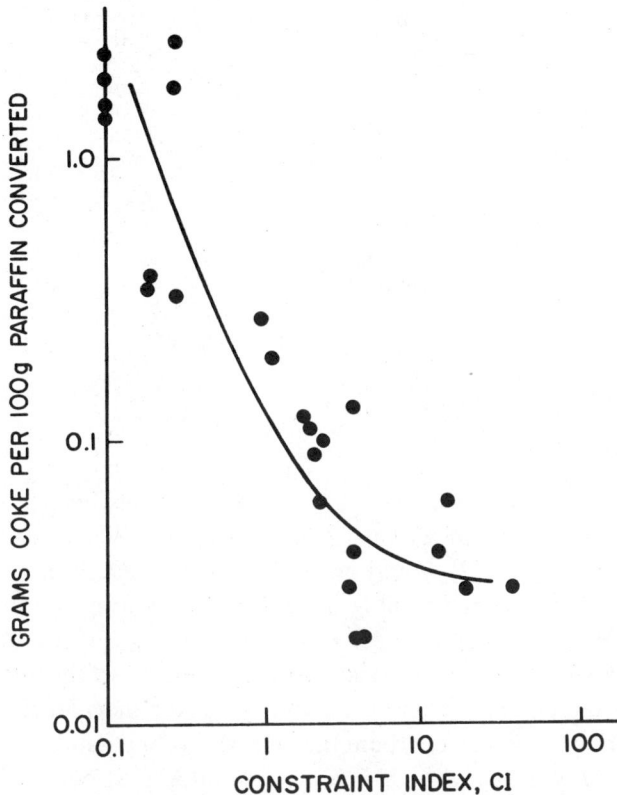

FIGURE 11. Dependence of coke selectivity on the pore size of zeolites as measured by their constraint index. From L. D. Rollmann and D. E. Walsh, *Journal of Catalysis* **56**, 140. Copyright © 1979, Academic Press.

site density (Derouane, 1985). It is unlikely that an effect exists other than that of raising the overall activity. However, in zeolites with very low Si–Al ratios (<5), the polar character is sufficiently large that it can lead to the preferential sorption of highly olefinic or aromatic coke precursors, and hence to faster coke formation.

5.3. Designing for the Edge

Optimal operating conditions and catalyst activity in commercial processes have usually been determined empirically. Yet it is interesting to note that many commercial processes are operating near the "edge" of diffusion

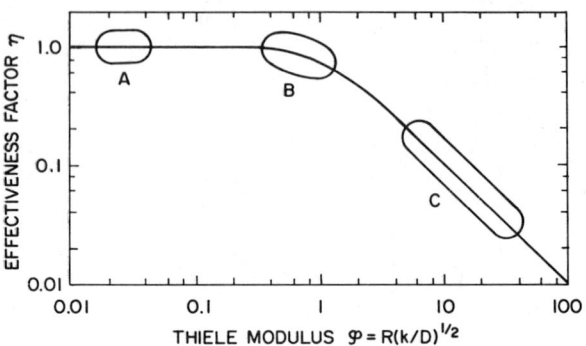

FIGURE 12. Effect of Thiele modulus parameters on catalyst design. Region B leads to optimal overall catalyst performance (see text).

limitation (Figure 12, point B). The practitioner has been guided to the optimum design point by the art, if not always the science. Operating in the region A (Figure 12) would not be advantageous: The catalyst size R is smaller than necessary leading to high pressure drop; or the activity k is low (more catalyst and a larger reactor is required); or the diffusivity D is higher than needed (the increased porosity means a lower catalyst density requiring a larger reactor and a catalyst of lower mechanical stability)—all this without any obvious benefits. On the other hand, while the rate advantages of working at high k values continue past point B, the occurrence of diffusion limitations in the C region leads to usually undesirable selectivity changes and often to faster catalyst deactivation, as mentioned above. However, low-aging shape-selective zeolite catalysts can take advantage of working in the C region of Figure 12, at least with regard to some feed or product molecules.

5.4. Shape Selectivity

Molecular-shape-selective catalysis in zeolites can be obtained when the dimensions of reactant or product molecules approach those of the intracrystalline pores. Several reviews describe the phenomenological aspects of shape selectivity (Csicsery, 1976, 1984).

Two different mechanisms operate to cause shape selectivity. In one, selectivity results from a large difference in the diffusivity of the participating molecules in the zeolite channels (mass transport selectivity). In the

5. GENERAL PRINCIPLES IN CATALYST DESIGN

other, selectivity is caused by steric constraints in the transition state of the catalytic transformation step (transition state selectivity). Recognition of these two different aspects of shape selectivity is more than a mechanistic detail. Since the origins are different, the catalyst features that are to be manipulated and optimized in catalyst design are different for the two cases as will be shown below.

Mass transport selectivity is responsible for the shape-selective reactions occurring with small-pore zeolites (Weisz et al., 1962; Chen and Weisz, 1967). It is also the origin of many selective reactions in medium-pore zeolites, such as the selective formation of para-substituted dialkylbenzenes with ZSM-5 catalysts (Olson and Haag, 1984). In these cases, the observed net rates of conversion or production of different molecular species is governed by the classical relationships for diffusion–reaction inhibition (Thiele, 1939; Weisz and Prater, 1954; Satterfield, 1970). For example, for two parallel reactions with an intrinsic selectivity $S = k_A/k_B$,

$$A \xrightarrow{k_A} P_1 \qquad\qquad M + A \xrightarrow{k_A} P_1$$
$$\text{or} \qquad\qquad\qquad\qquad\qquad (3)$$
$$B \xrightarrow{k_B} P_2 \qquad\qquad M + B \xrightarrow{k_B} P_2$$

the observed selectivity is

$$S_{obs} = \frac{k_A \eta_A}{k_B \eta_B} = S \frac{\eta_A}{\eta_B} \qquad (4)$$

The completely shape-selective conversion of linear paraffins (A) in the presence of branched molecules (B) with small-pore zeolites occurs when $D_B \sim 0$, that is, when the B molecules are essentially completely excluded from the zeolite interior (molecular sieving), as is the case with small-pore zeolites. With medium- and large-pore zeolites, both D_A and D_B are often finite. The observed selectivity in this case depends on the degree of diffusion limitation $\phi^2 = R^2 k/D$, of the more bulky, antiselective species B, as shown in Figure 13 for the case of $k_A = k_B$ (Weisz, 1980). Under conditions of severe diffusion limitation, $\eta = 1/\phi$ and Eq. (4) becomes

$$S_{obs} = \frac{k_A \phi_B}{k_B \phi_A} = \sqrt{\frac{k_A D_A}{k_B D_B}} \qquad (5)$$

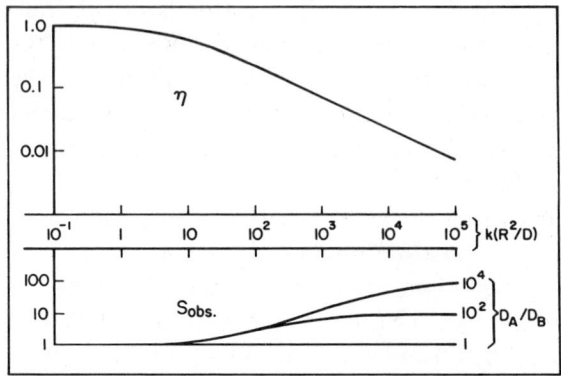

FIGURE 13. Observed selectivity in a parallel reaction as a function of Thiele modulus of the slower diffusing, antiselective species.

It is apparent that the degree of selectivity desired can be controlled by choosing the appropriate zeolite, which determines the individual diffusivities and their ratio. For a given zeolite (constant D_A/D_B), the selectivity can be adjusted by varying the crystal size (R) and the catalytic activity (k) via synthesis or steaming as shown in Section 4, to obtain the desired value of $\phi^2 = R^2k/D$. When both reactions are severely diffusion limited [Eq. (5)], a change in crystal size or activity does not change the observed selectivity (Figure 13).

An illustrative example in Table 7 shows how the selectivity S_{obs} for the cracking of n-hexane (H) and 2,2-dimethylheptane (DMH) can be varied

TABLE 7. Selective Cracking of n-Hexane(H)/2,2-Dimethylheptane(DMH)^a

Catalyst Sample:	A		B	C			D
CrystalSize, $R(\mu m)$:	0.025		0.10	1.35			1.8
k_H (= intrinsic activity)	0.85	29	25	1.8	29	36	43
$(k_{DMH})_{intr}$	1.9	63	53	4.0	64	79	95
$(k_{DMH})_{obs}$	1.9	63	53	2.5	8.4	7.6	6.6
$(\phi^2 = R^2k/D)_{DMH}$	4×10^{-4}	0.013	0.17	2.4	39	48	102
Selectivity $k_H/(k_{DMH})_{obs}$	0.45	0.46	0.47	0.72	3.4	4.7	6.5

^{a}With HZSM-5 catalyst, 811 K.

5. GENERAL PRINCIPLES IN CATALYST DESIGN

from the intrinsic value of 0.45 to 6.5, that is, by a factor of 14, by an increase in the crystal size and intrinsic activity (Haag, Lago, and Weisz, 1982). In the table, the intrinsic rate constant for DMH is calculated from the value $(k_{DMH}/k_H)_{intrinsic} = 2.20$ determined from the diffusion-free cases. The ϕ^2 values are calculated from the known crystal size R, the intrinsic k_{DMH}, and the experimentally determined diffusivity of 3×10^{-8} cm^{2}sec^{-1} (Haag, Lago, and Weisz, 1982). The data of Table 7 also illustrate the sometimes neglected fact that a catalyst that experiences a reduction in activity—for example, by repeated regeneration—can also lose selectivity as a direct consequence thereof. For example, an intrinsic activity loss of 20% of the most active C catalyst ($k_H = 36$) reduces the selectivity from 4.7 to 3.4.

Another example of catalyst design to obtain para-selective toluene disproportionation is given in Section 6.

The second mechanism for shape-selectivity control results from differences in the size of a reaction intermediate or of the transition state. This transition state (TS) selectivity (Csicsery, 1976; Haag, Lago, and Weisz, 1982) can be found in reactions whose reaction complex is larger than the feed or product molecules, such as in bimolecular reactions:

$$A + M \longrightarrow A^{\ddagger} \longrightarrow P_1$$
$$B + M \longrightarrow B^{\ddagger} \longrightarrow P_2$$

Csicsery (1971) has shown this phenomenon in the bimolecular reaction of transalkylation of dialkylbenzenes on mordenite catalysts.

Unlike mass transfer selectivity, TS selectivity is not defined by the variables of the Thiele parameter $\phi^2 = R^2k/D$; it is independent of crystal size and activity, but depends on the pore diameter and zeolite structure. A reaction showing TS selectivity is the cracking of n-hexane (H) and 3-methylpentane (MP). While cracking at high temperature (811 K) occurs partly by a monomolecular process (Haag and Dessau, 1984), its transition state at lower temperature involves a bimolecular hydride transfer step. The size of this transition state complex approaches the dimensions of the intracrystalline space in medium-pore zeolites and is subject to steric inhibition, which is greater for cracking of 3-methylpentane than of hexane (Haag, Lago, and Weisz, 1982). The observed selectivity k_H/k_{MP} = constraint index (C.I.) is a useful probe of the intracrystalline space (Frilette, Haag, and Lago, 1981) of medium-pore zeolites, as shown in Figure 14.

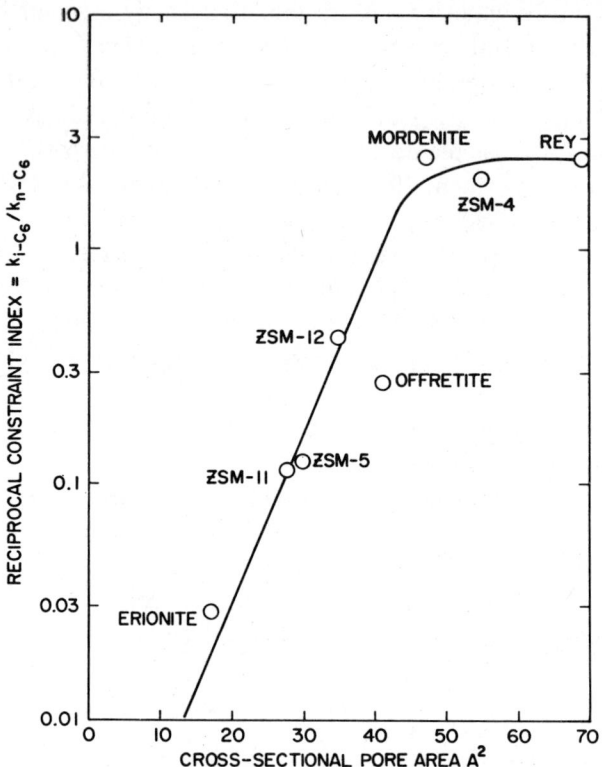

FIGURE 14. Dependence of constraint index on pore size.

It was found for HZSM-5 that a change in crystal size or in activity, each by two orders of magnitude, had no effect on the C.I.

Operation of TS selectivity in reactions of commercial importance will be discussed below.

6. DESIGN OF SHAPE-SELECTIVE CATALYSTS FOR COMMERCIAL PROCESSES

This section describes the use of zeolite catalysts in industrial processes and some of the important catalytic principles that are utilized and that underlie the matching of a catalyst to a particular process. A summary of

6. SHAPE-SELECTIVE CATALYSTS

TABLE 8. Processes Utilizing Zeolite Catalysts

Cracking	Gas oil to gasoline and distillates	REY, US-Y, ZSM-5
Hydrocracking	Heavy fractions to naphtha and distillates	Me + REX, REY, US-Y
Hysomer	Isomerization of pentane and hexane	Pt/Mordenite
Selectoforming	Post-reforming process	Erionite
M-forming	Upgrading of reformate	ZSM-5
MDDW, MLDW	Dewaxing of distillates and lube oil	ZSM-5
MLPI, MVPI, MHTI	Xylene isomerization	ZSM-5
MTDP	Toluene disproportionation	ZSM-5
MEB	Ethylbenzene synthesis	ZSM-5
MOGD	Conversion of olefins to gasoline and distillate	ZSM-5
MTG	Methanol conversion to gasoline	ZSM-5
MTO	Methanol conversion to olefins	ZSM-5
M2-forming	Formation of aromatics from paraffins and olefins	ZSM-5
Para-selective reactions	Synthesis of p-ethyltoluene synthesis of p-xylene from toluene	ZSM-5

processes in which zeolite catalysts have been used is given in Table 8. The important process of catalytic cracking of gas oil is treated separately in Section 7.

6.1. M-Forming

A new post-reforming process using ZSM-5 as the catalyst, known as the M-forming process, was developed in the early 1970s at Mobil. The catalyst performs two major functions, namely, to selectively crack paraffins and to alkylate benzene and toluene present in the reformate with the olefinic portion of the cracked products. The process improves liquid yield and reduces the concentration of benzene in the product.

The rate of cracking of paraffins over ZSM-5 increases with the length

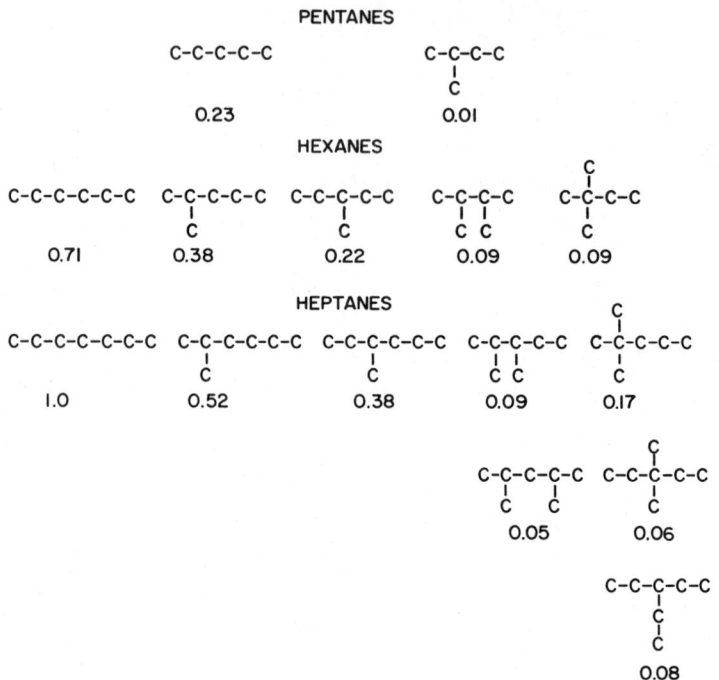

FIGURE 15. Relative rates of cracking of paraffins over ZSM-5 (1.4 LHSV, 35 atm pressure, 340°C).

of the molecules and decreases with the bulkiness of the molecules (Figure 15). For octane boosting, this is a most desirable feature: the more linear paraffins are the lowest octane components; they have the highest cracking rate.

The alkylation reaction over ZSM-5 favors benzene and toluene over the higher alkylaromatics in the reformate. These lighter alkylated products provide the additional liquid yield and retain the high octane rating of benzene and toluene, while heavy aromatics outside the gasoline boiling range are not produced to any significant extent because of shape-selective constraints.

Table 9 shows detailed compositional changes when a 85 clear octane reformate is converted over the M-forming catalyst at two different reaction severities (Chen, Garwood, and Heck, 1987). Aromatics alkylation is indicated by the disappearance of benzene and toluene and the appearance of C_8^+ aromatics.

TABLE 9. Compositional Change at Different M-Forming Severities[a]

C_5^+ (R + O)	Feed 84.5	Product 89.5	Δ	Product 92.7	Δ
Aromatics					
B	18.5	16.9	−1.6	16.3	−2.2
T	23.4	22.5	−0.9	21.7	−1.7
X	0.6	1.1	+0.5	1.5	+0.9
C_9	0.2	2.9	+2.7	3.7	+3.5
C_{10}	0	2.0	+2.0	3.3	+3.3
C_{11}	0	0.3	+0.3	0.4	+0.4
Total	42.7	45.8	+3.0	46.8	+4.1
Paraffins					
C_6	30.9	26.2	−4.7	23.4	−7.5
C_7	12.9	9.7	−3.2	8.7	−4.2
C_8	0.2	0	−0.2	0.3	+0.1
C_9	0	0	0	0	0
Total	44.0	35.9	−8.1	32.4	−11.6
Naphthenes	1.4	1.6	+0.2	1.1	−0.3
Pentanes	8.0	8.6	+0.6	9.0	+1.0
Butanes and ligher	3.9	8.1	+4.2	10.7	+6.8

[a] With C_6 − 80°C mid-continent naphtha, 28 atm pressure.

6.2. Distillate and Lube Oil Dewaxing (MDDW, MLDW)

Just as the octane rating of a paraffinic molecules is affected by the degree of branching of the molecule, the melting point of a paraffinic molecule is similarly determined by its degree of branching. Thus, the cold flow property (pour point, freeze point, cloud point, and cold filter plugging point) of distillate fuels and lubrication oils is largely determined by the composition of its paraffinic fraction. ZSM-5 has just the desired pore size to distinguish between the high-pour-point molecules from the low-pour-point bulkier molecules and shows high selectivity in pour-point reduction with minimum yield loss.

However, unlike the other applications of ZSM-5, the dewaxing catalysts were designed to operate without feed pretreatment and therefore had to tolerate nitrogen and sulfur compounds, which are known as catalyst poisons. Proper adjustment of the acid activity was found to improve dramatically its rate of deactivation in processing nitrogen-containing unpretreated feeds—a phenomenon not yet completely understood.

6.3. Xylene Isomerization

The commercial production of *p*-xylene involves its recovery from a C_8-aromatics stream by crystallization or sorption; the remaining para-depleted xylene fraction is isomerized to an equilibrium mixture and recycled.

Xylene isomerization over acid catalysts is accompanied by an undesirable disproportionation reaction to form toluene and trimethylbenzenes, leading to loss of valuable xylenes. Relative to larger-pore zeolites, this side reaction is greatly reduced with the use of smaller-pore zeolites. As discussed earlier (Haag, Olson, and Weisz, 1984; Olson and Haag, 1984), this is a result of transition state selectivity: The rate of xylene disproportionation is selectively inhibited because the bulky bimolecular transition state cannot be readily accommodated in the zeolite pores. Figure 16 shows how the ratio of disproportionation and isomerization rates depends on the size of the intracrystalline cavity.

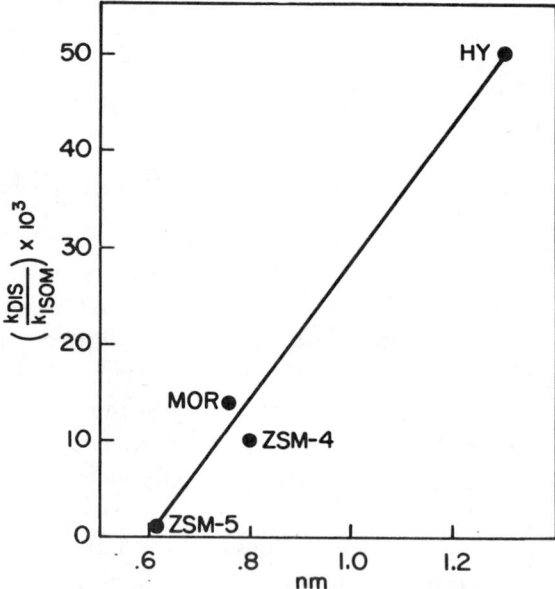

FIGURE 16. Effect of intracrystalline cavity diameter of several zeolites on selectivity in xylene isomerization. From D. H. Olson and W. O. Haag, *ACS Symposium Series* **248**, 280. Copyright © 1984, American Chemical Society.

6.4. Conversion of Olefins to Gasoline and Distillates

With increasing reaction severity, olefins undergo a series of acid-catalyzed reactions, starting with double-bond isomerization, cracking–polymerization, hydrogen transfer, cyclization, and aromatization. With medium-pore zeolites, these reactions are taking place under the constraints imposed by the size of the channels. As a result, unique products are formed.

Figure 17 shows an example of reacting propene over ZSM-5 at 200°C, 2.7 WHSV, and 35 atm. The products are predominantly trimers, tetramers, and pentamers (Garwood, 1983). At longer residence times, these oligomers undergo a series of cracking–polymerization steps (Haag, 1984a) to form a mixture of olefins, which approaches a pseudoequilibrium composition that is determined only by the temperature and pressure of the system and is independent of the feed. This effect is illustrated by the data in Figure 18 (Haag, Lago, and Rodewald, 1982); as the temperature is increased, the molecular weight distribution shifts to lighter products, as expected from thermodynamics. In each case, the activity of the catalyst (SiO_2–Al_2O_3 ratio R) and the contact time (WHSV) was adjusted to limit

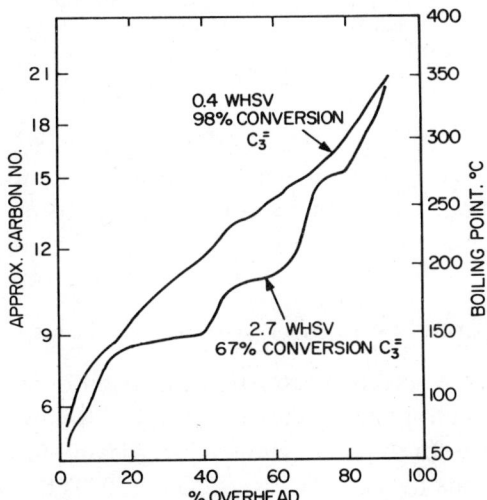

FIGURE 17. Effect of space velocity on propene conversion over HZSM-5 (35 atm pressure 200°C). From W. E. Garwood, *Intrazeolite Chemistry,* ACS Symposium Series, G. D. Stucky and F. G. Dwyer, Eds., Vol. 218, 386. Copyright © 1983, American Chemical Society.

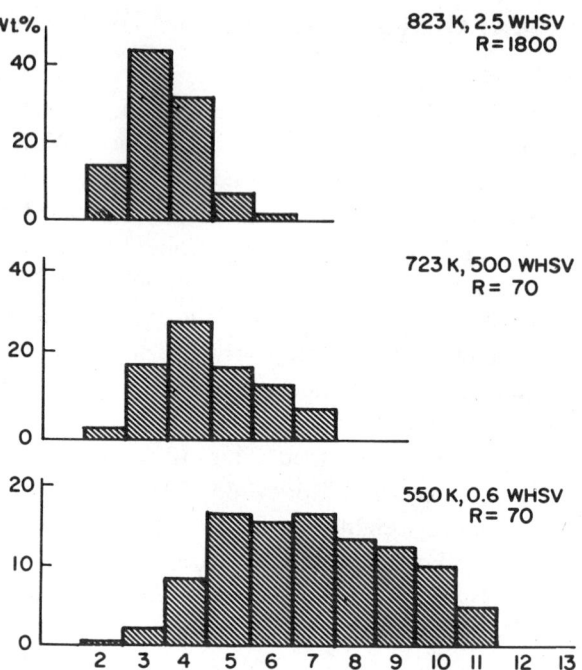

FIGURE 18. Olefin distribution from propylene. (ZSM-5, 1 bar pressure, $R = SiO_2/Al_2O_3$). Abscissa: carbon number. From W. O. Haag, R. M. Lago, and P. G. Rodewald, *Journal of Molecular Catalysis* **17,** 167 (1982).

the reaction to olefin interconversion. At higher temperatures, and/or longer contact time, hydrogen transfer reactions forming paraffins and aromatics become appreciable.

6.5. Methanol Conversion to Gasoline (MTG) and Olefins (MTO)

The discovery of the selective catalytic conversion of methanol to hydrocarbons over medium-pore zeolites led to the development of methanol conversion catalysts for the production of high-octane gasoline (MTG process) and light olefins (MTO process). Chang (1983) has reviewed the extensive literature describing the conversion and its mechanistic aspects.

The reaction path of this novel reaction has been elucidated by Chang and Silvestri (1977). The first step is the conversion of methanol/dimethyl ether to light olefins, which then react to form higher olefins and eventually aromatics and paraffins. The zeolite structure has a striking effect on the nature of the aromatics produced (Chang, Lang, and Bell, 1981) as can be seen from Figure 19. Large-pore zeolites (ZSM-4, mordenite), like other non-shape-selective acid catalysts, produce heavy aromatics rich in penta- and hexamethylbenzene by rapid alkylation of lower aromatics with methanol. With the medium-pore zeolites, ZSM-5 and ZSM-11, their formation

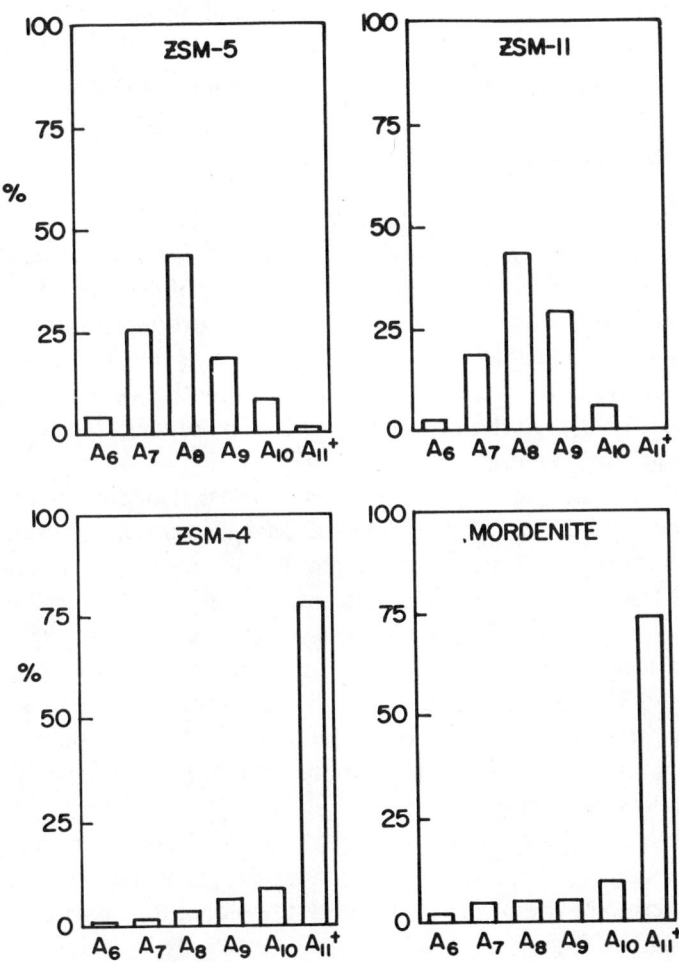

FIGURE 19. Aromatics distribution from methanol conversion over various zeolites. From C. D. Chang, W. H. Lang, and W. J. Bell, *Catalysis of Organic Reactions*, W. R. Moser, Ed., Chemical Industries Series, Vol. 5, p. 81 (1981) by courtesy of Marcel Dekker, Inc.

is prevented because of transition state control. The largest aromatic, formed in small amounts, is durene, the symmetrical 1,2,4,5-tetramethylbenzene. Thus, the success of this new process for the production of gasoline hinges critically on the fine control exerted by the precise dimension of the zeolite channels.

Instead of producing an aromatic gasoline, methanol can also give a high yield of light olefins. Preferential production of olefins is achieved by reducing the acidity of the zeolite and raising the operating temperature to above 500°C, where the relative rates of olefin formation reactions and aromatization reactions are sufficiently different so that C_2 to C_5 olefin yields as high as 80% have been obtained (Chang, Chu, and Socha, 1984).

6.6. Para-dialkylbenzenes

The design of catalysts to produce selectively para-dialkylbenzenes represents a prime example of how product selectivity may be achieved by the manipulation of the intrinsic chemical kinetics of the reactions involved and the diffusional transport inhibition of the product molecules. Such selectivity has been achieved for the production of p-xylene from toluene via disproportionation and via alkylation with methanol (Chen, Kaeding, and Dwyer, 1979) and the production of p-ethyltoluene from toluene and ethene (Kaeding, Young, and Chu, 1984).

In all these cases, large-pore catalysts give less than 50% selectivity for the desired para-isomer. Unmodified ZSM-5 also usually produces equilibrium mixtures of xylene, containing only 24% para-isomer. The reason for this is that the intrinsic rate constant for xylene isomerization is over 5000 times larger than that for toluene disproportionation. The catalyst has to be designed such that the rates of these two reactions are at least comparable. This has been accomplished by independently increasing some or all the variables of the Thiele modulus, $\phi = R\sqrt{k/D}$ for the isomerization reaction. A detailed discussion of this topic has been presented (Olson and Haag, 1984).

In practice, the adjustable range of each of these three independent variables extends over at least two or more orders of magnitude for each. The interaction of chemical kinetics and diffusional transport is illustrated schematically in Figure 20. High p-xylene selectivity is achieved when $D_P \gg D_{O,M}$, $k_I R^2/D_{O,M} \gg 1$, $k_T R^2/D_T \ll 1$, and (k_I/k_D) obs ≤ 1. The condition $k_I R^2/D_{O,M} \gg 1$ is most important and is subject to catalyst design. For constant activity k_I, the value $R^2/D_{O,M} = \tau_D$ represents a directly meas-

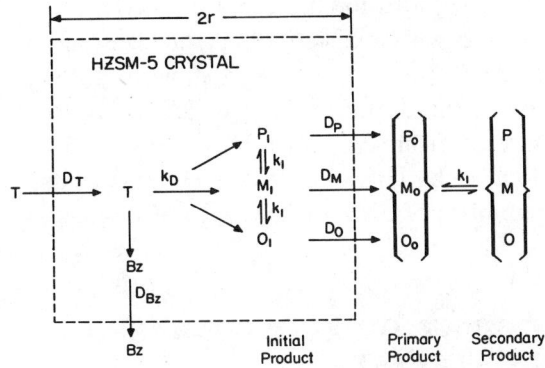

FIGURE 20. Model for selective toluene disproportionation. *Symbols: D* = diffusion coefficient; *k* = rate constant; *r* = crystal radius; Bz = benzene; T = toluene. P,M,O = para-, meta-, and ortho-xylene. *Subscripts:* I = isomerization; D = disproportionation; 1 = initial product; 0 = primary product; T = toluene; Bz = benzene. From D. H. Olson and W. O. Haag, *Catalytic Materials: Relationship Between Structure and Reactivity*, T. E. Whyte, R. A. Dalla Betta, E. G. Derouane and R. T. K. Baker, Eds., *ACS Symposium Series* **248**, 289. Copyright © 1984, American Chemical Society.

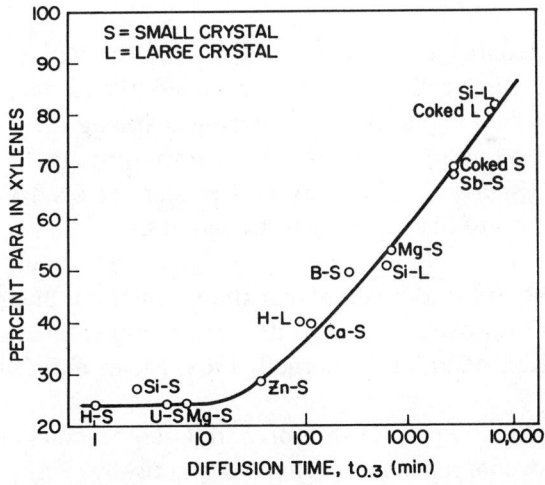

FIGURE 21. Relationship between diffusion parameter ($t_{0.3}$) and para-xylene selectivity in toluene disproportionation. TDP at 550 °C, P = 41 bar, 20% conversion; *o*-xylene sorption at 120°C, time to reach 30% of amount sorbed at infinite time. From D. H. Olson and W. O. Haag, *Catalytic Materials: Relationship Between Structure and Reactivity*, T. E. Whyte, R. A. Dalla Betta, E. G. Derouane and R. T. K. Baker, Eds., *ACS Symposium Series* **248**, 295. Copyright © 1984, American Chemical Society.

urable quantity. The diffusion time τ for o-xylene at 120°C has been measured for ZSM-5 with a variety of crystal sizes and different amounts and types of modifying agents. Figure 21 shows that the para-selectivity is a direct function of the diffusion time, which has been increased by four orders of magnitude from the small crystal HZSM-5 to a surface-coked large crystal, that is, by changing R and D. This example illustrates the power and capabilities of a scientific design of a selective catalyst.

7. DEVELOPMENT OF ZEOLITE-BASED CRACKING CATALYST

7.1. Principles Involved in the Early Discoveries

Plank, Rosinski, and Hawthorne at Mobil (1964) discovered that a commercially acceptable bead cracking catalyst could be formulated by dispersing 10% of a steam-stabilized rare earth-exchanged zeolite X in a silica-alumina matrix. This basic formulation, which has withstood the test of time, is truly a remarkable design encompassing a number of principles, worthy of some elaboration.

Rare Earth Exchange. Ion exchange studies show that for zeolite X all but 16 ions per unit cell are easily exchanged. These last 16 cations are located inside the sodalite cages, requiring stripping the hydration shells off the rare earth ions in order for them to move through the six-membered oxygen rings. (Sherry, 1968; 1971) This is best achieved by calcining the exchanged zeolite to 350°C between exchanges.

X-ray diffraction studies show that the thermal and hydrolytic stability of zeolite X is improved by having its sodalite cages filled with rare earth ions (Olson, Kokotailo, and Charnell, 1968; Meier and Olson, 1971).

Steaming. Steaming a freshly prepared zeolite cracking catalyst was found to improve its gasoline selectivity dramatically (Plank, Rosinski, and Hawthorne, 1964). The major function of steaming a zeolite catalyst is to reduce its acid site density so that its catalytic activity is made compatible with the rate of mass transport. Thus the gasoline selectivity is improved by reducing secondary cracking. Without polyvalent ion exchange, ammonium zeolite X, for example, is rendered amorphous by steaming because of uncontrolled loss of framework aluminum. The problem is solved by rare earth ion exchange, which reduces the number of protonic sites subject to hydrolytic aluminum removal.

Switch to Fluid-Bed Catalyst. The activity advantage of zeolite cracking catalysts over the amorphous catalyst has been well documented (e.g., Plank, Rosinski, and Hawthorne, 1964). Weisz and his coworkers (Weisz and Miale, 1965; Miale, Chen, and Weisz, 1966), using the α-test, demonstrated zeolitic activities 10^3–10^4 times greater than those of conventional catalysts. However, the availability of this high activity is of no immediate value to the catalytic cracking process. To take full advantage of the zeolite catalyst, the switch to the fluid-bed process was inescapable.

Zeolite–Matrix Formulation. The many roles served by the matrix have been documented (Oblad, 1963; Eastwood, Plank, and Weisz, 1971). The matrix provides shape and bulk to the particle and minimizes mechanical attrition. Some matrices may function as sodium sinks allowing solid–solid sodium ion exchange, contributing to the stability of the zeolite. The matrix also acts as a heat carrier facilitating heat transfer during regeneration. But seldom discussed are the basic principles underlying the empirically derived decision to put 10% zeolite in the matrix.

The catalytic cracking process couples the endothermic cracking reaction with the exothermic coke-burning reaction. A delicate heat balance is maintained between the cracking reactor and the catalyst regenerator. This heat balance is achieved by circulating solid catalyst particles between these two vessels. As shown schematically in Figure 22, the process has three independent variables, namely, the circulation rate of the solid catalyst, the feed rate, and the feed temperature. The temperature of both the regenerator and the reactor has stringent limits. The concentration of coke on the catalyst determines the maximum temperature of the catalyst leaving the regenerator, while the reactor temperature is controlled by the catalyst–oil ratio and the feed temperature. The feed rate and the catalyst–oil ratio must be adjusted to yield a coke concentration of about 1%, which would provide enough heat to raise the catalyst temperature to 740°C in the regenerator.

Laboratory studies using pure zeolites show that they are more active and selective than the amorphous catalyst. Coke selectivity is reduced to below 5% on feed. Furthermore, the catalyst becomes deactivated only after more than 10–15% of coke has deposited in the zeolite.

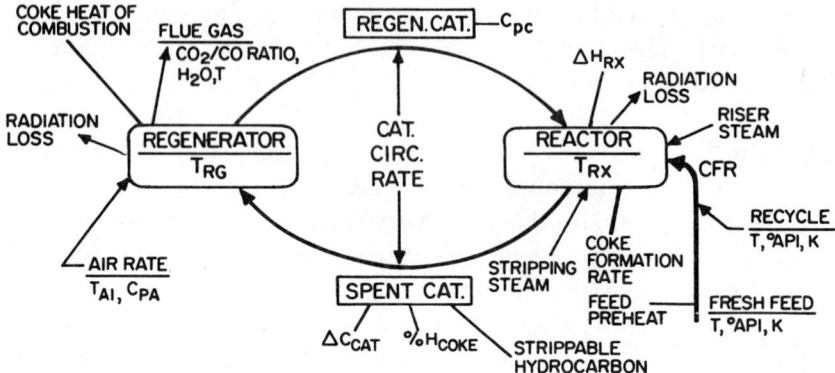

FIGURE 22. Coupling between reactor and regenerator in gas oil cracking. From P. B. Venuto and E. T. Habib, *Catalysis Reviews—Science and Engineering* **18**(1), 42 (1978) by courtesy of Marcel Dekker, Inc.

But when such a catalyst is to be used in the existing catalytic cracking units, it will be subjected to the same contraints applied to the amorphous catalyst. Thus, in order to operate the regenerator in the same temperature range, it would not be desirable to have more than 1% coke on the catalyst particles. In order for the catalyst particles to supply similar amounts of heat to the reactor, diluting the concentration of the active component to about 10% in the catalyst particle provides a simple and economical solution that fulfills all the process requirements.

Lower coke selectivity brings other benefits to an existing cracking unit. The same coke load on the regenerator can be translated into either higher throughput (lower catalyst–oil ratio) or higher single-pass conversion and still satisfy heat balance requirements.

Despite numerous advances made in improving the selectivity of the cracking catalyst and the development of the riser cracker and resid cracking over the past 30 years, the guiding principles underlying the concept of matrix formulation remain intact.

7.2. Principles Involved in Improved Activity and Selectivity

The high activity and the outstanding selectivity characteristics of zeolites were recognized early. For example, Table 10 (Blazek, 1968) shows the superiority of zeolites over silica-alumina in cracking a gas oil (namely, higher conversion and more gasoline at the expense of C_4's, dry gas and

TABLE 10. Comparison of Yield Structure for Fluid Catalytic Cracking of Waxy Gas Oil over Commercial Equilibrium Zeolite and Amorphous Catalysts[a]

Yields, at 80 vol% Conversion[b]	Amorphous, High Alumina	Zeolite, XZ-25	Δ Change from Amorphous
Hydrogen (wt%)	0.08	0.04	−0.04
$C_1 + C_2$'s (wt%)	3.8	2.1	−1.7
Propylene (vol%)	16.1	11.8	−4.3
Propane (vol%)	1.5	1.3	−0.2
Total C_3's	17.6	13.1	−4.5
Butenes (vol%)	12.2	7.8	−4.4
i-Butane (vol%)	7.9	7.2	−0.7
n-Butane (vol%)	0.7	0.4	−0.3
Total C_4's	20.8	15.4	−5.4
C_5−390 at 90% ASTM gasoline (vol%)	55.5	62.0	+6.5
Light fuel oil (vol%)	4.2	6.1	+1.9
Heavy fuel oil (vol%)[c]	15.8	13.9	−1.9
Coke (wt%)	5.6	4.1	−1.5
Gasoline octane number (R + 0)	94	89.8	−4.2

[a]Data from Blazek (1968); run at 950°F at 1 atm pressure in a fixed fluid bed. Reprinted from *Catalysis Reviews—Science and Engineering* **18**(1), 41 (1978) by courtesy of Marcel Dekker, Inc.
[b]No loss of basis on feed.
[c]HFO = 100 − conversion − LFO.

coke). The increased gasoline–(gas + coke) ratio reflects less secondary cracking reactions. As shown by the data in Figure 23, the gasoline fraction contained more paraffins and aromatics and less olefins than those from silica-alumina (Eastwood, Drew, and Hartzell, 1962), suggesting that the acidic zeolite catalysts have relatively higher reactivity for hydrogen transfer between hydrocarbon species than the amorphous catalysts (Weisz, 1970; Eastwood, Plank, and Weisz, 1971). However, our understanding of why zeolites are so active and selective in catalytic cracking evolved long after their discovery.

There is now a general consensus that the high catalytic activity is explained by the high concentrations of reactants in the very small pores; a

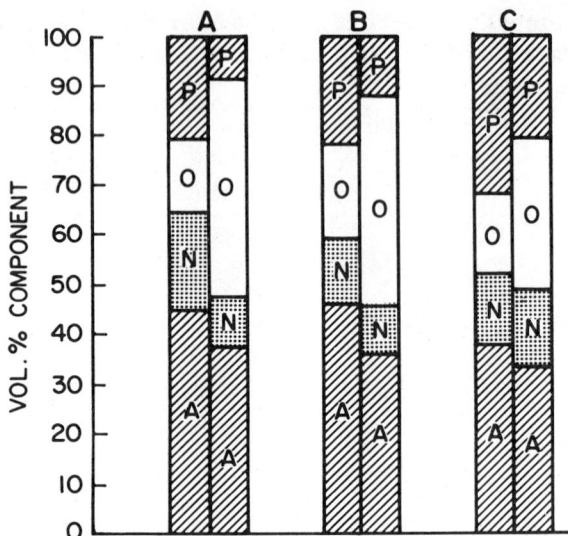

FIGURE 23. Comparison of paraffin, olefin, naphthene, and aromatic (P, O, N, A) contents in C_5 375–400°F A.S.T.M. 90% point TCC gasoline from catalytic cracking of (*A*) California virgin gas oil, (*B*) California coker gas oil, (*C*) Gachsaran virgin gas oil. Left and right members of bar graph pairs reflect zeolitic and amorphous cracking, respectively. From P. B. Venuto and E. T. Habib, *Catalysis Reviews—Science and Engineering* **18**(1), 55 (1978) by courtesy of Marcel Dekker, Inc.

higher reactant concentration would also favor hydrogen transfer reactions, a second-order reaction, over some of the first-order cracking reactions. Also, the number of catalytically active sites is much larger in zeolites.

Laboratory studies examining the composition of the gasoline fractions obtained with rare earth-exchanged cracking catalysts clearly reflect a quantitative shift of hydrogen according to

$$\text{olefins + naphthenes} \longrightarrow \text{paraffins + aromatics}$$

These hydrogen transfer reactions stabilize the gasoline product, as it is well known that olefinic products could undergo secondary cracking reactions. This also explains why with the zeolitic cracking catalyst, it is possible to increase single-pass conversion of the gas oil without losing product selectivity. However, the less olefinic gasoline produced by the zeolite catalyst is lower in octane than that produced by the amorphous silica-alumina catalyst (Table 10).

7.3. Recent Advances in Catalyst Design

The introduction of zeolitic catalytic cracking catalyst made possible a huge increase in the throughput of existing commercial cracking units. Further progress in catalytic cracking has been the result of having to race against a moving objective of more gasoline, higher octane, and lower coke-make with heavier feeds of lower hydrogen content. In fact, over the years the hydrogen content of the feed has deteriorated to the point that the refiners are facing the unchallengeable laws of conservation and stoichiometry, that is, unless an external source of hydrogen is added to the feed, such as hydrotreating, more carbon has to be rejected as coke.

Major advances in product selectivity have been achieved by replacing zeolite X with zeolite Y and more recently dealuminized ultrastable Y (Scherzer and Ritter, 1978). With the latter catalyst it is now possible to produce high-octane olefinic gasoline at high single-pass conversion without as much need as before to rely on hydrogen transfer reactions to maintain gasoline selectivity.

While there is general consensus that higher silica-to-alumina ratio zeolites have higher thermal and hydrothermal stability, there is much speculation on the structural factors contributing to product selectivity (Pine, Maher, and Wachter, 1984; Magee, Cormier, and Woltermann, 1985).

Besides the effect of lower site density on intrinsic activity, there is little evidence to support the contention that the gasoline olefinicity is related to its lack of hydrogen transfer activity. Those who contend that the change in product distribution is the result of having stronger acid sites as the population of framework aluminum is decreased are unable to differentiate its effect from the effect of lower site density (Pine, Maher, and Wachter, 1984).

It is our contention that the reduction in intrinsic activity resulting from the higher silica-to-alumina ratio, and with it, the changes in competitive sorption properties, that is, the selectivity between feed and product molecules, suffice to explain the experimental observations.

REFERENCES

M. W. Anderson, J. Klinowski, and X. Liu, *J. Chem. Soc. Chem. Commun.*, 1596 (1984).

R. J. Argauer and G. R. Landolt, U.S. Patent 3,702,886 (1972).

J. Bandiera, C. Hamon, and C. Naccache, in *Proceedings of the 6th International Zeolite Conference,* D. H. Olson and A. Bisio, Eds., p. 337, Butterworths, Guildford, (1984).

R. M. Barrer, *Chem. Ind. (London)*, 1203 (1968).

R. M. Barrer, *Zeolites and Clay Minerals as Sorbents and Molecular Sieves,* Academic Press, London, 1978.

R. M. Barrer, *Hydrothermal Chemistry of Zeolites,* Academic Press, London, 1982.

R. M. Barrer and L. V. C. Rees, *Trans. Faraday Soc.* **50,** 852, 989 (1954).

S. A. I. Barri, G. W. Smith, D. White, and D. Young, *Nature,* **312,** 533 (1984).

B. Beagley, J. Dwyer, F. R. Fitch, R. Mann, and J. Walters, *J. Phys. Chem.* **88,** 1744 (1984).

R. Beaumont and D. Barthomeuf, *J. Catal.* **26,** 218; **27,** 45 (1972).

H. A. Benesi and B. H. C. Winquist, *Advan. Catal.* **27,** 97 (1978).

H. K. Beyer and I. Belenykaja, in *Catalysis by Zeolites,* B. Imelik et al., Eds., p. 203, Elsevier, Amsterdam (1980).

J. J. Blazek, *Oil Gas J.* **66**(21), 112 (1968).

D. W. Breck, *Zeolite Molecular Sieves,* Wiley-Interscience, New York, 1974.

H. Bremer, K. P. Wendlandt, T. H. Chuong, U. Lohse, H. Stach, and K. Becker, *Geterog. Katal.* **5,** 435 (1983).

J. B. Butt, *Catalyst Deactivation,* Advances in Chemistry Series, R. F. Gould, Ed., Vol. 109, p. 259, American Chemical Society, Washington, D.C., 1972.

J. B. Butt, *Am. Chem. Soc. Symp. Ser.* **72,** 288 (1978).

C. D. Chang, *Hydrocarbons from Methanol,* Marcel Dekker, New York, 1983.

C. D. Chang and A. J. Silvestri, *J. Catal.* **47,** 249 (1977).

C. D. Chang, C. T.-W. Chu, and R. R. Socha, *J. Catal.* **86,** 289 (1984).

C. D. Chang, W. H. Lang, W. K. Bell, in *Catalysis of Organic Reactions,* W. R. Moser, Ed., p. 73, Marcel Dekker, New York, 1981.

C. D. Chang, C. T.-W. Chu, J. N. Miale, R. F. Bridger, and R. B. Calvert, *J. Am. Chem. Soc.* **106,** 8143 (1984).

C. D. Chang, S. D. Hellring, J. N. Miale, K. D. Schmitt, P. W. Brigandi, and E. L. Wu, *J. Chem. Soc. Faraday Trans. I,* **81,** 2215 (1985).

N. Y. Chen, *J. Phys. Chem.* **80,** 60 (1976).

N. Y. Chen, W. E. Garwood, and R. H. Heck, *Ind. Eng. Chem. Process Des. Dev.,* in press (1987).

N. Y. Chen, W. W. Kaeding, and F. G. Dwyer, *J. Am. Chem. Soc.* **101,** 6783 (1979).

N. Y. Chen and W. J. Reagan, *J. Catal.* **59,** 123 (1979).

N. Y. Chen and P. B. Weisz, *Chem. Eng. Prog. Symp. Ser.* **63,** 86 (1967).

C. T.-W. Chu and C. D. Chang, *J. Phys. Chem.* **89,** 1569 (1985).

S. M. Csicsery, *J. Catal.* **23,** 124 (1971).

S. M. Csicsery, "Shape Selective Catalysis," in *Zeolite Chemistry and Catalysis,* American Chemical Society Monographs, J. A. Rabo, Ed., Vol. 171, p. 680, American Chemical Society, Washington, D.C., 1976.

S. M. Csicsery, *Zeolites* **4,** 202 (1984).

G. Debras, A. Gourgue, J. B. Nagy, and G. DeClippeleir, *Zeolites* **5,** 369 (1985).

S. J. DeCanio, J. R. Sohn, P. O. Fritz, and J. H. Lunsford, *J. Catal.* **101,** 132 (1986).

B. Delmon and G. F. Froment, Eds., *Catalyst Deactivation,* Studies in Surface Science Catalysis, Vol. 6, Elsevier, Amsterdam, 1980.

REFERENCES

E. G. Derouane, *Catalysis by Acids and Bases,* Studies in Surface Science and Catalysis, B. Imelik et al., Eds., Vol. 20, p. 221, Elsevier, Amsterdam, 1985.

E. G. Derouane, J. P. Gilson, Z. Gabelica, C. Mousty-Desbuquoit, and J. Verbist, *J. Catal.* **71,** 447 (1981).

R. M. Dessau and G. T. Kerr, *Zeolites* **4,** 315 (1984).

V. Ducarme and J. C. Vedrine, *Appl. Catal.* **17,** 175 (1985).

F. G. Dwyer and E. E. Jenkins, U.S. Patent 3,941,871 (1976).

J. Dwyer, F. R. Fitch, F. Machado, G. Qin, S. M. Smyth, and J. C. Vickerman, *J. Chem. Soc. Chem. Comm.* 422 (1981).

J. Dwyer, F. R. Fitch, G. Qin, and J. C. Vickerman, *J. Phys. Chem.* **86,** 4574 (1982).

S. C. Eastwood, R. D. Drew, and F. D. Hartzell, *Oil Gas J.* **60**(44), 152 (1962).

S. C. Eastwood, C. J. Plank, and P. B. Weisz, *Proc. 8th World Pet. Cong.* **4,** 245 (1971).

P. F. Eberle, Jr., *J. Phys. Chem.* **67,** 2404 (1963).

G. Engelhard, U. Lohse, E. Lippmaa, M. Tarmak, and M. Magi, *Z. Anorg. Allg. Chem.* **482,** 49 (1981).

J. L. Figueiredo, Ed., *Progress in Catalyst Deactivation,* Martinus Nijhoff, The Hague, 1982.

V. J. Frilette, W. O. Haag, and R. M. Lago, *J. Catal.* **67,** 218 (1981).

Z. Gabelica, E. G. Derouane, and N. Blom, in *Catalytic Materials: Relationship Between Structure and Reactivity,* American Chemical Society Symposium Series, T. E. Whyte et al., Eds, Vol. 248 p. 219, American Chemical Society, Washington D.C., 1984.

W. E. Garwood, in *Intrazeolite Chemistry,* American Chemical Society Symposium Series, G. D. Stucky and F. G. Dwyer, Eds., Vol. 218, p. 383, American Chemical Society, Washington D.C., 1983.

R. Gramlich-Meier and W. M. Meier, *J. Solid State Chem.* **44,** 41 (1982).

W. O. Haag, in *Proceedings of the 6th International Zeolite Conference,* D. H. Olson and A. Bisio, Eds., p. 466, Butterworths, Guildford, 1984a.

W. O. Haag, in *Heterogeneous Catalysis,* B. L. Shapiro, Ed., p. 95, Texas A & M University Press, College Station, TX, 1984b.

W. O. Haag and R. M. Dessau, *Proceedings of the 8th International Congress on Catalysis,* Vol. II, p. 305, Verlag Chemie, Weinheim, 1984.

W. O. Haag, R. M. Lago, and P. G. Rodewald, *J. Molec. Catal.* **17,** 161 (1982).

W. O. Haag, R. M. Lago, and P. B. Weisz, *Faraday Disc.* **72,** 317 (1982).

W. O. Haag, R. M. Lago, and P. B. Weisz, *Nature,* **309,** 589 (1984).

W. O. Haag, D. H. Olson, and P. B. Weisz, in *IUPAC—Chemistry for the Future,* H. Grünewald, Ed., p. 327, Pergamon Press, Oxford, 1984.

L. L. Hegedus and R. W. McCabe, *Catalyst Poisoning,* Marcel Dekker, New York, 1984.

W. Hölderich , H. Eichhorn, R. Lehnert, L. Marosi, W. Mross, R. Reinke, W. Ruppel, and H. Schlimper, in *Proceedings of the 6th International Zeolite Conference,* D. H. Olson and A. Bisio, Eds., p. 545, Butterworths, Guildford, 1984.

K. G. Ione, L. A. Vostrikova, E. A. Paukshtis, E. Yurchenko, and V. G. Stepanov, *Akad. Nauk SSSR* **27,** 1160 (1981).

P. A. Jacobs, *Catal. Rev. Sci. Eng.* **24,** 415 (1982).

P. A. Jacobs and R. von Ballmoos, *J. Phys. Chem.* **86,** 3050 (1982).

W. W. Kaeding, L. B. Young, and C. Chu, *J. Catal.* **89,** 267 (1984).

J. Kärger, *Advan. Colloid Interface Sci.* **23,** 129 (1985).

G. T. Kerr, *J. Phys. Chem.* **71,** 4155 (1967).

J. Klinowski, S. Ramdas, J. M. Thomas, C. A. Fyfe, and J. S. Hartman, *J. Chem. Soc. Faraday Trans. II* **78,** 1025 (1982).

M. R. Klotz, U.S. Patent 4,268,420 (1981).

G. T. Kokotailo, J. L. Schlenker, F. G. Dwyer, and E. W. Valyocsik, *Zeolites* **5,** 349 (1985).

R. M. Lago, W. O. Haag, R. J. Mikovsky, D. H. Olson, S. D. Hellring and K. D. Schmitt, in *Proceedings of the 7th International Zeolite Conference,* Y. Murakami, A. Iijima and J. W. Wards, Eds., p. 677, Kodansha, Tokyo, Japan, 1986.

R. B. LaPierre, A. C. Rohrman, Jr., J. L. Schlenker, J. D. Wood, M. K. Rubin, and W. J. Rohrbaugh, *Zeolites* **5,** 346 (1985).

H. Lechert and W. Schweitzer, in *Proceedings of the 6th International Zeolite Conference,* D. H. Olson and A. Bisio, Eds., p. 210, Butterworths, Guildford, 1984.

J. H. Lunsford, *J. Phys. Chem.* **72,** 4163 (1968).

J. S. Magee, W. E. Cormier, and G. M. Woltermann, *Oil Gas J.* **81**(21), 59 (1985).

P. K. Maher and C. V. McDaniel, U.S. Patent 3,402,996 (1968).

C. V. McDaniel and P. K. Maher, in *Molecular Sieves,* p. 186, Society of the Chemical Industry, London, (1968).

W. M. Meier, in *Molecular Sieves,* p. 10, Society of the Chemical Industry, London, (1968).

W. M. Meier, *Z. Kristallogr.* **115,** 439 (1979).

W. M. Meier and D. H. Olson, *Advan. Chem. Ser.* **102,** 155 (1971).

W. M. Meier and D. H. Olson, *Atlas of Zeolite Structure Types,* International Zeolite Association Polycrystal Book Service, Pittsburgh, PA, 1978.

M. T. Melchoir, D. E. W. Vaughan, and A. J. Jacobson, *J. Am. Chem. Soc.* **104,** 4859 (1982).

J. N. Miale, N. Y. Chen, and P. B. Weisz, *J. Catal.* **6,** 278 (1966).

R. J. Mikovsky and J. F. Marshall, *J. Catal.* **44,** 170 (1976).

A. G. Oblad, *Oil Gas J.* **63**(13), 124 (1963).

D. H. Olson and W. O. Haag, in *Catalytic Materials: Relationship Between Structure and Reactivity,* American Chemical Society Symposium Series, Vol. 248, T. E. Whyte, R. A. Dalla Betta, E. G. Derouane and R. T. K. Baker, Eds., p. 275, American Chemical Society, Washington, D. C., 1984.

D. H. Olson, W. O. Haag, and R. M. Lago, *J. Catal.* **61,** 390 (1980).

D. H. Olson, G. T. Kokotailo, and J. F. Charnell, *J. Colloid Interface Sci.* **28,** 305 (1968).

D. H. Olson, G. T. Kerr, S. L. Lawton, and W. M. Meier, *J. Phys. Chem.* **85,** 2238 (1981).

J. Oudar and H. Wise, Eds., *Deactivation and Poisoning of Catalysts,* Marcel Dekker, NY, 1985.

A. Paravar and D. T. Hayhurst, in *Proceedings of the 6th International Zeolite Conference,* D. H. Olson and A. Bisio, Eds., p. 217, Butterworths, London, 1984.

C. J. Pereira, G. Kim, and L. L. Hegedus, *Catalysis and Surface Science,* H. Heinemann and G. A. Somorjai, Eds., p. 201, Marcel Dekker, NY, 1985.

L. A. Pine, P. J. Maher, and W. A. Wachter, *J. Catal.* **85,** 46 (1984).

C. J. Plank, R. J. Rosinski, and W. P. Hawthorne, *Ind. Eng. Chem. Prod. Res. Dev.* **3,** 165 (1964).

M. F. M. Post, J. van Amstel, and H. W. Kouwenhoven, in *Proceedings of the 6th International Zeolite Conference,* D. H. Olson and A. Bisio, Eds., p. 517, Butterworths, Guildford 1984.

A. C. Rohrman, Jr., R. B. LaPierre, J. L. Schlenker, J. D. Wood, E. W. Valyocsik, M. K. Rubin, J. B. Higgins, and W. J. Rohrbaugh, *Zeolites* **5,** 352 (1985).

L. D. Rollmann and D. E. Walsh, *J. Catal.* **56,** 139 (1979).

D. M. Ruthven and L. K. Lee, *AIChE J.* **27,** 654 (1981).

C. N. Satterfield, *Mass Transfer in Heterogeneous Catalysis,* MIT Press, Cambridge, MA, 1970.

J. Scherzer, *J. Catal.* **54,** 285 (1978).

J. Scherzer, in *Catalytic Materials: Relationship Between Structure and Reactivity,* American Chemical Society Symposium Series, T. E. Whyte, R. A. Dalla Betta, E. G. Derouane, and R. T. K. Baker, Eds., Vol. 248, p. 157, American Chemical Society, Washington, D.C., 1984.

J. Scherzer and J. L. Bass, *J. Catal.* **28,** 101 (1973).

J. Scherzer and R. E. Ritter, *Ind. Eng. Chem. Prod. Res. Dev.* **17,** 219 (1978).

W. Schirmer, H. Thamm, H. Stach, and U. Lohse, in *The Properties and Applications of Zeolites,* R. P. Townsend, Ed., Special Publication No. 33, p. 204, The Chemical Society, London, 1980.

J. L. Schlenker, W. J. Rohrbaugh, P. Chu, E. W. Valyocsik, nd G. T. Kokotailo, *Zeolites,* **5,** 355 (1985).

K. F. M. G. J. Scholle, A. P. M. Kentgens, W. S. Veeman, P. Frenken, and G. P. M. van der Welden, *J. Phys. Chem.* **88,** 5 (1984).

H. S. Sherry, *J. Colloid Interface Sci.* **28,** 288 (1968).

H. S. Sherry, *Advan. Chem. Ser.* **101,** 350 (1971).

D. S. Shihabi, W. E. Garwood, P. Chu, J. N. Miale, R. M. Lago, C. T.-W. Chu, and C. D. Chang, *J. Catal.* **93,** 471 (1985).

J. R. Sohn, S. J. DeCanio, P. O. Fritz, and J. H. Lunsford, *J. Phys. Chem.,* **90,** 4847 (1986).

H. Stach, H. Thamm, J. Jänchen, K. Fiedler, and W. Schirmer, in *Proceedings of the 6th International Zeolite Conference,* D. H. Olson and A. Bilio, Eds., p. 225, Butterworths, London, 1984.

S. Sundaresan and C. K. Hall, *Chem. Eng. Sci.* **41,** 1631 (1986).

K. Tanabe, *Solid Acids and Bases,* Academic Press, New York, 1970.

M. Taramasso, G. Perego, and B. Notari, in *Proceedings of the 5th International Conference on Zeolites,* L. V. Rees, Ed., p. 40, Heyden, London, 1980.

J. F. Tempere, D. Delafosse, and J. P. Contour, in *Molecular Sieves II,* American Chemical Society Symposium Series, J. R. Katzer, Ed., Vol. 40, p. 76, American Chemical Society, Washington, D.C., 1977.

H. Thamm, H. Stach, and W. Fiebig, *Zeolites* **3,** 95 (1983).

H. Thamm, H. Stach, W. Schirmer, and B. Fahlke, *Z. Phys. Chem. Leipzig,* **263,** 461 (1982).

D. Theodorou and J. Wei, *J. Catal.* **83,** 205 (1983).

E. W. Thiele, *Ind. Eng. Chem.* **31,** 916 (1939).

P. B. Venuto and P. S. Landis, *Advan. Catal.* **18,** 259 (1968).

P. B. Venuto, L. A. Hamilton, and P. S. Landis, *J. Catal.* **5,** 484 (1966).

R. von Ballmoos, *Collection of Simulated XRD Powder Patterns for Zeolites,* Butterworth Scientific, Guildford, 1984.

R. von Ballmoos and W. M. Meier, *Nature* **289,** 782 (1981).

D. E. Walsh and L. D. Rollmann, *J. Catal.* **49,** 369 (1977).

P. B. Weisz, *Ann. Rev. Phys. Chem.* **21,** 175 (1970).

P. B. Weisz, *Chemtech,* **3**(8), 498 (1973).

P. B. Weisz, *Pure Appl. Chem.* **52,** 2091 (1980).

P. B. Weisz and J. N. Miale, *J. Catal.* **4,** 527 (1965).

P. B. Weisz and C. D. Prater, *Advan. Catal.* **6,** 143 (1954).

P. B. Weisz, V. J. Frilette, R. W. Maatman, and E. B. Mower, *J. Catal.* **1,** 307 (1962).

7
MATHEMATICAL MODELS IN CATALYST DESIGN

R. ARIS, *Department of Chemical Engineering, University of Minnesota, Minneapolis, Minnesota*

1. INTRODUCTION

Mathematics can provide help to the would-be catalyst designer on two fronts, the theoretical and the practical. The latter might seem at first to be the more important in a design context, but, in many ways, the principal benefits that mathematics can bestow on any discipline lie not merely in providing formulas or design equations, but in clarifying its concepts and in making precise its underlying assumptions. We need to know exactly what we mean by the effectiveness of a catalyst particle and on what parameters, given the assumptions of the underlying mathematical model, the measures of effectiveness will depend. For a mathematical model we shall have to devise, as surely as M. Jourdain spoke prose, and it is well to do so as consciously as possible if we aspire to the wisdom implicit in the notion of design.

Though the primary benefit of mathematics is theoretical, its secondary benefits are practical enough. We shall see that important design quantities can be calculated or estimated, experimental data can be understood, and new areas of interest can be economically explored.

2. THE NOTION OF EFFECTIVENESS

The primary problem in the theory of diffusion and reaction, that of determining the distribution of the reacting species in a catalyst pellet, was solved by a German physicist, Ferencz Jüttner of Breslau in 1909 (Jüttner, 1909). His paper was remarkably far-sighted and touched on problems that have only been treated fully in recent times. It lay neglected until too late to be of any influence on the development of the theory of diffusion and reaction and it was not for 30 years that the problem appeared in the literature again. When it did so appear, it sprouted in three distinct places, almost simultaneously and quite independently. In Russia, Y. B. Zel'dovich (1939) recognized that the porous catalyst would present a region of internal diffusion limitation between the limiting cases of very low reaction rates, when the effect of mass transfer in negligible, and the other extreme of high reaction rate, when the slowness of mass transfer controls the rate of the process. Without actually solving the equations he saw that the asymptotic penetration depth would be proportional to the square root of $D/kS, D$ being the diffusion coefficient, k the rate constant per unit area, and S the catalytic area per unit volume of pellet. In Germany, G. Damköhler (1937) introduced his well-known dimensionless groups and recognized that the ratio of the diffusion rate to the reaction rate would be the critical parameter in determining the extent of diffusional limitation. He gave a criterion in terms of the pellet size, diffusion coefficient, and rate constant.

It was E. W. Thiele (1939) in the United States, however, who returned to the basic equations of diffusion and reaction. He did not solve them as fully or as generally as Jüttner (1909) had done, but he did something rather more useful. He expressed the overall rate of reaction in the pellet under the conditions of diffusional limitation as a fraction of the rate that would be obtained if there were no diffusional resistance at all. This wrapped up the results of the detailed calculations, in themselves of little interest to the engineer, in a factor known as the effectiveness (he coined the term "effectiveness factor" of which the second word, though really redundant, has stuck). This manner of presentation is both conceptually beautiful and practically useful and his analysis showed that this effectiveness would depend on a dimensionless parameter (now known as the Thiele modulus) which incorporates the size of the pellet with the rate constant and the diffusion coefficient. There may well be other dimensionless parameters expressing details of the kinetics but none need contain—indeed should not contain—any reference to the size of the pellet or the diffusion coef-

2. THE NOTION OF EFFECTIVENESS

ficient. The Thiele modulus is closely related to the characteristic parameters of Damköhler and Zel'dovich, though Thiele was not aware of their work at the time.

With the advantage of hindsight the matter can be expressed rather tersely by setting down the balances of mass and energy for the catalyst pellet. For ease of exposition we will suppose that a single reaction $\Sigma \, \alpha_j A_j = 0$ takes place in a porous pellet of arbitrary shape between reactants that diffuse independently. Thus if c_j denotes the concentration of A_j and D_{je} denotes its effective diffusion coefficient, we have

$$D_{je} \nabla^2 c_j + \alpha_j r(c_1,\ldots,c_p,T) = 0 \tag{1}$$

where r is the reaction rate per unit volume. From the energy balance we have

$$k_e \nabla^2 T + (-\Delta H) r(c_1,\ldots,c_p,T) = 0 \tag{2}$$

where k_e is the conductivity of the pellet and ΔH is the heat of reaction. We shall assume that there is no mass transfer resistance at the surface so that

$$c_j = c_{js} \quad \text{and} \quad T = T_s \quad \text{on} \quad S \tag{3}$$

Such equations cry out to be put in a nondimensional form, so we take a reference concentration and temperature, c_r and T_r, and set

$$u_j = \frac{c_j}{c_r}, \quad v = \frac{T}{T_r}, \quad R(u_1,\ldots,u_p,v) = \frac{r(c,T)}{r(c_r,T_r)} \tag{4}$$

Also we make the spatial variables dimensionless by dividing by a characteristic length, l_r. Then

$$\begin{aligned} \Delta_j \nabla^2 u_j + \alpha_j \phi^2 R(u,v) = 0 \quad \text{in } V, \quad u = U_j \quad \text{on } S \\ \nabla^2 v + \beta \phi^2 R(u,v) = 0 \quad \text{in } V, \quad v = 1 \quad \text{on } S \end{aligned} \tag{5}$$

where $\Delta_j = D_{je}/D_r$ (a reference diffusion coefficient), $\phi^2 = l_r^2 r_r / D_r c_r$ (the Thiele modulus), and $\beta = (-\Delta H) D_r c_r / k_e T_r$ (the heat of reaction in dimensionless form). There will still be dimensionless groups in the dimensionless reaction rate expression R (one that turns up all the time is the Arrhenius number E/RT or dimensionless activation energy), but the only one that contains any reference to the size of the pellet is the Thiele modulus.

The effectiveness is defined as the ratio of the actual rate of reaction to the rate obtaining when there is no diffusion limitation and the concentration and temperature are everywhere the same as at the surface S. By the adroit choice of the reference values we can make $R = 1$ at the surface conditions and then

$$\eta = \int\int\int R(u,v)\, dV \qquad (6)$$

Clearly this is a function of the Thiele modulus ϕ, the heat of reaction

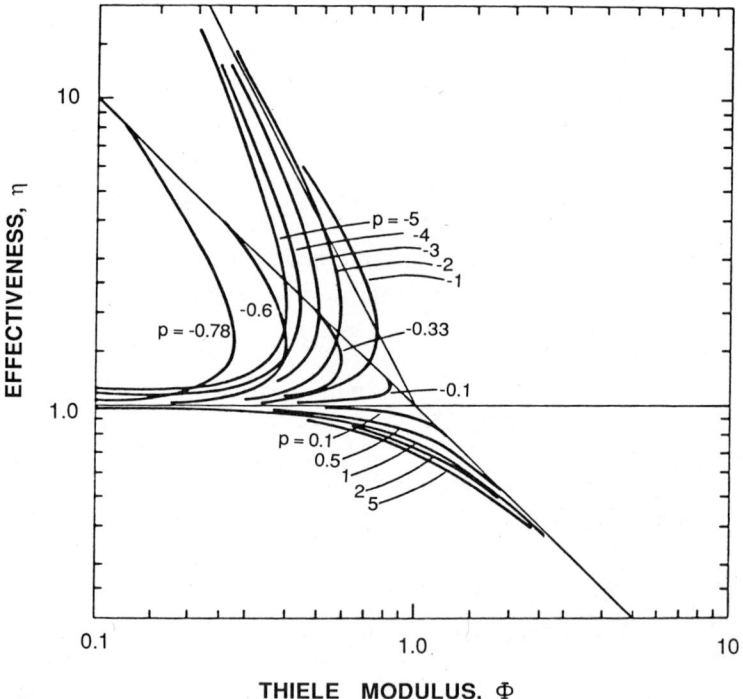

FIGURE 1. Isothermal effectiveness factor for pth-order reaction in a slab.

2. THE NOTION OF EFFECTIVENESS

parameter β, and the kinetic parameters. In Figure 1 the value of η for an isothermal pth-order reaction in a slab is shown as a function of $\Phi = \phi[(p+1)/2]^{1/2}$.

Furthermore it is clear that with a single reaction we need only one equation. The combinations $v - (\beta\Delta_j/\alpha_j)u$ are all potential functions constant on the boundary, and so by the maximum principle are constant throughout V. Substituting for each of the u_j gives

$$\nabla^2 v + \phi^2 F(v) = 0 \tag{7}$$

where

$$F(v) = \beta R[U_j + \frac{\alpha_j}{\beta\Delta_j}(v-1), v] \tag{8}$$

If there are mass and heat transfer resistances at the surface the boundary conditions must reflect this by being

$$k_{cj}(c_{jf} - c_{js}) = D_{je}(\mathbf{n} \cdot \nabla c_j)_s \tag{9}$$

and

$$h(T_f - T_s) = k_e(\mathbf{n} \cdot \nabla T)_s \tag{10}$$

where $\mathbf{n}$ is the outward unit normal to S, k_{cj} are mass transfer coefficients, h is the heat transfer coefficient, s (subscript) denotes a surface value, f (subscript) denotes the value far from the pellet. It is assumed that the relatively small area of catalyst exposed at the surface does not contribute significantly. In dimensionless form

$$\begin{aligned} \nu_j(U_j - u_{js}) &= \Delta_j(\mathbf{n} \cdot \nabla u_j)_s \\ \mu(1 - v_s) &= (\mathbf{n} \cdot \nabla v)_s \end{aligned} \tag{11}$$

where the Biot numbers are

$$\nu_j = \frac{k_{cj} l_r}{D_{je}}, \quad \mu = \frac{h l_r}{k_e} \tag{12}$$

By applying Green's theorem, it can be shown that

$$\eta = \frac{\sigma v_j(U_j - \bar{u}_{js})}{v(-\alpha_j)\phi^2} = \frac{\sigma\mu(\bar{v}_s - 1)}{v\beta\phi^2} \tag{13}$$

where

$$v = \frac{V_p}{l_r^3}, \quad V_p = \text{volume of pellet}$$

$$\sigma = \frac{S_x}{l_r^2}, \quad S_x = \text{external surface area of pellet}$$

and a bar denotes an average over the surface. If the pellet is spherically symmetrical in one, two, or three dimensions this average would be the surface value. If $l_r = V_p/S_x$, then $\sigma = v$ and the formulas simplify.

For a spherically symmetrical pellet the Laplacian is

$$\nabla^2 = \frac{1}{\rho^q}\frac{d}{d\rho}\left(\rho^q \frac{d}{d\rho}\right), \quad q = 0,1,2 \tag{14}$$

Then if we put

$$u_j(\rho) = u_{jf} + \frac{\alpha_j}{v_j}W + \frac{\alpha_j}{\Delta_j}w(\rho) \tag{15}$$

$$v(\rho) = 1 + \frac{\beta}{\mu}W + \beta w(\rho) \tag{16}$$

we have

$$\nabla^2 w + \phi^2 G(w, W, u_{jf}) = 0 \tag{17}$$

with

$$w = 0, \quad \frac{dw}{d\rho} = -W \quad \text{at} \quad \rho = 1 \tag{18}$$

and

$$\frac{dw}{d\rho} = 0 \quad \text{at} \quad \rho = 0$$

3. THE PARAMETERS

Before going on to look at the development of this model it would be advisable to pause and remember that mathematical models are implicit even in the physical constants that go to form the parameters. For example, we have used an effective diffusivity of reactant in the porous solid which could be measured by subjecting a slab of the material of area A and thickness L to a concentrations difference Δc. If J is the measured flux, the apparent diffusivity would be $JL/A(\Delta c) = D_e$. Now this depends on the structure of the porous material, the properties of the reactant, and the specific conditions; these effects can be disentangled only by conceiving a simplified model of the extremely complicated and particular structure of the porous solid. The effective diffusivity D_e can always be written as $D\epsilon/\tau$, where D is a suitable molecular diffusivity, ϵ the porosity and hence the fractional pore area, and τ the tortuosity. This really just puts the burden of fudging on the tortuosity.

The simplest, and still quite serviceable, model is the cylindrical pore model of Stewart and Johnson (1965), in which the pore volume is thought to consist of cylindrical tubes of a variety of lengths and radii but not interacting with one another. This makes use of the theory of diffusion in a long capillary in which the ratio of the radius to the mean free path is important. The mean free path greatly exceeds the tube radius in the regime of Knudsen diffusion, where the molecules interact with the walls rather than with one another. This yields a diffusion coefficient of

$$D_K = WL\left(\frac{RT}{2\pi M}\right)^{1/2} \qquad (19)$$

where M is the molecular weight, L is the length of the tube, r is its radius, and W is a factor that is a function of L/r falling from 1 to $8r/3L$ as L/r goes from 0 to 100 and is given by $8r/3L$ thereafter. D_K, the Knudsen diffusion coefficient, can be written as $2\bar{v}r/3$ where $\bar{v}$ is the average molecular speed and the molecular self-diffusion coefficient is $2\bar{v}\lambda/3$ where λ is the mean force path. Pollard and Present (1948) showed that Bosanquet's approximation,

$$\frac{1}{D} = \frac{1}{D_M} + \frac{1}{D_K} \qquad (20)$$

was a very good one over the whole range. Using it in the Stewart and Johnson model allows an effective diffusivity to be estimated, and this is a generally useful one. Smith and Wakao (1962) addressed the issue of the bimodal pore size distribution that is often found, and Butt and Foster (1966) used converging/diverging models. The dusty gas model has also been used (Mason, Evans, and Malinauskas, 1967) by considering the walls of the porous catalyst to be "immovable dust."

More sophisticated models that take account of the size and shape of the pores and molecules have also been developed. If the area of the pore does not vary drastically its harmonic mean can be used as the effective cross-sectional area. More precise calculations have been done by Petersen (1958) for hyperboloids and by Ballal and Zygourakis (1985) for cylindrical pores with spherical bulbs. When the size of a molecule is an appreciable fraction of the pore radius its diffusion is restricted, and this is sometimes accounted for by a restriction factor. Spry and Sawyer (1975) showed that this might be approximated by $(1 - r_m/r)^4$, where r_m is the "radius" of the molecule, and this has been used in studies of pore blocking by Rajagopalan and Luss (1979) and others. Do (1984) has used a formula attributed to Renkin (1954), in which the restriction factor is given by

$$1 - 2.104\left(\frac{r_m}{r}\right) + 2.09\left(\frac{r_m}{r}\right)^3 - 0.95\left(\frac{r_m}{r}\right)^5 \qquad (21)$$

There has also been a great deal of interest in the phenomenon of surface diffusion and its interaction with diffusion in pores. It is not hard to see that the combination of the two may produce a concentration-dependent diffusivity. For example, if the porosity of a pellet is ϵ, α is the surface area per unit volume, and $f(c)$ is the equilibrium adsorption relationship between the concentration in the pore and on the surface, then the flux will be $\epsilon D \nabla c + \alpha D f'(c) \nabla c$ giving an apparent diffusivity of $D_e = (\epsilon D/\tau)[1 + \alpha D f'(c)/\epsilon D]$, which now depends on the concentration. Smith (1968) has commented on Douglas and Gupta's (1967) experimental evidence for surface diffusion and his own work with Masamune (1965) on silica gel. It was found that surface diffusion was so important that D_e decreases with increasing temperature because of the decrease in surface coverage.

The diffusion coefficient is perhaps the trickiest component of the parameters. Levenspiel (1984) gives a range for D_e of 10^{-6}–10^{-5} m²/sec in commercial catalysts, with values lower by an order of magnitude or more

4. A PRIORI BOUNDS

for micropores. He also gives 10^{-1}–10 mole/m³ · sec as the range of catalytic reactions, and if we take 10^{-1}–10^2 mole/m³ as a range of gas concentrations and 1 mm to 1 cm as a size range of particles we see that the Thiele modulus is in the range 10^{-2}–10^1. Since asymptotic relations will undoubtedly be quite accurate beyond the central part of this range (say for Thiele moduli less than 10^{-1} or greater than 10) we see that the region of mathematical complexity is central to the realistic range.

The ranges of the Prater number, β, and of the Arrhenius number, γ, are linked in the sense that we are interested in cases where multiple steady states can occur. Values of β in the range 0.2–0.8 with γ in the range 23–27 have been reported for the hydrogenation of ethylene, whilst for its oxidation β is of the order of 0.13 and γ of 13. The hydrogenation of benzene has been credited with β ranging from 0.003 to nearly 1 and γ between 5 and 10. The high values $\beta = 0.25$, $\gamma = 65$ have been reported for the synthesis of vinyl chloride.

The Biot numbers for external mass transfer also have a fair range, say 10–150 for isolated particles and 5–500 in packed beds, with the Biot numbers for heat transfer about a hundredfold less.

The ratio of the effective diffusivities of heat and matter plays a role in stability problems and this, a Lewis number, has a range of about 10^{-3}–10^2. This does not have any influence on the form of the steady-state solution but tends, as it becomes large, to destabilize the steady state.

Further details on the range of parameters are given in Mercer and Aris (1971).

4. A PRIORI BOUNDS

The observation on the constancy of these linear combinations of concentration and temperature and the fact that no concentration can ever be negative immediately imposes a bound on the temperature at steady state. For an exothermic reaction this is $v \leq 1 + \beta$ or

$$T \leq T_s + \frac{(-\Delta H)D_r c_r}{k_e} \qquad (22)$$

For an endothermic reaction, we have

$$T \geq T_s - \frac{(\Delta H)D_r c_r}{k_e} \qquad (23)$$

As Wei (1966) has shown, the transient temperature may exceed this in arriving at the steady state and it is important to be aware of this, but the bound is nevertheless a usable and useful one. A. Wheeler (1951) obtained this bound for a sphere, but it was generalized for any shape by C. D. Prater (1958) and is sometimes known as the Prater number.

A priori bounds on w and W may be found but must be framed more carefully. Thus for exothermic reactions, we have

$$v_s + \beta w_- \leq v \leq v_s + \beta w_+ \tag{24}$$

where

$$w_- = \max\left(\frac{\Delta_j u_{js}}{-\alpha_j}, \alpha_j > 0\right), w_+ = \min\left(\frac{\Delta_j u_{js}}{-\alpha_j}, \alpha_j < 0\right)$$

whilst for endothermic reactions, we have

$$v_s - (-\beta w_+) \leq v \leq v_s - (-\beta)w_- \tag{25}$$

5. NORMALIZATIONS

The notion that there might be ways of bringing the results of calculations on many different shapes of particles together was first mentioned by A. Wheeler in a pioneering paper in 1951 (Wheeler, 1951) and was established by Aris in 1957 (Aris, 1957). By definition, the asymptotic value of the effectiveness at low values of the Thiele modulus is 1. When there is no external resistance the effectiveness tends to be inversely proportional to the Thiele modulus when it becomes large. Now at large Thiele modulus only a thin layer on the outside of the pellet is actively engaged in the reaction and the rate will be proportional to the surface area. But the effectiveness is calculated as a volume average so that there will be a factor of (S_x/V_p). This suggests that we should use the length (V_p/S_x) as the reference length so that the asymptotes for large Thiele modulus should come together. This turns out to be a great help in bringing the curves together. Various ad hoc improvements have been made to this scheme (see Miller and Lee, 1983), and some rather more subtle questions come up when the activity is variable and zero at the surface (see Morbidelli and Varma,

5. NORMALIZATIONS

1982). The basic normalization of the Thiele modulus was discovered independently and almost simultaneously by Petersen (1965), Bischoff (1965), and Aris (1965).

Another important normalization arises from the recognition that under conditions of severe diffusion limitation there is only a thin region in which the reaction is taking place to any significant degree. The thinness of this region implies that its shape is not of any importance and that the results for the slab geometry are asymptotically valid for any reaction. But the equation for the slab geometry,

$$v'' + \phi^2 F(v) = 0, \quad v(1) = 1, \quad v'(0) = 0 \qquad (26)$$

can be solved by quadrature as follows. Multiply the equation by $2v'$ and integrate, denoting the value of v at the center of the slab by v^*, then

$$[v'(x)] = -\left[\int_{v(x)}^{v^*} 2F(w)\, dw\right]^{1/2} \phi \qquad (27)$$

Rearrangement and a further integration gives

$$\phi = \int_1^{v^*} dw \left[\int_w^1 2F(w')\, dw'\right]^{-1/2}, \quad \eta = \left[\int_1^{v^*} 2F(w)\, dw\right]^{1/2} \bigg/ \phi \qquad (28)$$

Instead of regarding one as an equation to be solved for v^* which would then be substituted in the other, let us use v^* as a parameter in defining η as a function of ϕ. In particular $v^* \to 1 + \beta$ as $\phi \to \infty$ and hence in an asymptotic sense

$$\eta\phi \sim \left[\int_1^{1+\beta} 2F(w)\, dw\right]^{1/2} \qquad (29)$$

It follows that by redefining the Thiele modulus as

$$\Phi^2 = \left(\frac{V_p}{S_x}\right)^2 \left[\frac{r_r}{D_r c_r \int_1^{1+\beta} 2F(v)\, dv}\right]$$

the relation $\eta\Phi \sim 1$ will be asymptotically valid for any kinetics. This gives a criterion for diffusional effects to be less than 1%, namely,

$$\Phi^2 < 10^{-2} \frac{3v}{3+v} \frac{1}{|R'(1)|} \tag{30}$$

A serious objection to this criterion is that it involves quantities that cannot be observed because the observed reaction rate is the effectiveness multiplied by the true rate. A Thiele modulus based on the observed rate could be written

$$\phi_o^2 = \eta\phi^2 = \frac{V_p^2 \, r_o}{S_x^2 D_e c_r} \tag{31}$$

Bischoff (1967) has applied this to the irreversible reaction for which $r_r = k(T_r)g(c_r)$ and

$$\Phi_o^2 = \left(\frac{V_p^2}{S_x^2}\right) r_o g(c_r) \left[2 \int_{c_e}^{c_f} D_e(c) g(c) \, dc\right]^{-1} \tag{32}$$

Then Bischoff's criterion for no serious diffusional limitation is

$$\Phi_o < 1 \tag{33}$$

In this context "serious diffusional limitation" means $\eta < 0.7$.

Carberry and Cassiere (1973) have treated the problem with external mass transfer by defining an observable quantity

$$\chi = r_o/S_x k_c c_r = \frac{\eta\phi^2}{v} \tag{34}$$

For the flat plate and an isothermal first-order reaction, we have

$$\eta = \left(\phi \coth \phi + \frac{\phi^2}{v}\right)^{-1} \quad \text{and} \quad \chi = \left(1 + \frac{v \coth \phi}{\phi}\right)^{-1} \tag{35}$$

It is a simple matter to use ϕ as a parameter along a curve of constant η

5. NORMALIZATIONS

and determine ν and χ by

$$\nu = \frac{\eta\phi^2}{1 - \eta\phi \coth \phi}, \quad \chi = 1 - \eta\phi \coth \phi \tag{36}$$

Other cases with more complicated geometry or kinetics have to be solved numerically (see Carberry and Cassiere, 1973). Hudgins (1972) has given a general criterion for avoiding film diffusion in heterogeneous catalysis.

A variety of curves normalized by shape and kinetics is shown in Figures 2 and 3. These are for a first-order nonisothermal reaction in the slab and sphere, respectively. Thus from Eq. (7) we get

$$F(v) = \beta^{-1}(1 + \beta - v) \exp\left[\frac{\gamma(v - 1)}{v}\right] \tag{37}$$

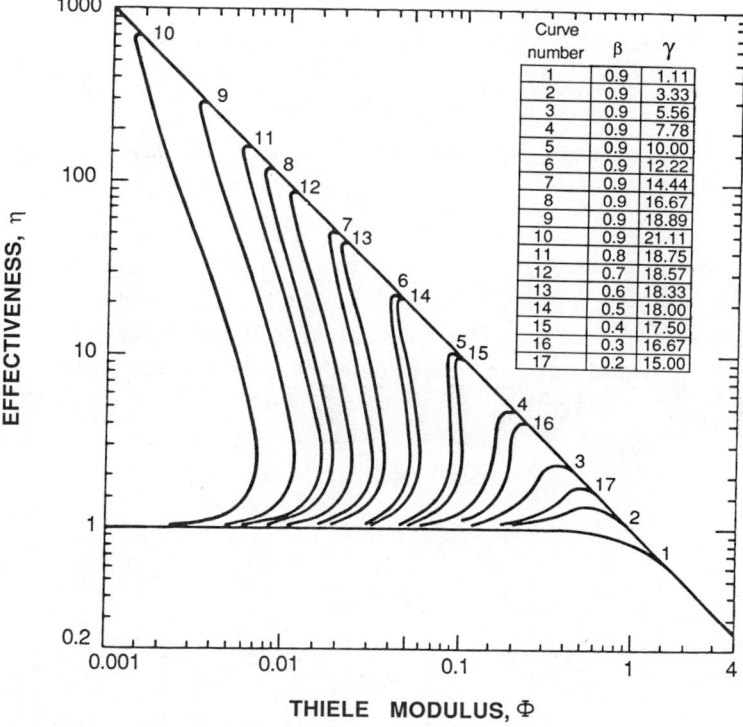

FIGURE 2. Nonisothermal effectiveness factor for an exothermic first-order reaction in a slab.

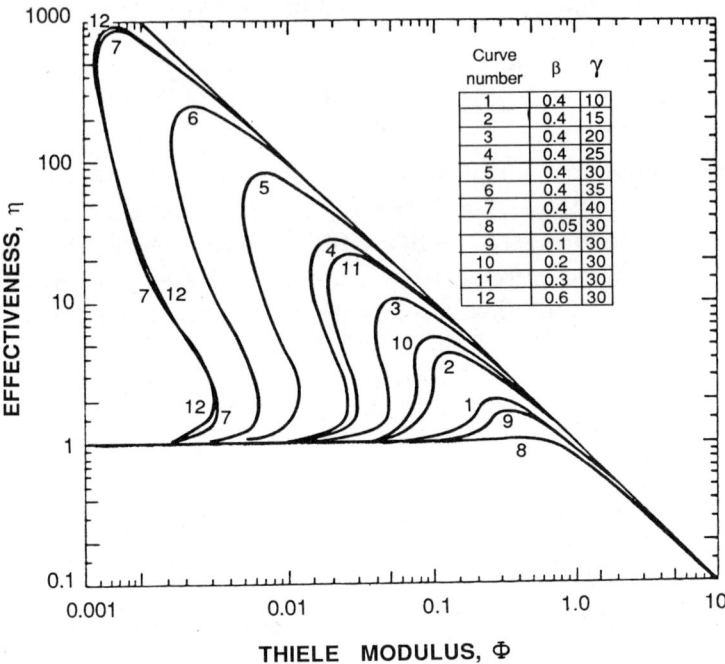

FIGURE 3. Nonisothermal effectiveness factor for an exothermic first-order reaction in a sphere.

and $\Phi = \phi M$, $\phi^2 = a^2 k(T_s)/D_e$, and a = half thickness of slab or 3 times the radius of the sphere. The constant M is the integral of the kinetics from Eq. (29) and this can be evaluated in terms of the parameters and the exponential integral

$$E_n(z) = \int_1^\infty e^{-zt} t^{-n} dt \tag{38}$$

It is

$$M^2 = 2 \frac{e^\gamma}{\beta^2} \left\{ (1 + \beta)^2 \left[E_2\left(\frac{\gamma}{1+\beta}\right) - E_3\left(\frac{\gamma}{1+\beta}\right) \right] \right. \\ \left. - (1 + \beta) E_2(\gamma) + E_3(\gamma) \right\} \tag{39}$$

where

$$\beta = \frac{(-\Delta H)D_e c_s}{k_e T_s}, \quad \gamma = \frac{E}{RT_s} \tag{40}$$

6. OPTIMIZATION

Various optimization problems have been considered in connection with the choice of catalyst. One of the first was for multifunctional catalysts that contain two components, one promoting the first step of a reaction and the other the second (see work by Gunn and Thomas, 1965). Langmuir kinetics with bifunctional catalysts have also been considered (Mendiratta, Prabhu, and Davidson, 1971). Here it is pleasant to note that the authors found that "the optimal profiles displayed a bang-chatter bang-bang policy with multiple switching times." Fortunately some of the bang-bang and bang-straight-bang suboptimal policies were nearly as good as the bang-chatter bang-bang policy.

Bailey (1971) considered the influence of the difference that might be obtained between the diffusivities of two reactants participating in a second-order reaction. He found that it was optimal to have somewhat less than an equal concentration of the reactant with the greater diffusivity. A later discussion of Bailey's problem is given by DeVera and Varma (1980).

More recently, with the advent of layered catalyst pellets the question of the optimal distribution of activity through the pellet has been raised. From this follows a nice optimization problem for the tubular reactor. The external composition along the bed varies as the reactant is used up and thus the position within the pellet of the optimal location of the concentrated activity varies also. The very neat solution that Morbidelli and Varma (1982) obtain is complicated by the fact that more than one stable steady state can exist within some of the pellets. Moreover it is not very practical to vary the composition of the catalyst continuously along the length of the bed so that a suboptimal piecewise constant distribution needs to be considered. Also the sharp distribution of activity within the pellet can only be approximately achieved. But here again a mathematical theory may be of help.

The reasoning of Morbidelli, Servida, and Varma (1984) may be illustrated by one of their examples using the isothermal kinetics expression

$$R(u) = \frac{(1 + \sigma)^2 u}{(1 + \sigma u)^2} \tag{41}$$

For this expression there is a value of u, say u_m, at which $R(u)$ is greatest: $R(u_m) > R(u)$, $u \neq u_m$. If the total activity of the pellet is constant but it may be distributed throughout the volume at will, then it is clear that it cannot be better disposed than at such a level where u has fallen to u_m, because the rate will then be at its greatest and

$$\eta = R(u_m) = \frac{(1 + \sigma)^2}{4\sigma} \tag{42}$$

since

$$u_m = 1/\sigma, \quad \sigma > 1 \tag{43}$$

But for a slab, we have

$$\nu(1 - u_s) = \frac{u_s - u_m}{d} = \phi^2 R(u_m)$$

where u_s is the surface concentration and d is the fractional depth at which the activity is concentrated. Thus

$$d = \frac{4(\sigma - 1)}{\phi^2(\sigma + 1)^2} - \frac{1}{\nu} \tag{44}$$

If $\nu < \phi^2 (\sigma + 1)^2/4(\sigma - 1)$ it is best to have the activity on the surface, whereas for $\phi^2 > 4\nu(\sigma - 1)/(\nu + 1)(\sigma + 1)^2$ it is concentrated at the center. Similar formulas may be used for some other symmetric shapes and simple kinetics, but certain cases demand computation (see Morbidelli and Varma, 1982; Morbidelli, Servida, and Varma, 1984).

Another optimal problem that has been considered is the conjecture that the sphere is the worst possible shape for a first-order reaction (Aris, 1957). This is fairly obvious in the sense that the sphere has the least surface of all shapes of given volume and so may be felt to have its inside buried the least accessibly. Amundson and Luss (1967) were able to prove this using the variational formulation and Steiner symmetrization. The variational inequality

$$E(U) = \frac{1}{v\phi^2} \int \int \int [(\nabla U)^2 + \phi^2 U^2]\, dV \geq E(u) \quad (45)$$

holds for all suitably differentiable functions U in Ω, the space of the pellet, with $U = 1$ on its surface $\partial\Omega$, when u is the solution,

$$\nabla^2 u = \phi^2 u \quad \text{in } \Omega, \, u = 1 \quad \text{on } \partial\Omega \quad (46)$$

Thus

$$\eta = \min E(U), \quad U = 1 \quad \text{on } \partial\Omega \quad (47)$$

If $\mathbf{j} = \nabla u$, then $u = (\nabla \cdot \mathbf{j})\phi^2$ is the vector field associated with the solution u of (39). Then, if $\mathbf{J}$ is any other vector field with $\nabla \cdot \mathbf{J} = \phi^2$ on $\partial\Omega$, we have

$$G(\mathbf{J}) = \frac{2}{v\phi^2}\left\{\int\int (\mathbf{n}\cdot\mathbf{J})\, dS - \int\int\int\left[\mathbf{J}^2 + \frac{(\nabla\cdot\mathbf{J})^2}{\phi^2}\right]dV\right\} \quad (48)$$
$$\leq G(\mathbf{j}) = \eta$$

This gives bounds on the effectiveness which, by adroit choice of trial functions, can be made quite tight. An immediate extension to the case of external mass transfer resistance can be made. Extension to nonlinear kinetics is less easy and satisfactory (see Arthurs, 1970; Strieder and Aris, 1973; Aris, 1975; Rester and Aris, 1972).

7. THE STRUCTURE OF THE CATALYST PELLET

The idea of controlling the distribution of catalytic activity within the pellet arose when it was realized that the greatest reaction rate did not necessarily correspond to the highest, or surface, concentration. This was in fact something that might have been considered as soon as an effectiveness greater than unity had been discovered by Amundson and Schilson (1961) in the mid-1950s, and indeed some work was done by Maatman and Prater (1957), but its real impact came in the 1970s with the tremendous increase in demand for catalysts for the automobile exhaust system. This is not the place to review the development of the subject, especially because an extensive review has appeared elsewhere (see Lee and Aris, 1985). The manufacture of egg-shell, egg-white, and egg-yolk catalysts (well-chosen and self-explanatory names) was discussed by Becker and Wei (1977). Techniques of manufacture and the advantages of this type of formulation both in performance and in resistance to poisoning are described by Hegedus and others (Hegedus and Summers, 1977; Hegedus, Oh, and Baron, 1977; Hegedus et al., 1979; Hegedus and McCabe, 1981; Hegedus, 1980). The optimization problem associated with the distribution of activity has just been described, and the practical attainment of nonuniform distribution is the subject of the next paragraph. But there is also the physical structure of the pellet to be considered, and the approach to the optimal design of structure which Hegedus has achieved using the complex search method is worth noting.

Hegedus (1980) looked for the best distributions of macropores and micropores. If these have volumes V and v per unit mass and the ρ_s is the solid density, the apparent density of the pellet is

$$\bar{\rho} = \left(\frac{1}{\rho_s} + V + v\right)^{-1} \tag{49}$$

giving macro- and microporosities

$$E = V\bar{\rho} \quad \text{and} \quad \epsilon = v\bar{\rho} \tag{50}$$

If R and r are the average radii of macro- and micropores, respectively, the surface area per unit mass is

$$\xi = \frac{2V}{R} + \frac{2v}{r} \tag{51}$$

For the two types of pore the Knudsen diffusion coefficients are given by

$$D_K = \frac{2}{3} R \left(\frac{8\mathbf{R}T}{\pi M}\right)^{1/2}, \quad D_k = \frac{2}{3} r \left(\frac{8\mathbf{R}T}{\pi M}\right)^{1/2} \tag{52}$$

7. THE STRUCTURE OF THE CATALYST PELLET

where M is the molecular weight of the diffusing species. If D is its molecular diffusivity, the pore diffusivities in macro- and micropores are given by

$$D_P = \left[\left(\frac{1}{D}\right) + \left(\frac{1}{D_K}\right)\right]^{-1} \quad \text{and} \quad D_p = \left[\left(\frac{1}{D}\right) + \left(\frac{1}{D_k}\right)\right]^{-1} \quad (53)$$

respectively. These can be combined into an effective diffusivity,

$$D_e = E^2 D_P + \frac{\epsilon^2(1 + 3E)}{1 - E} D_p \quad (54)$$

This effective diffusivity must be worked out for each species or for representative species (such as propylene) representing hydrocarbons of reactant and poison precursor.

The fractional distance ξ to which the poison penetrates after a time t is given implicitly by

$$\frac{3D_{ep}c_p}{\bar{\rho}s\sigma L^2} t = \xi^3 \left(\frac{1}{v_p} - 1\right) + 3\xi^2 \left(\frac{1}{2} - \frac{1}{v_p}\right) + 3\xi \frac{1}{v_p} - 1 \quad (55)$$

where c_p is the external concentration of the poison, v_p is its Biot number for mass transfer, L is the radius of catalyst particle, and σ is the saturation constant for the poison. The penetration of poison and hence the effectiveness of each catalyst particle can thus be calculated. Hegedus expressed this as the conversion after aging,

$$C = 1 - \exp\left\{-\left[\frac{4\pi D_e N_p L}{Q}\left(\frac{1}{v} + \frac{\xi}{1-\xi}\right)^{-1}\right]\right\} \quad (56)$$

where now D_e and v are calculated for the reactant rather than the poison, N_p is the number of pellets in the reactor, and Q the flow rate.

The optimal structure was sought by varying V, v, R, and r so that the conversion after 1600 hr would be greatest. As well as placing specific constraints on the four design variables, there were implicit constraints on $\bar{\rho}$ and s, since it is not feasible to make pellets of too low a density or to stabilize very high surface areas. Box's search method COMPLEX was used and led away from standard formulations to a bimodal, low-density structure. This is not the place to quote the results in toto because they were specific and called for further experimental proof, but a comparison between the reference catalyst (a current low-density automobile formulation) and Hegedus' best case is informative:

Quantity	V	v	R	r	s	$\bar{\rho}$	C
Standard	.37	.63	7000	110	114	.78	93.2
Optimized design	1.15	.37	99,600	50	150	.55	99.0
Units	cm³/g	cm³/g	Å	Å	m²/g	g/cm³	%

The rather large macropore radii can be obtained by embedding polymeric microspheres in the pellet and subsequently removing them by calcination (Bedford and Berg, 1977).

Do (1984) has also looked at an interesting problem in connection with catalytic cracking of large molecules where restricted diffusion and surface access play a role. If the pores are too small the diffusion will be too restricted and, if too large, there will be insufficient catalytic area. He used the restriction coefficient given in Eq. (21) and also adapted an observation of Glandt's on the partition coefficient for large molecules (Glandt, 1981). If $\lambda = r_m/r^p$ is the ratio of the molecular radius to the pore radius and $\gamma = 8Nr_m^3$, where N is Avogadro's number, the concentration at the pore mouth as a fraction of the concentration c in the bulk phase is

$$(1 - \lambda)^2 + \lambda L(\lambda)\gamma c + \lambda M(\lambda)\gamma^2 c^2$$

where L and M are fifth-order polynomials in λ. Do's results show that as the Thiele modulus decreases the optimal value of λ falls, that is, the more reactive molecules do better in larger pores as one might expect. In the absence of the partition effect the optimal λ falls from about $\frac{1}{3}$ for Thiele modulus <1 to around $\frac{1}{8}$ for Thiele modulus >10. When $\gamma c = 0.2$ the decrease is from 0.48 to 0.2 and takes place over two orders of magnitude. However, when $\gamma c = 0.5$ the optimal λ falls from 0.57 for small ϕ_o to 0.48 at $\phi = 4$ and there drops to 0.33 before falling more gradually to 0.2 as ϕ tends to infinity. This remarkably sensitive behavior is a possibility that the catalyst designer should be aware of. It also implies that the pore size should vary along the reactor, and while an optimal continuous variation might not be practical the sudden jump suggests that in a reactor with two

8. PREPARATION BY IMPREGNATION

When the preparation of the catalyst involves the impregnation of the pellet with a solution of a precursor such as hexachloroplatinic acid, this impregnation may be done either by diffusion into a wet pellet or by capillary imbibition. The precursor is adsorbed onto the surface and, after drying, the pellet is treated to form the active catalytic surface. When coimpregnated with a strongly adsorbed species, such as citric acid, the precursor is driven into the interior and an egg-white, or even an egg-yolk, catalyst is formed. We will briefly outline the mathematics of dry imbibition into a spherical pellet.

Let $c_j(r,t)$ denote the concentration of species j in the pores of the pellet at a radius r and time t, and $n_j(r,t)$ be the adsorbed concentration. Then a mass balance neglecting diffusion gives

$$4\pi r^2 \frac{\partial}{\partial t}(c_j + n_j) = \dot{V}(t) \frac{\partial}{\partial r} c_j \tag{57}$$

and

$$\frac{\partial}{\partial t} n_j = k^+ c_j \left(N - \sum n_i\right) - k^- n_j \tag{58}$$

These equations can be transformed into those of column chromatography by taking the volume imbibed as the time variable,

$$\tau = \left(\frac{3}{4\pi R^3}\right) \int \dot{V}(t)\, dt \tag{59}$$

and, in place of r, the volume outside the radius r, namely,

$$\rho = 1 - \left(\frac{r}{R}\right)^{1/3} \tag{60}$$

If in addition we assume that the adsorption equilibrium is attained very rapidly there is a functional relationship between n_j and c_j. For example, a Langmuir adsorption isotherm is given by

$$n_j = \frac{Nk_j c_j}{1 + \sum k_i c_i} \tag{61}$$

Then we might take $u_j = k_j c_j$ and $n_j/N = u_j/(1 + \sum u_i) = g_j(u)$ with $f_j(u) = u + g_j(u)$ to give

$$\frac{\partial u_j}{\partial \rho} = \frac{\partial f_j}{\partial \tau} \tag{62}$$

These equations can be solved subject to the initial and boundary conditions

$$u_j(\rho, 0) = 0, \quad u_j(0, \tau) = U_j \tag{63}$$

For a single substance ($j = 1$) the equation may be written

$$\frac{1}{f'(u)} \frac{\partial u}{\partial \rho} = \frac{\partial u}{\partial \tau} \tag{64}$$

and this can be interpreted to mean that a point of concentration u moves with a speed of $1/f'(u)$ or "slowness" $f'(u)$. With an isotherm such as the Langmuir, $f'(u)$ decreases as u increases so the higher concentrations go faster. This means that an advancing front would steepen and eventually break like a wave. This is physically unacceptable and a discontinuity must form. A mass balance for the discontinuity shows that if u drops from u_1 to u_2 across it then it must move with speed $(u_1 - u_2)/(f(u_1) - f(u_2))$. Another way of putting this is to say that the slowness of the discontinuity is $[f(u)]/[u]$, where the square brackets denote the finite differences in f and u. As the discontinuity approaches continuity the ratio $[f]/[u]$ approaches $f'(u)$. Now impregnation with a single substance starts with a sharp discontinuity from U to 0 and hence, since $f(0) = 0$, it advances with a slowness $s = f(U)/U$ or speed $U/f(U)$. If the imbibition ceases when $\tau = T$ the sharp front has reached $\rho = P = TU/f(U)$. The thickness of the layer on the outside of the sphere as a fraction of the radius is $\Gamma = 1 - (1 - P)^{1/3}$. If the imbibition is perfectly symmetrical the advancing

8. PREPARATION BY IMPREGNATION

front of liquid traps and compresses the air in the pores. Even so these are small enough that the capillary pressure is very high and so T is close to 1.

When there are two solutes obeying the Langmuir isotherm

$$n_i = \frac{Nk_ic_i}{1 + k_1c_1 + k_2c_2}, \quad i = 1,2 \tag{65}$$

we may set

$$n_i = k_ic_i, \quad k = \frac{k_1}{k_2}$$
$$\Delta = 1 + u_1 + u_2, \quad x = \rho, \quad y = \tau - \rho \tag{66}$$

Then

$$\frac{\partial u_1}{\partial x} + \frac{1 + u_2}{\Delta^2}\frac{\partial u_1}{\partial y} - \frac{u_1}{\Delta^2}\frac{\partial u_2}{\partial y} = 0$$
$$-\frac{u_2}{\Delta^2}\frac{\partial u_1}{\partial y} + k\frac{\partial u_2}{\partial x} + \frac{1 + u_1}{\Delta^2}\frac{\partial u_2}{\partial y} = 0 \tag{67}$$

and the boundary conditions are

$$u_1(x,0) = u_2(x,0) = 0 \tag{68}$$
$$u_1(0,y) = U_1, \quad u_2(0,y) = U_2 \tag{69}$$

The solution of these equations is more difficult than for the single solute in imbibition, but the form of the solution is easily understandable and is shown in Figure 4. A mixture of the two species fills the outside of the pellet while the less strongly adsorbed species is pushed ahead into the interior. Without loss of generality we can call the less strongly adsorbed species 1 (so that $k < 1$) by methods described in books on partial differential equations (see Amundson and Aris, 1973). Remarkably enough the formulas can be given in closed form. Thus $\hat{U}$, the concentration of the less strongly adsorbed species in the interior, is

$$\hat{U} = \tfrac{1}{2}\{k - 1 + U_1 + kU_2 + [(k - 1 + U_1 + kU_2)^2 + 4kU_1U_2]^{1/2}\} \tag{70}$$

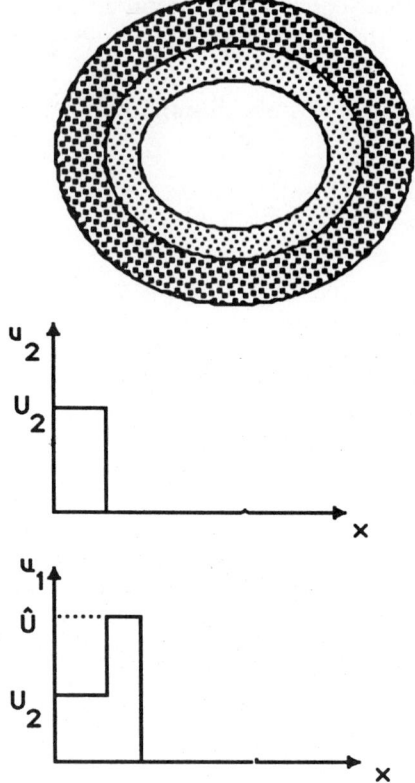

FIGURE 4. Distribution of coimpregnated precursors.

The egg-white formed at this concentration is buried beneath an egg-shell of thickness

$$\xi = \frac{\chi}{1 + v/(1 + U_1 + U_2)} \tag{71}$$

where $\chi = k_1 \alpha NT$ and $v = k_2 \alpha N$. The egg-white is found from $\rho = x = \xi$ to $\rho = x = \Xi$ where

$$\Xi = \frac{\chi}{1 + kv/(1 + \hat{U})} \tag{72}$$

The redistribution of solute during the drying process is more difficult to calculate particularly as thermal effects are involved. The sharp discontinuities formed during imbibition become blurred but it is still possible to get a general idea of what is involved. More work is needed, both experimental and theoretical, to see how far this model conforms to reality. If done well, then there would be a real incentive to rethink Morbidelli and Varma's work on the optimal design in terms of distributions that could actually be achieved through the imbibition process. Then a catalyst design problem would start with the kinetics, ask whether and where a maximum rate might be attained, and then tailor the manufacturing process to achieve a design that would have the best structure and distribution.

9. MULTIPLICITY, STABILITY, AND OTHER MODELS

Much of the work of the last 10 years has been on questions of multiplicity and stability. In the comparatively humdrum world of actual catalysts some of the more exotic phenomena that have been discovered would be rare and, if present, to be avoided. Nevertheless this work is important in two ways. Firstly, although he may not wish to make use of them, the designer should be aware of some of the possibilities that exist lest he stumble into a region of parameter space where the behavior is unusual. Secondly, phenomena such as spontaneous oscillations may not always be undesirable, for it has been shown that in some cases they lead to improvements.

Great advances in the multiplicity problem have been made in recent years and if Aris' two volumes (1975) were to be brought up to date, the second, on multiplicity, stability, and transients, would be longer than the first. But the real insights into the question for partial differential equations are only just beginning to be gained (Neilsen and Villadsen, 1985; Hu, Balakotaiah, and Luss, 1985) and there will undoubtedly be much more to come. Nielsen and Villadsen have compared the various models that can be propounded. In their nomenclature, Ia is the model with external heat transfer resistance and internal diffusional resistance ($\beta = 0, \nu = \infty$) while Ib has an external mass transfer resistance. These are isothermal since $\beta = 0$ arises from an assumption of infinite conductivity. The corresponding models IIIa and IIIb with nonzero β are nonisothermal. Models II and IV are similar to I and III, respectively, but allow a discontinuity as a result of adsorption at the surface. All are for the geometry of the flat plate and all give a maximum of five steady states. In the isothermal model of the flat plate the equation can be solved quite simply for the internal distribution but the temperature has to be determined by an overall energy balance. Thus, in their model Ia the reaction is first order and $\theta = T_p/T_f$ is the temperature of the pellet T_p as a fraction of the bulk temperature far from the pellet,

$$\frac{d^2u}{d\rho^2} = \phi_e^2 \gamma^{(1-1/\theta)} u = \phi_p^2 u \tag{73}$$

$$\frac{du}{d\rho} = 0 \text{ at } \rho = 0, u = 1, \quad \frac{du}{d\rho} = \frac{\mu}{\beta}(\theta - 1) \text{ at } \rho = 1 \tag{74}$$

where ϕ_p is a Thiele modulus based on the pellet temperature and ϕ is a Thiele modulus using the bulk temperature. Thus θ is determined by

$$\phi_p \tanh \phi_p = \frac{\mu}{\beta}(\theta - 1) \tag{75}$$

a nonlinear equation that can be analyzed in the usual way though it makes more demands than usual on the manipulative virtuosity and computational skill of the analyst. Hu, Balakotaiah, and Luss (1985) looked at the same model with a zeroth-order reaction using the methods of singularity theory. The fact that there can be a "dead" core within which the reactant is entirely absent introduces a discontinuity of a kind not considered in the mathematical literature. They also point out that this is the first example in the chemical engineering literature where the number of solutions exceeds the codimension of the most singular point by more than one. In a second paper they look at reaction order and show that for an order of irreversible reaction greater than 1.053 only three (rather than five) steady states can exist. Another striking result is obtained by letting the geometry be governed by an index s. The slab, long cylinder, and sphere are governed by the same set of equations,

$$\frac{1}{\rho^s} \frac{d}{d\rho}\left(\rho^s \frac{du}{d\rho}\right) = \phi^2 u \tag{76}$$

$$\frac{du}{d\rho} = 0, \rho = 0; \quad \frac{du}{d\rho} = \nu(1 - u), \rho = 1 \tag{77}$$

9. MULTIPLICITY, STABILITY, AND OTHER MODELS

with $s = 0$ for the slab, $s = 1$ for the cylinder and $s = 2$ for the sphere. Hu, Balakotaiah, and Luss (1985) find that if s is treated as a continuous variable there are five solutions for a first-order reaction if $s < 0.0548$, but otherwise only three. This result is the more interesting because it is known that for $\beta > 0$, that is, the nonisothermal case, there can be a very large number of solutions when $s = 2$ though probably only five for $s = 0$ and 1 (Copelowitz and Aris, 1970; Villadsen and Michelsen, 1972).

On the question of stability the designer should also be aware that even a unique steady state can become unstable for small ratios of the thermal and material diffusivities (Luss and Lee, 1970). In general the maximal and minimal steady states are likely to be stable, though, in the case of five steady states induced by the external resistances to mass and heat transfer, the middle one may also sometimes be stable (Jackson, 1973). Little has been done to try and map out the stability picture in a comprehensive way, which is scarcely surprising considering the immensity of the task and the difficulty of some of the computations.

Finally it should be mentioned that work is beginning on different classes of model for the catalytic surface. Thus Mukesh et al. (1984) have discussed island models that may be more suitable for the description of some oxidation reactions than those in which it is assumed that the adsorbates are homogeneously intermingled. If a species clumps into an island its reactivity is proportional to the length of its perimeter rather than its area, that is, to $(N\theta)^{1/2}$, where θ is the fractional area coverage and N is the number of islands. If, for example, CO forms islands and oxygen does not, the rate of reaction would be governed by $\theta' \sqrt{N\theta}$, where θ' is the oxygen coverage and θ and N refer to CO. If N is constant then the reaction rate is proportional to $\theta'\theta^{1/2}$. If the island size is constant, $N \propto \theta$ and the rate is proportional to $\theta'\theta$. It is also possible for the number of islands to depend parabolically on θ, that is, $N = N_{max} \theta(2\Theta - \theta)/\Theta^2$. This gives rise to some interesting nonlinear kinetics that have been successfully fitted to experimental data (Mukesh et al., 1984). Diffusion between regular or random catalytic archipelagos has also been considered (Kuan, Davis, and Aris, 1983), and it is found that the effectiveness is closely modeled by an annulus with the same perimeter and adsorption area ratio.

Takoudis has shown that certain models positing further surface requirements can lead to oscillatory reactions (Takoudis, Schmidt, and Aris, 1982). For example, A and B may be adsorbed on the surface with coverages θ_A and θ_B, respectively, but require two adjacent sites to react,

$A^* + B^* + 2S \to C + 4S$. Then arise such equations as

$$\begin{aligned}\theta_A &= k_A p_A(1 - \theta_A - \theta_B) - k'_A \theta_A - k\theta_A \theta_B (1 - \theta_A - \theta_B)^2 \\ \theta_B &= k_B p_B(1 - \theta_A - \theta_B) - k'_B \theta_B - k\theta_A \theta_B (1 - \theta_A - \theta_B)^2\end{aligned} \quad (78)$$

where p_A and p_B are the partial pressures of the reactants above the catalytic surface. Spontaneous oscillations are found in some regions and a number of striking phenomena are seen when one of the partial pressures is forced periodically (Kevrekidis, Schmidt, and Aris, 1986).

Finally, some of the results showing multiplicity of hysteresis loops that have been reported (Barelke, 1973; Hegedus, Oh, and Baron, 1977) have stimulated a domain model of heterogeneous catalysis by Chang (1983). He envisages a distribution of domains each with a steady state x_s that satisfies

$$f(x_s; r, \lambda(t)) = 0 \quad (79)$$

where r is a distribution parameter and λ is a scanning variable. Then the observed x is

$$\hat{x}(\lambda) = \int x_s(\lambda, r) \psi(r) \, dr \quad (80)$$

and Chang goes on to show that this is capable of showing the hysteresis phenomena observed.

An area that has not yet come into its own but which, with the rise of supercomputing, may have a future is the direct Monte Carlo calculation of problems in diffusion and reaction. Abbasi, Evans, and Abramson (1983) simulated a porous solid by a set of random spheres, determined the mean pore size and its standard deviation, and then computed the courses of many molecular trajectories in the porous space. By comparing the results with the standard formulation they are able to correlate the tortuosity with the pore structure. Nothing seems to have been done in this line, however, for a reactive system. If this could be set up to take full advantage of the parallel processing which is the forte of the new computers, it might be possible to include a number of effects. The prospect is at once attractive and repulsive; repulsive, in the sense that it seems to be giving in to the number-crunching spirit of the age; attractive, in that it is clearly so big a task that without a good deal of ingenuity it will be impossible.

These are some of the currently interesting theoretical phenomena; they will eventually filter down to the world of praxis and perhaps be found to solve some problem that has long resisted solution. Meanwhile there is plenty to do on all fronts.

NOMENCLATURE

A	Area for diffusion
A_j	jth chemical species
c	Conversion in Hegedus' model; Eq. (56)
c_j	Concentration of A_j
D_{je}	Effective diffusivity of A_j
D_k	Knudsen diffusion coefficient
D_M	Molecular diffusion coefficient
d	Depth of active zone; Eq. (44)
E_n	Exponential integral; Eq. (38)
$E(U)$	Variational integral; Eq. (45)
$F(v)$	Dimensionless reaction rate
$f_j(u)$	Column isotherm; Eq. (62)
$G(\mathbf{J})$	Variational integral; Eq. (48)
$g_j(u)$	Adsorption isotherm; Eq. (61)
h	Heat transfer coefficient
J	Flux
k_c	Mass transfer coefficient
k_e	Effective conductivity
l_r	Reference length
M	Molecular weight
p	Order of reaction
q	Dimension parameter; 0 = slab; 1 = cylinder; 2 = sphere
R	Dimensionless reaction rate; gas constant
r	Radius of pore
r_m	Molecular radius
S_x	External surface area

T	Temperature
U_j	Bulk value of u_j
u_j	Dimensionless concentration of A_j
V	Bulk value of v
V_p	Particle volume
v	Dimensionless temperature
W, w	Variables defined by Eqs. (15) and (16)
α_j	Stoichiometric coefficient
β	Dimensionless adiabatic temperature rise
γ	Arrhenius number
Δ_j	Dimensionless diffusion coefficient
E, ϵ	Voidage of macro-, micropores
η	Effectiveness
θ	Fractional coverage
μ	Biot number for heat transfer
ν_j	Biot number for mass transfer
Ξ	Egg-white penetration; Eq. (72)
ξ	Poison penetration; Eq. (55)
ρ	Dimensionless radius
σ	Dimensionless surface area
υ	Dimensionless volume
Φ	Normalized Thiele modulus
Φ_0	Normalized Thiele modulus based on observable quantities
ϕ	Thiele modulus
ϕ_o	Thiele modulus based on observable quantities

Suffixes

e	Effective
f	Bulk
j	Of species A_j
s	Surface
r	Reference

REFERENCES

M. H. Abbasi, J. W. Evans, and I. S. Abramson, *AIChE J.* **29,** 617 (1983).

N. R. Amundson and R. Aris, *First Order Partial Differential Equations with Applications,* Prentice-Hall, Englewood Cliffs, N.J., 1973.

N. R. Amundson and D. Luss, *AIChE J.* **13,** 759 (1967).

N. R. Amundson and R. E. Schilson, *Chem. Eng. Sci.* **13,** 226, 237 (1961).

R. Aris, *Chem. Eng. Sci.* **6,** 262 (1957).

R. Aris, *Ind. Eng. Chem. Fundam.* **4,** 227 (1965).

R. Aris, *Chem. Eng. Ed.* **8,** 19 (1974).

R. Aris, *The Mathematical Theory of Diffusion and Reaction in Permeable Catalysts* (2 vols.), Clarendon Press, Oxford, 1975.

A. M. Arthurs, *Complementary Variational Principles,* Clarendon Press, Oxford, 1970.

J. E. Bailey, *Chem. Eng. Sci.* **26,** 991 (1971).

G. Ballal and K. Zygourakis, *Chem. Eng. Sci.* **40,** 1477 (1985).

V. V. Barelke, *Kinet. Katal.* **14,** 196 (1973).

E. R. Becker and J. Wei, *J. Catal.* **46,** 365 (1977); **46,** 372 (1977).

R. E. Bedford and M. Berg, U.S. Patent 4 051 072, September 1977.

K. B. Bischoff, *AIChE J.* **11,** 351 (1965).

K. B. Bischoff, *Chem. Eng. Sci.* **22,** 525 (1967).

J. B. Butt and R. N. Foster, *AIChE J.* **12,** 180 (1966).

J. J. Carberry and G. Cassiere, *Chem. Eng. Ed.* **7,** 22 (1973).

H.-C. Chang, *Chem. Eng. Sci.* **38,** 535 (1983).

I. Copelowitz and R. Aris, *Chem. Eng. Sci.* **25,** 906 (1970).

G. Damköhler, *Chemieinginieur* **3,** 359 (1937).

A L. DeVera and A. Varma, *Ind. Eng. Chem. Fundam.* **19,** 320 (1980).

D. D. Do, *AIChE J.* **30,** 849 (1984).

W. J. M. Douglas and V. P. Gupta, *AIChE J.* **13,** 883 (1967).

E. D. Glandt, *AIChE J.* **27,** 51 (1981).

D. J. Gunn and W. J. Thomas, *Chem. Eng. Sci.* **20,** 89 (1965).

L. L. Hegedus and J. C. Summers, *J. Catal.* **51,** 185, 1978; **48,** 345 (1977).

L. L. Hegedus and R. W. McCabe, *Catal. Rev.* **23,** 377 (1981).

L. L. Hegedus, *I&EC Proc. Des. Dev.* **19,** 533 (1980).

L. L. Hegedus, S. H. Oh, and K. Baron, *AIChE J.* **23,** 632 (1977).

L. L. Hegedus, T. C. Chou, J. C. Summers, and N. M. Potter, in *Preparation of Catalysts II,* Proceedings of the 2nd International Symposium on the Scientific Bases for the Preparation of Heterogeneous Catalysts, Elsevier, Amsterdam, 1979, p. 171.

R. Hu, V. Balakotaiah, and D. Luss, *Chem. Eng. Sci.* **40,** 589, 599 (1985).

R. R. Hudgins, *Can. J. Chem. Eng.* **50,** 427 (1972).

R. Jackson, *Chem. Eng. Sci.* **28,** 1355 (1973).

F. Jüttner, *Z. Phys. Chem.* **65,** 595 (1909).

I. G. Kevrekidis, L. D. Schmidt, and R. Aris, *Chem. Eng. Sci.* **41,** 1546 (1986).

D.-Y. Kuan, H. T. Davis, and R. Aris, *Chem. Eng. Sci.* **38,** 719, 1563 (1983).

S.-Y. Lee and R. Aris, *Catal. Rev.* **27,** 207 (1985).

O. Levenspiel, *The Chemical Reactor Omnibook,* Oregon State University Bookstore, Corvallis, OR, 1984.

D. Luss and J. C. M. Lee, *AIChE J.* **16,** 620 (1970).

R. W. Maatman and C. D. Prater, *Ind. Eng. Chem.* **49,** 253 (1957).

S. Masamune and J. M. Smith, *AIChE J.* **11,** 41 (1965).

E. A. Mason, R. B. Evans, and A. P. Malinauskas, *J. Chem. Phys.* **46,** 3199 (1967).

A. K. Mendiratta, A. V. Prabhu, and B. Davidson, *Chem. Eng. Sci.* **26,** 885 (1971).

M. Mercer and R. Aris, *Latin Amer. J. Chem. Eng. Appl. Chem.* **1,** 149 (1971).

D. J. Miller and H. H. Lee, *Chem. Eng. Sci.* **38,** 363 (1983).

M. Morbidelli and A. Varma, *I&EC Fundam.* **21,** 278, 284 (1982).

M. Morbidelli, A. Servida, and A. Varma, *Biotech. Bioeng.* **26,** 1508 (1984).

D. Mukesh, W. Morton, C. N. Kenney, and M. B. Cutlip, *Surf. Sci.* **138,** 237 (1984).

P. H. Nielsen and J. Villadsen, *Chem. Eng. Sci.* **40,** 571 (1985).

E. E. Petersen, *AIChE J.* **4,** 343 (1958).

E. E. Petersen, *Chem. Eng. Sci.* **20,** 587 (1965).

W. G. Pollard and R. D. Present, *Phys. Rev.* **73,** 762 (1948).

C. D. Prater, *Chem. Eng. Sci.* **8,** 284 (1958).

K. Rajagopalan and D. Luss, *Ind. Eng. Chem. Process Des. Dev.* **18,** 459 (1979).

E. M. Renkin, *J. Gen. Physiol.* **28,** 225 (1954).

S. Rester and R. Aris, *Chem. Eng. Sci.* **27,** 347 (1972).

C. N. Satterfield and G. W. Roberts, *Ind. Eng. Chem. Fundam.* **4,** 288 (1965).

J. C. Spry and W. H. Sawyer, Paper 16–20, AIChE 68th Annual Meeting, Los Angeles (1975).

J. M. Smith, *AIChE J.* **14,** 650 (1968).

J. M. Smith and N. Wakao, *Chem. Eng. Sci.* **17,** 825 (1962).

W. E. Stewart and M. F. L. Johnson, *J. Catal.* **4,** 248 (1965).

W. Strieder and R. Aris, *Variational Methods Applied to Problems of Diffusion and Reaction,* Springer-Verlag, Heidelberg, 1973.

C. G. Takoudis, L. D. Schmidt, and R. Aris, *Chem. Eng. Sci.* **37,** 69 (1982).

R. Tartarelli, S. Cioni, and M. Capovani, *J. Catal.* **18,** 212 (1970).

E. W. Thiele, *Ind. Eng. Chem.* **31,** 916 (1939).

W. J. Thomas and R. M. Wood, *Chem. Eng. Sci.* **22,** 1607 (1967).

J. Villadsen and M. L. Michelsen, *Chem. Eng. Sci.* **27,** 751 (1972).

J. Wei, *Chem. Eng. Sci.* **21,** 1171 (1966).

P. B. Weiss and J. S. Hicks, *Chem. Eng. Sci.* **17,** 265 (1962).

A. Wheeler, *Adv. Catal.* **3,** 249 (1951).

Y. B. Zel'dovich, *Acta Phys-chim. USSR.* **10,** 583 (1939).

8

TOWARD THE DESIGN OF HYDRODEMETALLATION CATALYSTS

J. WEI, *Department of Chemical Engineering, Massachusetts Institute of Technology, Cambridge, Massachusetts*

1. INTRODUCTION

Crude oil contains a number of metal compounds in small quantities. The most important metals are vanadium and nickel, which may be above 1000 ppm by weight in Venezuelan oil. These compounds must be removed before refining and use, because they are potent poisons for cracking catalysts, and vanadium oxides can erode furnace linings and turbine blades.

The removal of metals from oil should be considered together with the removal of other heteroelements, especially sulfur, nitrogen, and oxygen. The most important process for the removal of these elements is catalytic hydrogen treatment under high pressure and temperature, and the heteroatoms are removed as gaseous species sulfur dioxide, ammonia, and water. However, the metal elements do not form gaseous hydrides; they deposit on, and thus shorten the useful lifetime of the hydrotreating catalysts (Nelson, 1976). When the metals content exceeds 200 ppm, catalyst consumption exceeds 1 lb per 10 barrels of feed, so that hydrodemetallation cannot be carried out economically. The oil would have to be demetallized first, or processed by other methods such as coking, thermal cracking, and deasphalting. If we use the strict definition of a catalyst as "the substance that causes reactions without itself being affected," then the hydrodemetallation catalyst is not a catalyst at all but a reagent. There may even be a stoichiometric relation between metals and catalyst before the catalyst ceases to function. It should be the goal of catalytic scientists to develop a true HDM catalyst.

An ideal strategy for the removal of metals from oil would involve the separation of the metals from the organometallic compounds, followed by complexing of the metals in a soluble form that can be extracted. An analogy would be with the metabolic removal of iron from mammalian hemoglobin. In the human, hemoglobin in the red blood cell is taken apart after 120 days in the liver and spleen: There is enzymatic cleavage of the α-methine bridge to a biliverdin–iron complex in room temperature and pressure, the heteroatom iron remains in solution and combines with plasma transferrin and is transported for reuse, globin is liberated, and the tetrapyrrole is excreted (White, Handler, and Smith, 1964). There is no fouling of the spleen with iron deposits. The cost of such biochemical demetallation could be much lower than industrial catalytic hydrodemetallation, since it does not need high-pressure reactors and pumps, and the byproducts nickel and vanadium can be valuable resources instead of nasty catalyst poisons.

The metals can also be removed by a number of chemical means such as extraction by strong acids to form water-soluble salts (Kukes and Aldage, 1985). However, the strong acids would also cause undesirable effects on the oil. Deasphalting with pentane or heptane would cause the precipitation of the asphaltenes together with a large fraction of metals. However, currently the only commercially viable process for a high degree of metals removal is hydroprocessing, which is primarily designed to remove sulfur and nitrogen.

The two main types of metal compounds in oil are the porphyrins and the asphaltenes. The former are closely related in structure to hemoglobin in mammalian blood and chlorophyll in green plants, and are probably derived from long-buried dinosaurs and ferns by clay zeolitic cracking and ion exchange of the iron and magnesium ions by nickel and vanadyl ions. There is a great deal of knowledge about the porphyrins because of the interests of the biochemists (Baker and Palmer, 1978). The basic porphyrin has four pyrrole rings and four N atoms chelated to a central metal or hydrogen ion and may have substitution groups around the ring and a long tail. The basic molecular weight of a porphyrin is about 300–600, with a

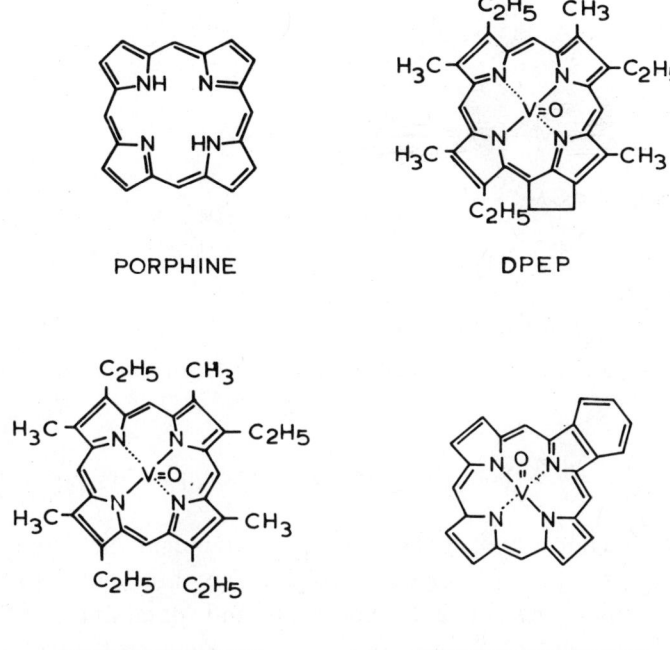

FIGURE 1. Porphyrin skeletal structure and metalloporphyrins representative of those in petroleum.

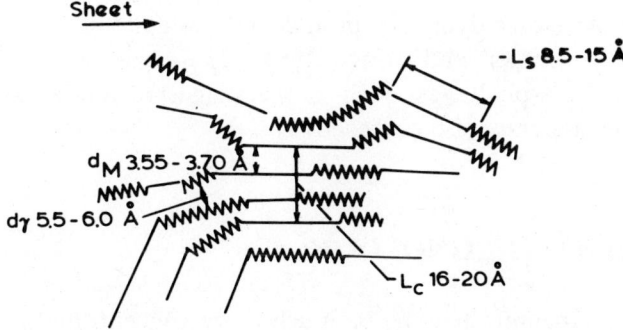

FIGURE 2. Schematic view of asphaltene micelles. —— represents the edge of flat sheets of condensed aromatic rings, ⋙ represents paraffinic chains or napthenic rings. (Adapted from Yen, Erdman, and Pollack, 1961.)

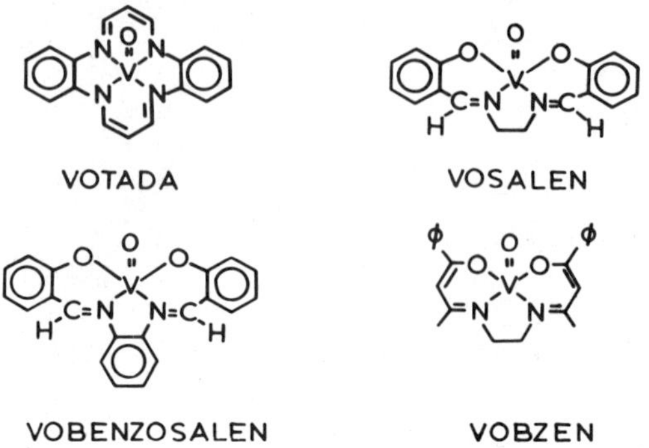

VOTADA

VOSALEN

VOBENZOSALEN

VOBZEN

FIGURE 3. Nonporphyrinic model compounds. (Adapted from Fish and Komlenic, 1984.)

diameter of 12–20 Å (Scheidt, 1978). Some examples are given in Figure 1.

There is much less known about the more abundant asphaltenes, which may be up to 400,000 in molecular weight. They are known to be much more polar and aromatic than hydrocarbons and contain much of the heteroatoms S, N, and O as well as Ni and V. They probably form large micelles of 20–80 Å in diameter at room temperature, with a core of asphaltenes solubilized by the less polar resins, shown in Figure 2 (Bunger and Li, 1981; Spry and Sawyer, 1975; Yen, Erdman, and Pollack, 1961). Recent evidence points to the concept that these micelles are held together by rather weak forces that are easily broken by shear or thermal motion at the processing temperatures of more than 700°F, so that the individual molecules or organometallic compounds, chelated by S and O, as well as N, may be no larger than 1000 in molecular weight and 12 Å in diameter (Tamm, Harnsberger, and Bridge, 1981). There is a need to study model compounds of asphaltenes, such as the suggested model compounds in Figure 3, to understand their reactivity.

2. PRESENT TECHNOLOGY

Nickel and vanadium tend to be much more concentrated in the heavier portions in the crude oil, particularly in the residuum after distillation—atmospheric resid boils above 650°F and vacuum resid boils above 1020°F. Refineries have concentrated on the problem of the removal of sulfur,

2. PRESENT TECHNOLOGY

which is a powerful environmental pollutant and is strongly regulated by law. The process of hydrodemetallation (HDM) is often part of the process of catalytic hydrotreating or hydrodesulfurization (HDS). Many of the important crude oils have sulfur content 1–6%, nitrogen 0.1–0.6%, nickel 5–90 ppm, and vanadium 10–1000 ppm.

The process of HDS of distillate gas oil takes place under more mild conditions than HDS of resids. The reactor is usually a trickle bed reactor, where hydrogen and oil trickle through a packed bed of catalysts. The reactor may have a diameter of 3–6 ft, and a length of 6–30 ft. The operating conditions are typically 750–850°F in temperature, 500–3000 psig in hydrogen pressure, and 0.5–2 hr^{-1} in liquid hourly space velocity. Since the distillate oil contains much less metal, the catalysts can be regenerated by burning off coke, and catalyst life may be up to 10 yr. For the resids that have much higher metal content, the catalysts are irreversibly fouled by the metal deposits and cannot be regenerated by combustion, so that catalyst life may be in the range of 0.5–1.0 yr. An oil with 100 ppm of metals, operating at a space velocity of 1 hr^{-1}, would take only 0.2 yr to accumulate 18% metals on the catalyst. The oils with the highest metal content cannot be treated with current technology because the catalyst replacement rate would be too high.

The catalyst contains usually 2–5% cobalt and 7–15% molybdenum over γ-alumina. Sometimes tungsten is used instead of molybdenum, and nickel is used instead of cobalt. The catalysts are extrudates of 1.5–3 mm in diameter, with surface area 150–300 m^2/g, pore diameters 60–120 Å, and pore volume 0.3–0.7 cm^3/g. The structure of commercial catalysts was reviewed by Gates, Katzer, and Schuit (1979). MoS_2 and WS_2 are not the most active metal sulfides for the HDS reactor. Pecoraro and Chianelli (1981) have shown that group VIII metal sulfides RuS_2, Rh_2S_3, OsS_2, and IrS_2 have 20–40 times the activity of MoS_2 (see Figure 4). Among the nonprecious metals, Cr_2S_3 approaches MoS_2 for activity.

Under operating conditions, the coke concentration in the catalyst quickly builds in 1–2 days up to 15–25% and remains stable for a long time (Beuther and Schmid, 1963; Inoguchi et al., 1971a, 1971b). The metal concentration gradually builds up to 20–40%. The catalyst activity declines steadily with time. Since a constant throughput of oil at a constant level of conversion is desirable, the reactor temperature is raised as the catalyst ages to maintain a constant level of activity. The aging, measured by temperature required to maintain a given level of activity, is fast at the beginning, slows down to a linear rate for a long time, and speeds up toward

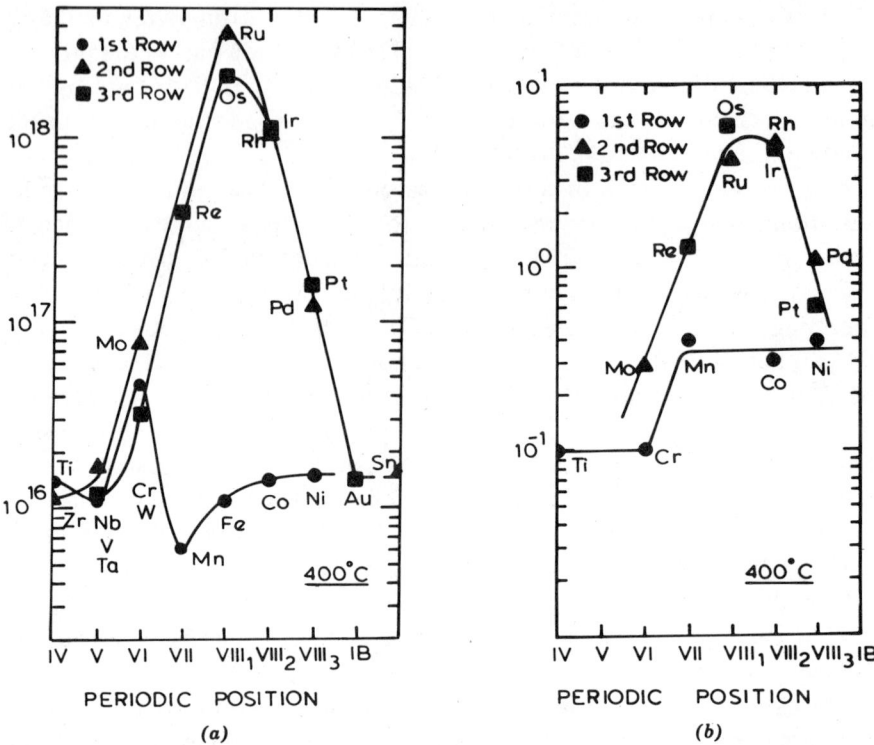

FIGURE 4. The activity of various metals on hydrodesulfurization of dibenzothiophene. (a) Activity per millimole of metal. (b) Activity per square meter of metal. (Adapted from Pecararo and Chianelli, 1981).

the end (see Figure 5) (Tamm, Harnsberger, and Bridge, 1981). The usual interpretation is that during the first phase, aging is caused by coke deposition (Chang and Silvestri, 1976); the second phase, linear aging, is caused by metal deposition to reduce catalyst surfaces and pore diffusivity; and the third phase, rapid aging, is caused by pore mouth plugging (Oxenreiter et al., 1972; Newson, 1975).

Metal deposition occurs with sharp gradients within a catalyst pellet and with gentler gradients along the reactor axis. Usually, vanadium is highly concentrated on the outside of catalyst pellets, forming a U-shaped distribution. Nickel deposition tends to be more uniform inside the catalyst (see Figure 6) (Audibert and Duhaut, 1970; Kwan and Sato, 1970; Sato

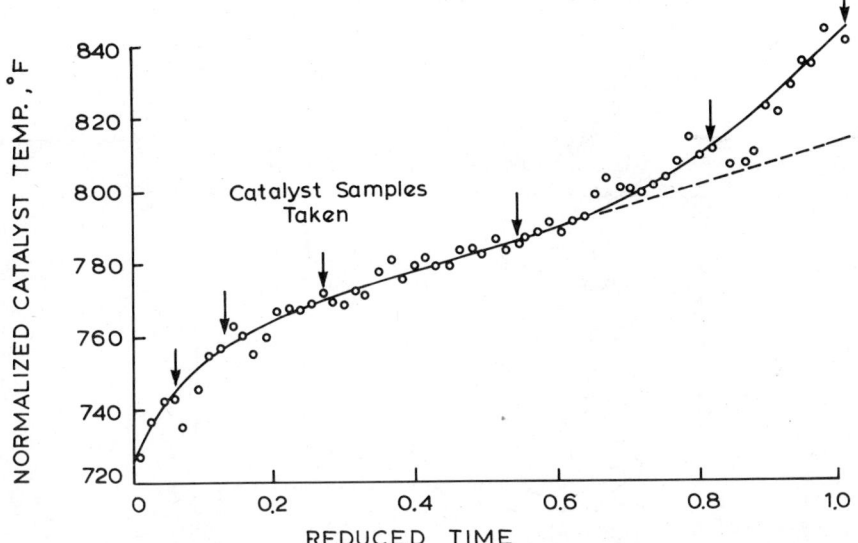

FIGURE 5. Typical deactivation curve for residuum hydroprocessing catalysts: Temperature required for an activity level versus time. Arabian heavy atmospheric residuum desulfurization, product sulfur = 1.0 wt%, $\frac{1}{16}$-in. extrudate catalyst. (Adapted from Tamm, Harnsberger, and Bridge, 1981.)

et al. 1971). Some authors reported that metal deposition rises from the edge to a maximum inside the catalyst before declining to the center (Tamm, 1981). The relation of this U-shaped distribution to the Thiele analysis was pointed out by Sato et al. (1971). The sharpness of U-shaped distribution is quantified by Tamm et al. by the Θ factor, which is defined in a cylindrical geometry as

$$\Theta = \frac{\int Mr\, dr}{M_{max} \int r\, dr}$$

Here M_{max} is the maximum metal concentration before pore plugging, which depends on pore diameter and catalyst porosity. The Θ factor is the ratio of average metal concentration to the maximum concentration, and should

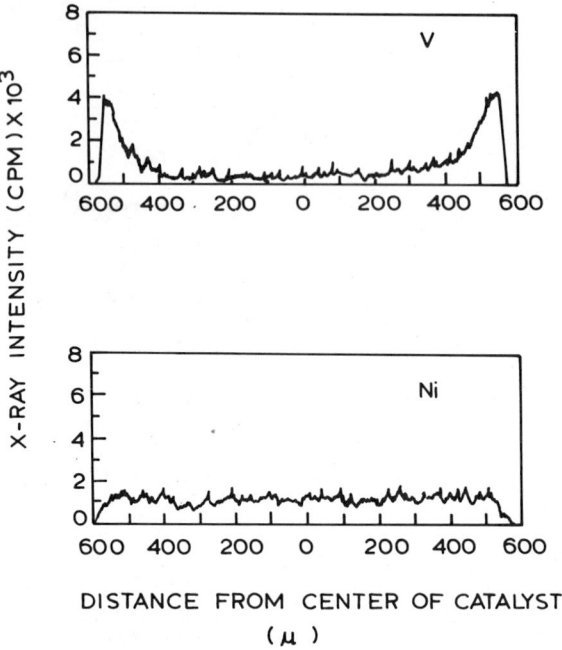

FIGURE 6. Concentration profiles of vanadium and nickel deposition on the hydrodesulfurization catalyst after 50 h of reaction. (Adapted from Sato et al., 1971.)

equal the effectiveness factor as long as the concentration maximum is at the edge of the catalyst. A low Θ factor would mean that when pore plugging condition is reached, only a small fraction of pore volume is filled with metals. Therefore, there is a great deal of catalyst potential that is now beyond reach and wasted. The Θ factor decreases when reaction rates are increased by higher catalyst loading, higher temperature, or higher hydrogen pressure, quite consistent with an increase in Thiele modulus (Pazos, Gonzalez, and Salazar-Guillen, 1983).

Beuther and Schmid (1963) found that it is much easier to remove vanadium than nickel in Kuwait resid; they attributed this difference to the projection of the vanadyl complex above the porphyrin plane, which makes it more accessible to the catalyst. Nickel is a more planar and less polar compound, which makes it less reactive. Vanadium compounds are also more adsorptive over clays than nickel compounds.

3. MECHANISM AND KINETICS

Crude oils and residual oils contain a large variety of metalloorganic compounds of unspecified nature and quantity, with the possibility of a wide range of reactivities. The interactions of these complex mixtures, and the presence of sulfur and nitrogen compounds, would make it very difficult to study the kinetics of HDM in crude oils and resids. Some authors found that the hydrodemetallation reaction of petroleum residue can be described as either a second-order kinetics or a first-order kinetics. There may be several "lumps" of reactive species of different reactivity, each obeying first-order kinetics, so that the aggregate appears to obey second-order kinetics. Beuther and Schmid (1963) found that the HDS of Kuwait residue can be described by a simple second-order kinetics with an activation energy of 27 kcal/mole; they also found a strong dependency of the rates on hydrogen pressure from 250 to 1000 psig, but the rates stayed fairly constant up to 3000 psig. On the other hand, Chang and Silvestri (1976) found a first-order kinetics that sharply increases with hydrogen pressure up to 2000 psig. Model compounds dissolved in clean oils would form the basis for scientific studies to resolve some of these questions.

Porphyrins are good model compounds, since they are easy to obtain and well characterized. There is much less known about asphaltenes in oil; future studies should reveal more information. The most widely used porphyrin compounds are nickel and vanadyl forms of etioporphyrin, tetraphenylporphyrin (TPP), and tetra(3-methyl-phenyl) porphyrin (T3MPP) (see Figure 7).

It is now well established that the mechanism of HDM of porphyrin proceeds by a consecutive step through one or more hydrogenated intermediates (Agrawal, 1980; Agrawal and Wei, 1984; Rankel, 1981; Rankel and Rollman, 1983; Kameyama et al., 1981, 1982, 1985; Galiasso et al., 1985; Morales, Garcia, and Prada, 1984) (see Figure 8):

$$\text{metal-porphyrin} \rightleftharpoons \text{metal-chlorin} \longrightarrow \text{metal deposition and ring breakup}$$

The metal-porphyrin is first reversibly hydrogenated to form a metal-chlorin, which is usually green in color and hence its name. This intermediate has been isolated and identified by mass spectrometry. It is not a

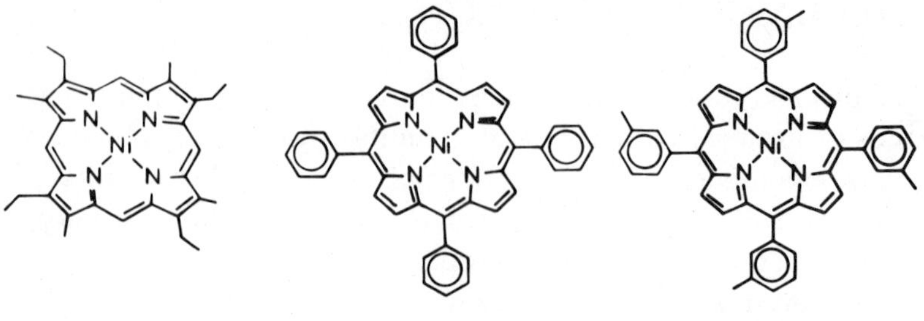

Ni-ETIOPORPHYRIN

(Ni-EP)

Ni-TETRAPHENYLPORPHY-RIN

(Ni-TPP)

Ni-TETRA(3-METHYL-PHENYL) PORPHYRIN

(Ni-T3MPP)

FIGURE 7. Structure of model nickel porphyrin compounds.

planar molecule such as porphyrin and is regarded as less stable. The second step is irreversible hydrogenolysis of the ring, followed by deposition of the metal on the catalyst. Under reactor conditions, hydrogen forms of the porphyrin or chlorin are not found. There is a very low ratio of nitrogen/carbon deposit on the spent catalyst surface, so the destination of the porphyrin ring must be in tetrapyrroles or further fragments dissolved in the oil.

For the tetraphenylporphyrines, several authors (Ware, 1983; Ware and Wei, 1985; Weitkamp, Gerhardt, and Scholl, 1984) have found a further intermediate beyond chlorin, termed the isobacteriochlorin (see Figure 9):

$$M-P \rightleftharpoons M-PH_2 \rightleftharpoons M-PH_4 \longrightarrow \text{deposit and breakup}$$

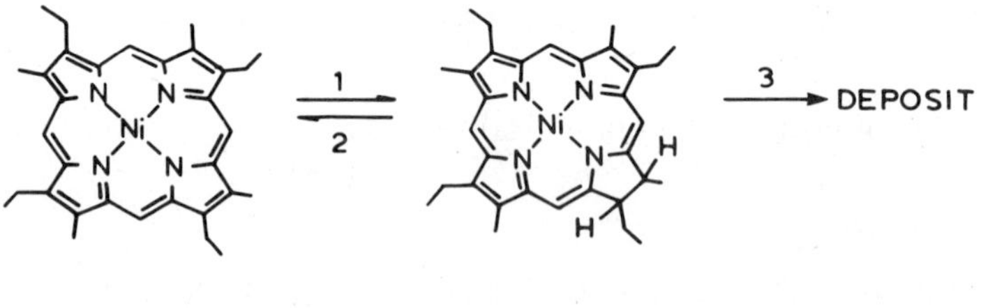

Ni-EP

Ni-EPH$_2$

FIGURE 8. Reaction sequence for the hydrodemetallation of Ni-etioporphyrin. (Adapted from Agrawal, 1980.)

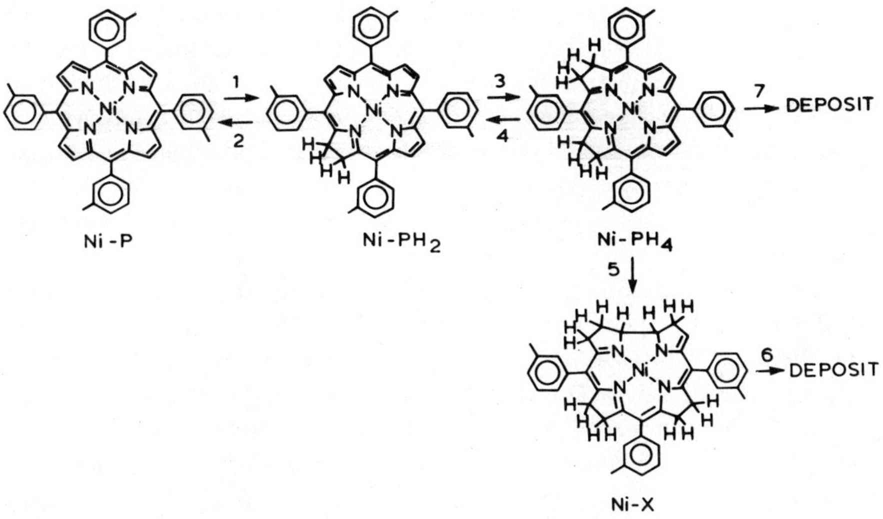

3. MECHANISM AND KINETICS

FIGURE 9. Reaction sequence for the hydrodemetallation of Ni-tetra(3-methylphenyl) porphyrin. (Adapted from Ware, 1983.)

It is possible that this longer sequence is also followed for etioporphyrins, except that the ratio k_5/k_3 is very large so that the intermediate M–PH$_4$ cannot be detected.

For nickel etioporphyrin, the kinetics are first-order with respect to the porphyrin species. The values of the three preexponential factors, activation energies, and dependencies on hydrogen pressure are given below:

Reaction	$\ln k_0$	E	n
1	15.4	17	1
2	23.6	23	0
3	27.4	33	2

For vanadyl etioporphyrin, the values are:

Reaction	$\ln k_0$	E	n
1	16.9	18	1
2	17.8	17	0
3	20.8	25	2

Thus the reactivity of the vanadyl porphyrins is not significantly greater than that of the nickel porphyrins, which contradicts the Sato observation of deeply U-shaped distribution (with a Thiele modulus greater than 8) of vanadium deposits on catalysts versus the more uniform distribution (with a Thiele modulus less than 1) of nickel deposits. It is possible that vanadyl compounds interfere with the kinetics of the nickel compounds. Preliminary results show that vanadyl compounds adsorb less strongly on the catalyst surface than nickel compounds.

These kinetics results are totally consistent with the metal deposition profiles found on spent catalysts. In a complex first-order reaction system, the steady-state diffusion and reaction rates are balanced in a pore $\mathbf{D}\nabla^2 \mathbf{a}(r) = \mathbf{K}\mathbf{a}(r)$, where $\mathbf{D}$ is the diffusion matrix, $\mathbf{K}$ is the reaction matrix, and $\mathbf{a}(r)$ is the vector of concentrations. Let the matrix $\mathbf{D}^{-1}\mathbf{K}$ have eigenvectors $\mathbf{Y}_i$ and eigenvalues μ_i: $\mathbf{D}^{-1}\mathbf{K} = \mathbf{Y}\tilde{\boldsymbol{\mu}}\mathbf{Y}^{-1}$; let $\mathbf{b}(r) = \mathbf{Y}^{-1}\mathbf{a}(r)$ represent the eigenconcentrations, then $\nabla^2 \mathbf{Y}^{-1}\mathbf{a}(r) = \tilde{\boldsymbol{\mu}}\mathbf{Y}^{-1}\mathbf{a}(r)$ or $\nabla^2 \mathbf{b}(r) = \tilde{\boldsymbol{\mu}}\mathbf{b}(r)$. The solution is each $b_{i(r)} = (\cosh \sqrt{\mu_i}\, r / \cosh \sqrt{\mu_i}\, R)[b_i(R)]$, where R is the total length of the pore in a slab geometry. The concentrations are $a_j(r) = \Sigma_i Y_{ji} b_i(r)$

$$\mathbf{a}(r) = \mathbf{Y}\left\{\frac{\cosh \mu_i r}{\cosh \mu_i R}\right\} \mathbf{Y}^{-1} \mathbf{a}(R)$$

where { } indicates a diagonal matrix. We define $\phi_i = R\sqrt{\mu_i}$, the Thiele moduli, so that $\mathbf{a}(r) = \mathbf{Y}\{\cosh(\phi_i r/R)/\cosh \phi_i\}\, \mathbf{Y}^{-1}\mathbf{a}(R)$ is the concentration profile in the pore. The Θ factor of Tamm in the geometry is for the depositing species:

$$\Theta = \sum_j \int_0^R a_j(r)\, dr / [\sum_j a_j(r)]_{\max} \int_0^R dr$$
$$= \sum_j \frac{[\mathbf{Y}\{\eta_i\}\, \mathbf{Y}^{-1}\mathbf{a}(R)]_j}{[\sum_j a_j(r)]_{\max}}$$

where $\eta_i = (\tanh \phi_i)/\phi_i$, the effectiveness factors (Agrawal, 1980; Wei and Wei, 1982). The appropriate diffusivity to use is that of the metalloporphyrins in the pores during reaction temperatures, which should be on the order of 10^{-6}–10^{-8} cm^2/sec (Chang and Silvestri, 1976; Newson, 1970). As metal deposition increases with catalyst age, this diffusivity can be expected to drop according to some law of restricted diffusion in pores, such as

3. MECHANISM AND KINETICS

Renkin's equation (Renkin, 1954) or the Spry and Sawyer rule (Spry and Sawyer, 1975).

Tamm, Harnsberger, and Bridge (1981) observed in their study with Arab heavy crude oil that in the reactor inlet, the maximum in metal deposition does not occur at the edge of the catalyst but in the interior. For catalyst samples removed in the reactor middle and exit, the maximum in metal deposition moved to the edge (see Figure 10). Agrawal and Wei (1984) and Ware and Wei (1985) have found this interior maximum at reactor inlets with model compounds in clean oil without the presence of sulfur (see Figure 11). This result is consistent with the consecutive mechanism, since at the reactor inlet there is a very small concentration of metallochlorin. The metalloporphyrin must diffuse inside the catalyst, then hydrogenate to metallochlorin, before it can react to form deposit. In the middle and exit of the reactor, metallochlorin concentration builds up and can form deposits at the edge of the catalysts. Tamm, Harnsberger, and Bridge (1981) suggested that this interior maximum is caused by the in-

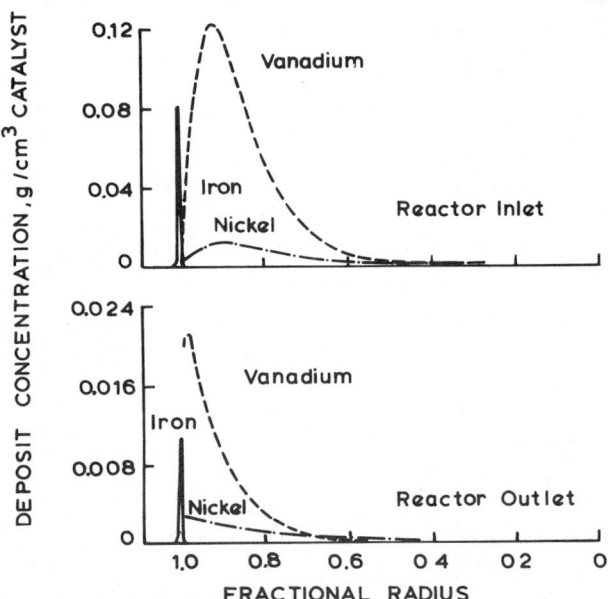

FIGURE 10. Intraparticle vanadium, nickel, and iron concentration profiles of aged residuum hydrotreating catalysts from inlet and outlet reactor positions. (Adapted from Tamm, Harnsberger, and Bridge, 1981.)

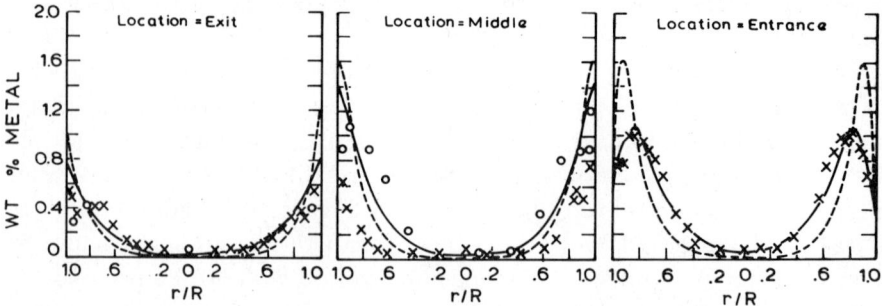

FIGURE 11. Vanadium deposition profiles of catalysts with model compounds, at reactor inlet (entrance), middle, and exit. (Adapted from Agrawal and Wei, 1984.)

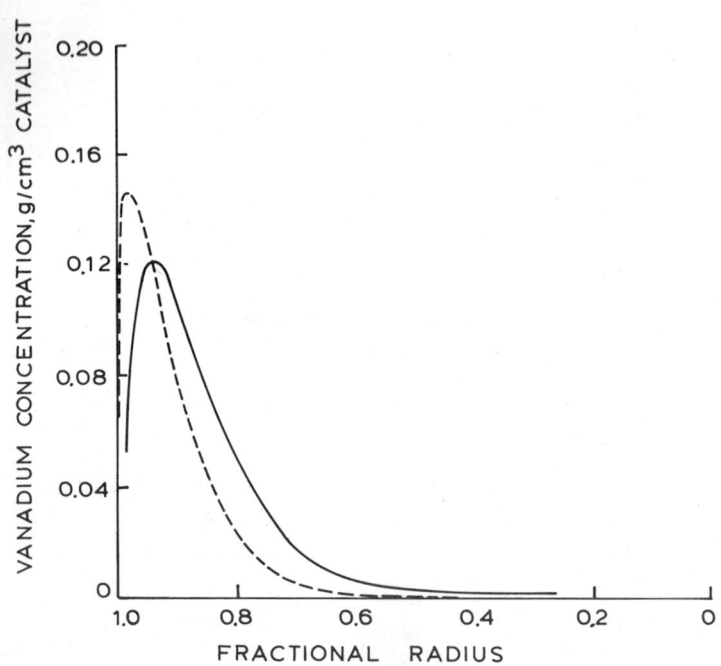

FIGURE 12. The effect of reaction temperature on vanadium deposition with Arabian heavy atmospheric residuum. —— reaction temperature 700°F, $\Theta_v = 0.33$. --- reaction temperature 750°F, $\Theta_v = 0.24$. (Adapted from Tamm, Harnsberger, and Bridge, 1981.)

3. MECHANISM AND KINETICS

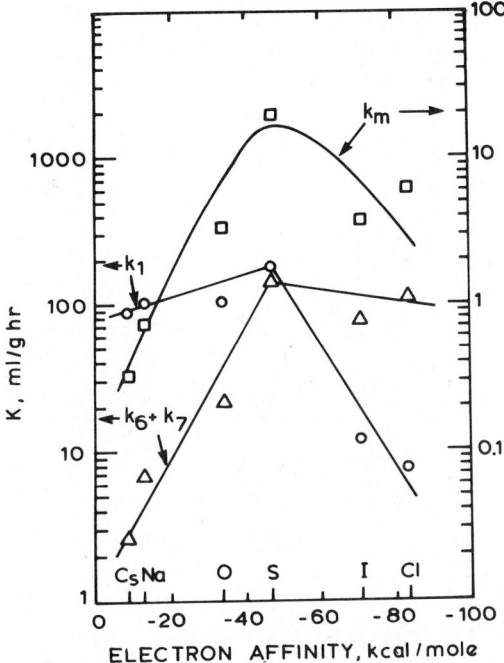

FIGURE 13. Selectivity of the hydrodemetallation of Ni-T3MPP kinetic rate parameters versus electron affinity of dopants on the modified catalysts. k_1, hydrogenation; $k_6 + k_7$, hydrogenolysis; k_m, overall reaction rate. (Adapted from Ware and Wei, 1985.)

termediate hydrogen sulfide, which is low at the reactor inlet; but the data of Agrawal (1980) was obtained under sulfur-free conditions, and this interior maximum was still observed. As the reaction rate increases, so would the Thiele moduli and the profiles cosh $(\Phi_i r/R)$; the interior maximum would move toward the edge of the catalyst and the Θ factor would decrease. This has been demonstrated by Tamm, Harnsberger, and Bridge (1981) by increasing the reaction temperature (see Figure 12). Pazos, Gonzalez, and Salazar-Guillen (1983) have also demonstrated this shift by increases in temperature and pressure.

The pool of intermediate chlorins that can be found during the reaction depends on the hydrogenolysis–hydrogenation ratio. When this ratio is small, the intermediate chlorins would build up to large values. This ratio is decreased at low hydrogen pressure and low temperature because of the higher dependency on H_2 and greater activation energy of hydrogenolysis.

For cobalt–molybdenum catalysts on alumina, this ratio can be depressed by an order of magnitude by the addition of alkali metals such as sodium and cesium and can be increased by two orders of magnitude by the addition of acid elements such as sulfur, iodine, and chlorine (see Figure 13). For the chlorided catalyst, the concentration of reaction intermediates is very small, so that the reaction sequences highly resemble the simple reaction

$$\text{M–P} \longrightarrow \text{deposit and breakup}$$

Webster (1984) showed that group VIII metals vary a great deal with regard to the k_3/k_1 ratio.

The data of Fleisch et al. (1984) also supports the consecutive demetallation mechanism. In a fixed bed of 12 ft length, they found that the maximum deposition of vanadium and nickel occurs not at the bed entrance, but at 17% of bed depth; the coke profile is relatively uniform. This agrees well with the calculations of Agrawal (1980).

4. SURFACE OF WORKING CATALYST

Since the HDM catalyst works mostly in a state where the surface is covered with coke and metal deposits, an understanding of the surface of aged catalysts is necessary. It is generally agreed that molybdenum sulfide is the catalyst for HDM, while cobalt acts as a promotor (Gates, Katzer, and Schuit, 1979).

For a typical catalyst with 12 wt% molybdenum and 3 wt% cobalt, the Mo and Co sulfides would cover a surface of about 110 and 30 m^2/g of catalyst, respectively, which is not quite a monolayer on a 200 m^2/g catalyst. On top of this, 20 wt% coke would occupy 400 m^2/g, 20 wt% vanadium pentasulfide would occupy 200 m^2/g, and 10 wt% nickel sulfide would occupy 100 m^2/g. How do these five types of materials, sufficient to form four monolayers, coexist on the surface? When the essential molybdenum is covered with another material, does it lose its catalytic activity or can it migrate up to the surface again? Do vanadium and nickel have enough activity to carry on?

If two monolayers worth of graphite come down uniformly, and if the molybdenum atoms cannot migrate to the surface, then the catalyst would not survive the first period of coke buildup. In fact, the loss of activity

4. SURFACE OF WORKING CATALYST

after a coke buildup is modest, thus coke deposition is not uniform. It is also known that new coke contains a considerable amount of hydrogen and can act as hydrogen donor in the hydrogenation of metal compounds. After a period of time, the hydrogen content decreases and the coke "hardens" to become graphitelike and loses the hydrogen donor ability.

Another model of deposition is the purely random Poisson processes, where any surface is equally likely to be the next deposition site, regardless of whether it is bare alumina or covered by previous depositions. For such a model, the probability of a surface covered with n number of deposits would be

$$p_n(m) = \frac{m^n e^{-m}}{n!}$$

where m is the average number of deposits on a surface site (Feller, 1957). Thus the probability of a bare site is

$$p_0(m) = e^{-m}$$

and the probability of a site with a single deposit is

$$p_1(m) = me^{-m}$$

If the catalyst activity is directly related to the probability of bare catalyst surfaces, then the catalyst activity after a single monolayer's worth of deposition would decline to 36.8% of its original value and to 13.5% of its original value after two monolayers' worth of deposition (see Figure 14). This would agree very well with the observation that during the second period of catalyst aging, an approximately linear temperature increase with time is needed to maintain conversion level,

$$k = be^{-Q/RT} = b_0 e^{-m} e^{-Q/RT} = \text{a constant}$$

or

$$T = \frac{Q/R}{\ln(b_0/k)^{-m}}$$

We give here a table for the relation between $p_n(m)$, the probability that

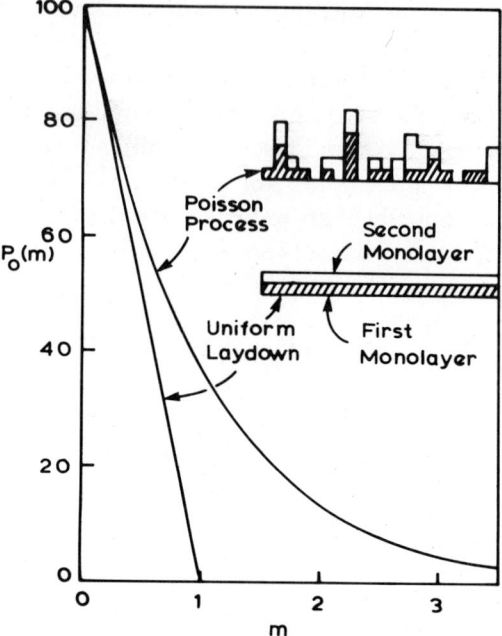

FIGURE 14. Bare surfaces remaining $P_0(m)$ versus number of monolayers of coverage m, under the uniform laydown and Poisson process assumptions.

a surface has precisely n deposits upon it, and m, the average number of deposits on the entire surface:

			m		
P_n	0	1	2	3	4
0	1	0.368	0.135	0.050	0.018
1	0	0.368	0.271	0.149	0.073
2	0	0.184	0.271	0.224	0.146
3	0	0.061	0.180	0.224	0.195
4	0	0.015	0.090	0.168	0.195
5	0	0.003	0.036	0.101	0.156

It is seen that after the average coverage reaches two monolayers, the surface resembles the Grand Canyon, with a large number of hills of various

4. SURFACE OF WORKING CATALYST

heights but very little of the original surface exposed. We do not know how realistic this Poisson model is, since we do not know whether the next deposit is more likely to fall on a clean surface or on a pile. An equal deposition probability would imply the final step is not surface-specific, even though the overall rate is controlled by the amount of bare molybdenum left. The data of Webster (1984) indicated that even at very high values of m, bare alumina and molybdenum signals can still be detected.

On the other hand, Takeuchi et al. (1985) found that when vanadium deposited on bare carriers reached 15 wt%, the surface is just as active as 15 wt% vanadium on metal carriers for the purposes of vanadium removal and asphaltene cracking, although not as active for HDS (see Figure 15). This would imply that CoMo are used only as "starters," and the deposits of V and Ni became in-situ-formed catalysts for HDM.

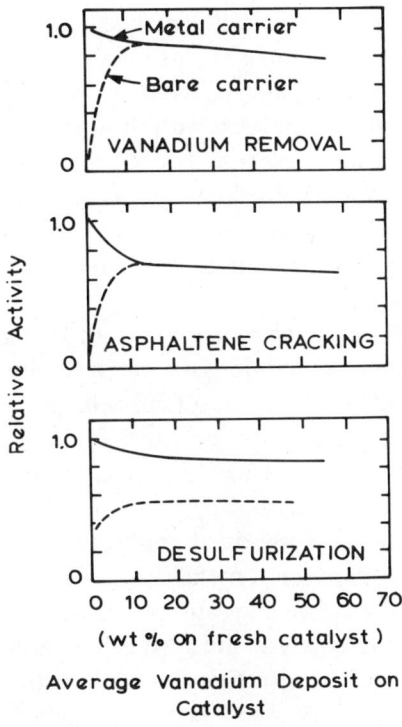

FIGURE 15. Catalytic activities versus vanadium deposition, for initially bare carrier and for initially metal/carrier. (Adapted from Takeuchi et al., 1985.)

5. CATALYST DESIGN AND OPTIMIZATION

Since the catalyst for HDM must also serve to remove sulfur and nitrogen compounds from oil, the optimal design must take the various requirements into account. The main classes of sulfur compounds are the sulfides, mercaptans, and thiophenes, namely, molecules that are smaller in diameter than asphaltenes and porphyrins. The destruction of the more resistant benzothiophenes and dibenzothiophenes also involves hydrogenation and hydrogenolysis steps. A much larger fraction of sulfur compounds is in the middle boiling point and molecular weight ranges and much easier to remove. The removal of nitrogen compounds tends to be more difficult.

The normal goals of catalyst design are greater activity, selectivity, and stability. If we have a catalyst with greater activity, how would we exploit it? If we operate the new catalyst at the same reactor conditions, it would increase the Thiele modulus ϕ, decrease the Θ and η factors, and shorten catalyst life. Unless we can work with catalyst diameters much smaller than 1 mm, we cannot take advantage of this greater activity. Another scheme is to lower the temperature and pressure to maintain the same level of activity, which can lead to a process with thinner reactor walls, less pumping and consumption of hydrogen, but more coking. A goal of better selectivity would be to minimize hydrocracking, hydrogenation of double bonds, and hydrogen consumption per barrel of resid. A goal of better stability would be the adsorption of more metals before shutdown and replacement.

Since the reaction mechanism involves both hydrogenation and hydrogenolysis reactions, one should design a catalyst with the balance of these two functions in mind. Pecoraro and Chianelli (1981) showed that group VIII metals are far more active than Mo and Co. Not enough is known about the specific actions of the active catalyst molybdenum (or tungsten) and the promoter cobalt (or nickel) to provide a guide on the optimal quantities to be used and methods of application.

Since the relentless barrage of metals raining on the catalyst surface will eventually cover up the active sites and block the pore openings, the physical arrangement of the catalysts must be designed to absorb this barrage to delay total deterioration of activity. Another strategy is to protect the HDS functions behind very small pores to exclude the larger metallo-organic compounds but to admit the smaller sulfur compounds and to provide a separate guard chamber with cheap throw-away catalysts to absorb the metals. Howell et al. (1984) demonstrated that smaller-pore cat-

5. CATALYST DESIGN AND OPTIMIZATION

alysts are better at HDS than for HDM. Graded catalyst beds with the larger-pore catalysts in front means a lower initial activity but longer life (see Figure 16). Ultrafiltration would be another method to separate the larger metallo-organic compounds for special treatments.

The first strategy requires the design of pores and a topology that represent a compromise between large surface area and rapid access. A small pore would mean a high surface area per gram of catalyst, which is favorable for initial catalyst activity and acceptance of metals barrage. It would also mean low diffusivity of the metallo-organic compounds to the active sites and rapid pore mouth plugging (see Figure 17) (Rajagopalan and Luss, 1979). Rajagopalan and Luss have proved that if the pores all behave as parallel pores without connections, then given a fixed surface area and porosity, the largest initial activity is obtained from uniform pore size. This proof is valid only for parallel bundles and not necessarily valid for bundles that are interconnected. Inoguchi et al. (1971b) also showed that small-pore catalysts give high initial activity, but the activity falls rapidly with time; the larger-pore catalysts have less initial activity, but maintain them for long periods of time.

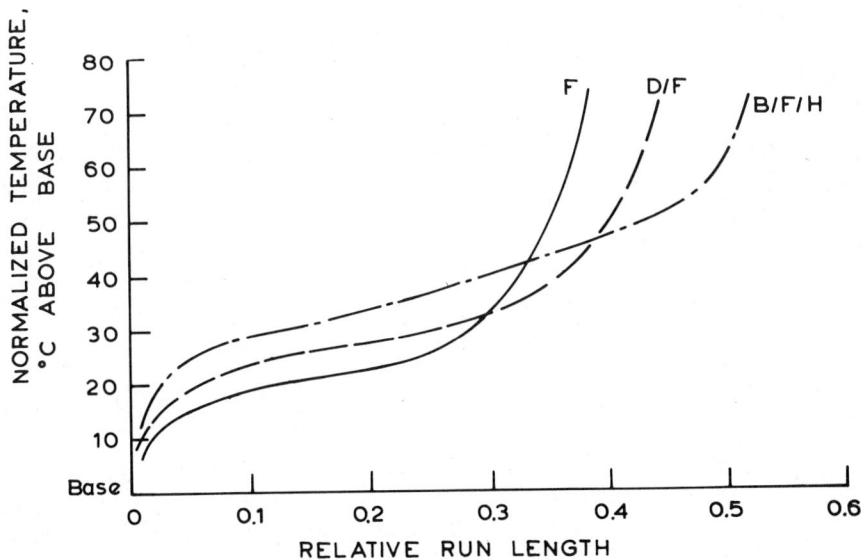

FIGURE 16. The effect of graded catalysts in beds on initial activity and relative run lengths. The catalysts H, F, D, and B are in the order of increasing pore sizes. (Adapted from Howell et al., 1984.)

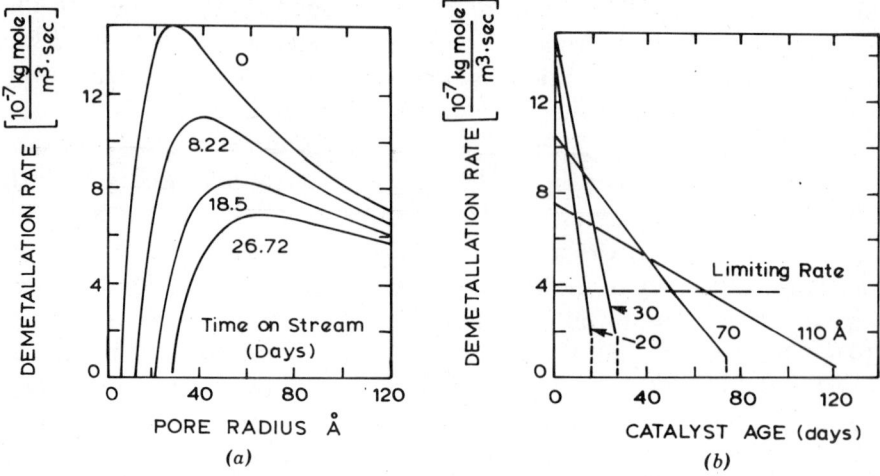

FIGURE 17. The effects of pore sizes and catalyst age on demetallation rates. (a) Dependence of demetallation rate on pore size and time on stream. (b) Dependence of demetallation rate on catalyst age for several pore sizes in angstroms. (Adapted from Rajagopalan and Luss, 1979.)

The access problem would be greatly improved for smaller catalyst particles. In a packed bed, the particles cannot be much smaller than 1 mm in diameter, to prevent excessive pressure drops across the bed and plugging of the bed. In a slurry bed, the particles can be much smaller but must be larger than the micron range, since the separation of oil and catalyst would be very difficult. For a catalyst with a single pore size, the optimal size can be computed under a variety of assumptions of diffusivities and their dependence on pore size narrowing by deposition. Spry and Sawyer (1975) suggested $D/D_0 = (1 - d_m/d_p)^4$. For particles in the range of 1–3 mm in diameter, the optimal pore size is in the range of 80–120 Å. Smaller particles have smaller optimal pore diameters (Bridge and Green, 1979). After metal deposition, we find many aged catalysts with deep U-shaped metal profile with Θ factors much less than 0.2. The Thiele modulus can be decreased by lowering the reaction rates: less loading of cobalt and molybdenum, lower temperature and lower hydrogen pressure. However, a lower-activity catalyst would also entail a lower space velocity and a larger reactor volume (which are also unfavorable) to effect the same conversion and throughput. The best catalyst should have both a high reactivity and a high factor.

5. CATALYST DESIGN AND OPTIMIZATION

If we are allowed to have two pore sizes, then we can design a catalyst with pore structure that resembles a highway map of the United States. There would be a network of superhighways or large pores for rapid transportation but it would not provide much surfaces for catalysis; this would be complemented by numerous local streets or fine pores that lead from the superhighways to each hamlet and neighborhood, which provide the surfaces needed for catalysis and metal deposition. Such a two-pore catalyst can be manufactured by pressing many microspheres lightly together, so that the space around the microspheres would become the macropores and the pores inside the microspheres would be the micropores. Typical pore sizes are given in Figure 18. A computation of the best two-pore strategy by the grain model of Sohn and Szekely (1972) would involve the search for the optimal size of the microspheres and the optimal sizes of the micropores and the macropores. The best two-pore catalyst would be significantly better than the best one-pore catalyst. One can come closer to the goal of catalysts with a uniform deposition of metals ($\Theta = 1$) and yet be consistent with the goal of adequate overall rate (Pazos, Gonzalez, and Salazar-Guillen, 1983). When there are two pore sizes, the optimal micropore would be smaller than the optimal micropore for a monopore catalyst. Very good performances can be attained by using micropores of 40 Å and microspheres of 40 μm in a particle of 3-mm diameter. Agrawal (1980) showed that as the grains become larger, the optimal pore radius would increase, as shown in Figure 19. The best strategy would involve a trade-

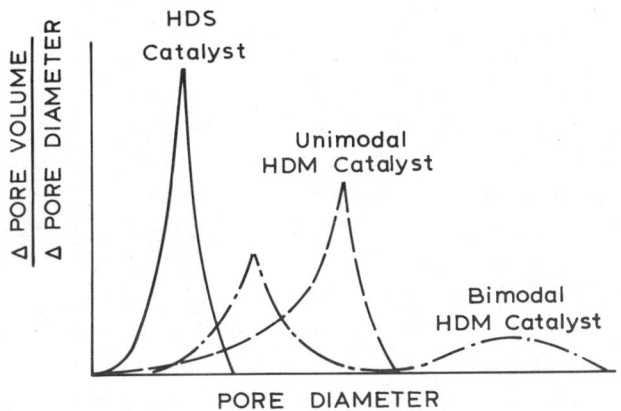

FIGURE 18. Pore size distributions for typical residuum hydroprocessing catalysts. (Adapted from Howell et al., 1984.)

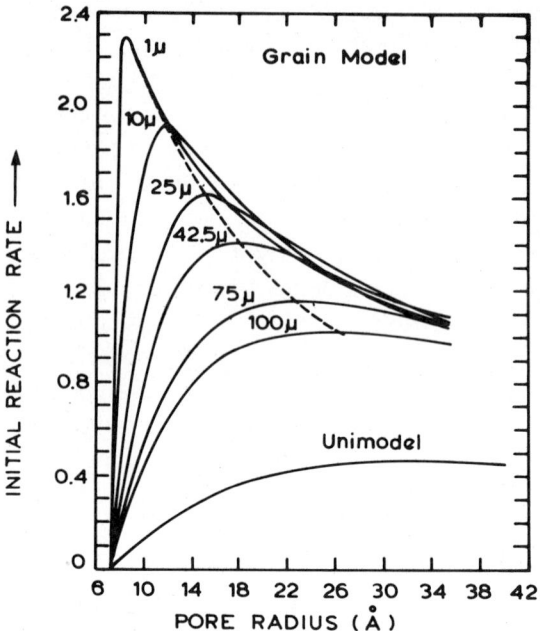

FIGURE 19. Initial reaction rate versus micropore radius and grain sizes at a fixed porosity. (Adapted from Agrawal, 1980.)

off between good initial activity, which is favored by small pores, and long life, which is favored by larger pores.

For a particular refinery, the favorite source of crude oil would have its own metal concentrations distributed among a characteristic blend of various porphyrins and asphaltenes of different reactivity and molecular sizes. The most economic frequency of catalyst replacement would depend on many factors. The best pore sizes and microsphere sizes should be designed to fit that particular application.

Is there any reason to believe that the best three-pore catalyst would be significantly better than the best two-pore catalyst? What should be the connectivity of the best three-pore catalysts? Should it be analogous to a hierarchy of primary–secondary–tertiary roads to serve decreasing domains with decreasing road widths? When would the point of diminishing return set in when we consider the best four or five pore catalysts? What about the following far more challenging questions: What is the best catalyst with

5. CATALYST DESIGN AND OPTIMIZATION

a continuous pore size distribution? What connectivity should these pores have? Can such fancy catalysts ever be manufactured?

The second strategy attempts to shield the good catalysts for desulfurization from the barrage of raining metals. One method involves an initial guard chamber with large-pore catalysts for the removal of metals through HDM, followed with a small-pore HDS catalyst that would exclude the larger metallo-organic compounds (Hastings, James, and Mounce, 1975).

A knowledge of the consecutive nature of the HDM mechanism and an ability to manipulate the ratio of hydrogenolysis–hydrogenation activities leads to a different two-bed strategy. Alkali doping would lead to catalysts in the first bed that produce a large concentration of intermediate metallochlorins in dynamic equilibrium with the metalloporphyrins, and this could be accomplished without the deposition of metals ruining the first bed catalyst. The second bed can be filled with large-pore and high-void sacrifice catalysts, such as chlorine on alumina, which would convert the metallochlorins into metal deposits. The first bed would be kept clean and the second bed would be discarded periodically.

To allow more external surface per volume of catalyst without employing very small particles, which would lead to high pressure drops, a number of new shapes have been proposed recently. Figure 20 shows the "minilith"

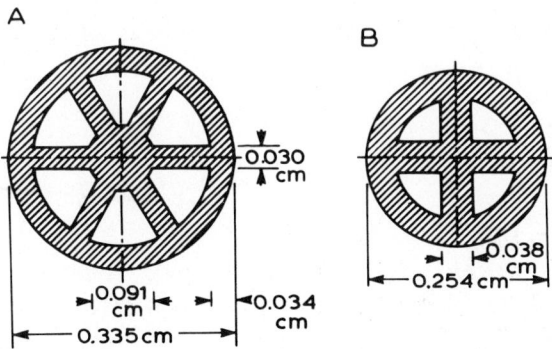

FIGURE 20. Minilith HDM catalysts for increasing surface area per volume. (Adapted from Pereira et al., 1985.)

of Pereira et al. (1985), which resembles miniature Raschig rings. It is important to ensure that oil and water will flow through these narrow slots inside the miniliths. Toulhoat et al. (1985) invented a "chestnut bur" material that resembles a star burst, or a compact heat exchanger with radiating fins. This shape has the potential of very high porosity and a very high metals loading. Finally, Dautzenberg and de Deken (1985) suggested a small porous basket filled with small spheres. If Takeuchi et al. (1985) are correct that vanadium and nickel form self-sustained in-situ catalysts for HDM, a "fractal" catalyst could be designed as a growing Brillo pad that could be periodically split into two parts and expanded to fit the original volume. These catalysts would be both everlasting and resource-generating. All of these measures would not be needed if we could keep the metals in solution, instead of them depositing on the catalyst surface.

6. CONCLUSIONS

Catalyst design is traditionally performed as an evolutionary process, exploring the regions around a successful design and making improvements incrementally. Good theoretical understanding and fundamental data can be very helpful in this process of evolutionary change, followed with evaluations of the performance of the new design, adjustments in the theory and parameter value, and another round of design. Rational design of the physical configurations of HDM catalysts has been fruitful since the dominant factors of transport and access to surfaces are fairly well defined and understood. Rational design of the chemical constituents of HDM catalysts waits for greater understanding on the chemistry of surfaces, reactivity, and selectivity.

REFERENCES

R. Agrawal, Sc.D. Thesis, MIT (1980).

R. Agrawal and J. Wei, *I&EC Proc. Des. Dev.* **23**, 505–515 (1984).

F. Audibert and P. Duhaut, Paper given at the American Petroleum Institute meeting in Houston, May 1970.

E.W. Baker and S.E. Palmer, in *The Porphyrins*, D. Dolphin, Ed., Vol 1, Chapter 2, Academic Press, New York, 1978.

H. Beuther and B.K. Schmid, *Sixth World Petroleum Congress, Frankfurt*, Section III, Paper 20, p. 297 (1963).

REFERENCES

A.G. Bridge and D.C. Green, *Div. Pet. Chem. Am. Chem. Soc. Prepr.* **24**(3), 791 (1979).

J.W. Bunger and N.C. Li, *Chemistry of Asphaltenes,* Advances in Chemistry Series, No. 195, American Chemical Society, Washington, D.C., 1981.

C.D. Chang and A.J. Silvestri, *I&EC Proc. Des. Dev.* **15**(1), 161 (1976).

F.M. Dautzenberg and J.C. de Deken, *Div. Pet. Chem. Am. Chem. Soc. Prepr.* 8 (February 1985).

W. Feller, *An Introduction to Probability Theory and Its Applications,* Vol 1, 2nd ed., John Wiley & Sons, New York, 1957.

R. Fish and J.J. Komlenic, *Anal. Chem.* **56**, 510–517 (1984).

T.H. Fleisch, B.L. Meyers, J.B. Hall, et al., *J. Catal.* **86**, 147–157 (1984).

R. Galiasso, J. Garcia, L. Caprioli, et al., *Div. Pet. Chem. Am. Chem. Soc. Prepr.* 50 (February 1985).

B.C. Gates, J.R. Katzer, and G.C.A. Schuit, *Chemistry of Catalytic Processes,* p. 411, McGraw-Hill, New York, 1979.

K.H. Hastings, L.C. James, and W.R. Mounce, *Oil Gas J.* 122 (June 30, 1975).

R.L. Howell, C. Hung, K.R. Gibson, and H.C. Chen, paper presented at the Japan Petroleum Institute meeting, October 1984.

M. Inoguchi, K. Inaba, et al., *Bull. Jpn. Pet. Inst.* **13**(1), 12 (1971a).

M. Inoguchi, H. Kagaya, K. Daigo, et al., *Bull. Jpn. Pet. Inst.* **13**(2), 153 (1971 b).

H. Kameyama, M. Sugishima, M. Yamada, et al., *J. Jpn. Pet. Inst.* **24**, 317 (1981).

H. Kameyama, M. Sugishima, M. Yamada, et al., *J. Jpn. Pet. Inst.* **25**, 118 (1982).

H. Kameyama, M. Sugishima, M. Yamada, et al., *J. Jpn. Pet. Inst.* **28**, 83 (1985).

S.G. Kukes and A.W. Aldage, *Div. Pet. Chem. Am. Chem. Soc. Prepr.* p. 119 (Fall 1985).

T. Kwan and M. Sato, *Nippon Kagaku Zasshi* **91**(1), 1 (1970).

A. Morales, J.J. Garcia, and R. Prada, in *Eighth International congress on Catalysis,* p. II-341, Verlag Chemie, Frankfurt am Main, 1984.

W.L. Nelson, *Oil Gas J.,* p. 72 (November 15, 1976).

E. Newson, *Div. Pet. Chem. Am. Chem. Soc. Chicago Meeting* p. A141 (September 1970).

E. Newson, *I&EC Proc. Des. Dev.* **14**(1), 27 (1975).

M.F. Oxenreiter, C.G. Frye, G.B. Hoekstra, et al., Desulfurization of Khafji and Gach Saran Resids, paper presented at Japan Petroleum Institute meeting, November 1972.

J.M. Pazos, J.C. Gonzalez, and A.J. Salazar-Guillen, *I&EC Proc. Des. Dev.* **22**(4), 653 (1983).

T.A. Pecoraro and R.R. Chianelli, *J. Cat.* **67**, 430 (1981).

C.J. Pereira, J.W. Beeckman, W.C. Cheng, R.G. Donnelly, and L.L. Hegedus, *Div. Pet. Chem. Am. Chem. Soc. Prepr.* **30**(1), 74 (April–May 1985).

K. Rajagopalan and D. Luss, *I&EC Proc. Des. Dev.* **18**(3), 459 (1979).

L.A. Rankel, *Div. Pet. Chem. Am. Chem. Soc. Prepr.* **26**(3), 689 (1981).

L.A. Rankel and L.D. Rollman, *Fuel* **62**, 44 (1983).

E.M. Renkin, *J. Gen. Physiol.* **38**, 225 (1954).

M. Sato, N. Takayama, S. Kurita, et al., *Nippon Kagaku Zasshi* **92**(10), 834 (1971).

W.R. Scheidt, in *The Porphyrins,* D. Dolphin, Ed., Vol. 3, Chapter 10, Academic Press, New York, 1978.

H.Y. Sohn and J. Szekely, *Chem. Eng. Sci.* 27, 763 (1972).

J.C. Spry and W.H. Sawyer, 68th AIChE Meeting, Los Angeles, November 16–20, 1975.

C. Takeuchi, S. Asaoka, S. Nakata, et al., *Div. Pet. Chem. Am. Chem. Soc. Prepr.*, p. 96 (February 1985).

P.W. Tamm, H. F. Harnsberger, and A.G. Bridge, *I&EC Proc. Des Dev.* **20,** 262 (1981).

H. Toulhoat, J.C. Plumail, G. Martino, et al., *Div. Pet. Chem. Am. Chem. Soc. Prepr.*, p. 85 (February 1985).

R.A. Ware, Sc.D. Thesis, MIT (1983).

R.A. Ware and J. Wei, *J. Catal.* **93,** 100–121, 122–134, 135–151 (1985).

I.A. Webster, Sc.D. Thesis, MIT (1984).

J. Wei and R.G. Wei, *Chem. Eng. Commun.* **13,** 251 (1982).

J. Weitkamp, W. Gerhardt, and D. Scholl, *Proc. Eighth Int. Cong. Catal.* **2,** 269 (1984).

A. White, P. Handler, and E.L. Smith, *Principles of Biochemistry,* p. 796, McGraw-Hill, New York, 1964.

T.F. Yen, J.G. Erdman, and S.S. Pollack, *Anal. Chem.* **33,** 1587 (1961).

INDEX

Ab initio design of catalysts, 7
Acetic acid, by carbonylation of methanol, 74
Acid sites, 23
Acidity, of oxide surfaces, 105–109
Adsorption:
  heat of, 18
  of metal ions on supports, 109–112
Aerosol, propellants in ozone destruction, 142
AES (Auger electron spectroscopy), 5, 12
Aggregation of supported organometallics, 91–92
Aging, catalyst, 231, 250
Alkali metals, in hydrotreating catalyst, 260
Alkanes:
  alkylation of, 163
  aromatization of, 184
  C-H bond activation in, 7
  cracking of, 172, 193, 194
  dehydrogenation of, 133
  hydrogenolysis of, 96, 131, 133
  isomerization of, 11, 133, 158, 163
  melting point of, 195
  M-forming of, 193
Alkenes:
  acid-catalyzed reactions of, 197
  aromatization of, 184, 197
  conversion of, 197, 198
  cracking of, 184, 197
  cracking-polymerization of, 197, 198
  cyclization of, 197
  dismutation of, 154
  double-bond isomerization of, 197
  hydroformylation of, 73, 77, 90
  hydrogenation of, 96, 97, 133
  hydrogen-transfer in, 197
  isomerization of, 163
  metathesis of, 96
  polymerization of, 71, 98–99
Alkylation, *see under individual reactants and catalysts*
Alloys, 7, 40–42, 64
Alpha-test, 203
Alumina:
  in automotive catalysts, 135
  as binder with zeolites, 178, 184
  halogenated, 6, 12, 22
  in hydrotreating catalyst, 260
  as a structural promoter, 64
  surface acidity of, 106–110
Alumina-boria, surface acidity of, 108
Alumina-chromia, surface acidity of, 108
Aluminosilicates, 57
Aluminum, tetrahedral, 24
Aluminum oxide, *see* Alumina
Ammine complexes, impregnation by, 110, 112, 114
Ammonia decomposition, on molybdenum, 149–151
Ammonia synthesis:
  activation energy of, 18
  catalysts for, 11, 13, 43
  commercial catalysts for, 53
  effect of potassium on, 51
  kinetics of, 150
  model catalysts and, 161
  modifiers in, 43

Ammonia synthesis (*Continued*)
  product inhibition in, 151
  role of potassium in, 53
  structure-sensitivity in, 25–28, 151
Ammonium-Y-zeolite, ammonia desorption from, 179
Anchoring of complexes, 89
Antibonding orbitals, 53
Aromatics:
  alkylation of, 163, 194, 199
  hydrogenation of, 133
  isomerization of, 163
Aromatization:
  catalysts for, 11
  structure-sensitivity in, 51
  see also under individual reactants and catalysts
Arrhenius number, 216, 221
Arsenic, in hydrodesulfurization of dibenzothiophene, 250
Asphaltenes:
  cracking on vanadium surface, 263
  in oil, 246
  micelles of, 247, 248
Auger electron spectroscopy, 5, 12
Automobile emissions, 1, 4, 8, 135
Automobile exhaust catalysts, *see* Automotive catalysts
Automotive catalysts, 1, 4, 135, 230
A-zeolite:
  silica-to-alumina ratio of, 170
  structure of, 165–167
  zoning in, 183

Back-bonding, 53
Bang-chatter, bang-bang, 227
Benzene:
  alkylation of, 193
  hydrogenation of, 221
  M-forming of, 193
  on platinum, 37, 39
  production of, 46
  on rhodium, 37–39
Benzothiophenes, in crude oil, 264
Beryllium, in zeolites, 184
Beta-zeolite, constraint index of, 169
Bifunctional catalysts, 94, 95, 159, 227
Biliverdin-iron complex, 246
Bimetallic catalysts, 5, 40, 131, 132

Bimodal hydrotreating catalyst, 267
Bimolecular reactions, 172, 191
Bimolecular transition state, 196
Biot number, 217, 221
Boron, in zeolites, 165, 184
Bosanquet's approximation, 219
Bronsted acid(s):
  sites in cracking, 181
  sites on supports, 105–109
  sites in zeolites, 178–180, 182, 184
  solid acids as, 24
Butane, hydrogenolysis of, 47
n-butane, diffusion of, 176
Butene, on molybdenum, 40
t-butylbenzene, diffusion of, 176
B-ZSM-5:
  aromatization with, 184
  cracking with, 184
  dealkylation with, 184
  isomerization with, 184
  methanol conversion with, 184

Calcination, 114, 115
Carbonaceous deposit, 48–51, 61
Carbonium-ion-like intermediates in cracking, 24
Carbon monoxide:
  coadsorption with potassium, 51–54
  dissociation of, 52, 53, 55
  emission of, 4
  hydrogenation of, 13, 35, 51, 53–56, 58, 85, 96–97, 123–130
  on nickel, 51, 53, 59
  oxidation of, 13, 135
  on palladium, 18, 19
  on platinum, 37, 39, 51
  on rhodium, 37, 39, 53
  on ruthenium, 51
  suppression of chemisorption, 118, 120, 121
Catalyst-oil ratio, in cracking, 203
Catalytic converter, 4
Cation hydrolysis, 110, 111, 114
Ceria:
  in automotive catalysts, 135
  as a surface modifier, 117
Chabazite, structure of, 166
Chain reactions, 147, 155, 157
Charge fluctuations, 22

INDEX

C-H bond activation in alkanes, 7
Chemisorption, for metal surface-area determination, 60
Chestnut bur, catalyst shape, 270
Chloride complexes, impregnation by, 112, 114
Chlorine:
  as an electron acceptor, 64
  in ozone destruction, 141, 142
Chlorine-alumina, as sacrifice catalyst, 269
Chlorins, metal, as intermediates in hydrodemetallation, 253, 257, 259, 269
Chloroplatinic acid, 113, 233
Chromia, surface acidity of, 110
Chromium:
  complexes anchored to metal oxides, 95
  in hydrodesulfurization of dibenzothiophene, 250
  in zeolites, 184
Chromium sulfide, in hydrotreating catalyst, 249
Citric acid, 233
Clays, as acid catalysts, 163
Clinoptilolite, stacking faults in, 168
Cloud point, 195
Clusters:
  bimetallic, 7, 60, 91
  carbonyl, 75, 77
  as catalysts, 75
  heteropolynuclear, 113
  iron-osmium on silica, 91
  osmium, supported, 83, 90–92, 96
  rhodium-osmium on alumina, 91
  ruthenium osmium on alumina, 91
  and structure of the catalytic entity, 146
  in supported metal catalysts, 80
  in zeolites, 97
Cobalt:
  CO hydrogenation on, 13
  in formic acid decomposition, 122
  in hydrodesulfurization of dibenzothiophene, 250
  in hydrotreating catalyst, 249, 260, 263, 264
  methanol carbonylation by complexes of, 78
Cobalt-alumina:
  CO hydrogenation on, 125

  in formic acid decomposition, 122
Cobalt-silica:
  alkane hydrogenolysis on, 134
  CO hydrogenation on, 125
Cobalt-titania:
  alkane hydrogenolysis on, 134
  CO hydrogenation on, 125
Cocatalyst, 94
Coke:
  deposition profiles of, 260
  precursors to, 187
Coke buildup in hydrotreating catalyst, 249, 250
Coke selectivity, 171, 203, 204
Coking, 245
  in HZSM-5, 175, 177, 186
  on zeolites, 186, 187
Cold filter-plugging point, 195
Column chromatography, equations of, 233
Compositional zoning, 183
Configurational diffusion, 174
Constraint index, 169, 187, 191, 192
Consumption of catalysts, 2,3
Cooperativity, 151–153
Copolymers, 161
Copper-nickel, ethane hydrogenolysis on, 131, 132
Coupled cycles, 154–160
Coupling:
  kinetic, 147–154
  thermodynamic, 147
Cracking:
  carbonium-ion-like intermediates in, 24
  catalyst-oil ratio in, 203
  development of zeolite catalysts for, 202
  in fluidized bed, 203
  gas-oil, 163, 186, 204–206
  rare-earth exchange in catalysts for, 202, 206
  regeneration in, 203
  resid, 204
  steam, 155
  see also under individual reactants and catalysts
Crude oil:
  metals in, 245, 249
  sulfur content, 249
Cumene, dealkylation of, 181
Cyclization, of hydrocarbons, 16, 197

Cyclohexane, dehydrogenation of, 121, 123, 131, 132, 134
Cyclopropane:
  hydrogenolysis of, 97
  isomerization of, 180

Dachiardite, stacking faults in, 168
Damköhler number, 215
Deasphalting, 246
Decoration, of metals by oxides, 116, 117, 130
De Donder relation, 148, 149
Degeneracy of electronic states, 20
Dehydrocyclization:
  of hydrocarbons, 13, 16, 45
  requirements of, 44
  see also under individual reactants and catalysts
Dehydrogenation:
  of hydrocarbons, 13
  see also under individual reactants and catalysts
d-electron metals, 20, 22
Demanding reactions, see Structure-sensitivity
Deprotonation, of hydrido complexes, 80
Dewaxing, see MDDW; MLDW
p-dialkylbenzenes, formation of, 189, 200–202
Dibenzothiophene:
  in crude oil, 264
  hydrodesulfurization of, 14, 250
Diesel fuel, from heavy petroleum, 22
Diffraction:
  low-energy electron, 5, 12
  X-ray, 6
  X-ray, grazing angle, 12
Diffuse reflectance spectroscopy, 6
Diffusion limitation, 188–190, 214
Dimensionless groups, 214
2,2 Dimethylbutane, diffusion of, 174, 176
2,2 Dimethylheptane, cracking of, 174, 190
Dinosaurs, oil derived from, 246
Dismutation of alkenes, 154
Dispersion, of metals on supports, 105–109
Dissolution of metal, into support, 115, 117
Distillate dewaxing, see MDDW
Dodecane, cracking of, 181
DPEP, structure of, 247

DRS (Diffuse reflectance spectroscopy), 6
Durene, 200
Dusty gas model, 220

EELS, see Electron energy-loss spectroscopy
Effectiveness, 214–218
Effectiveness factor, 173, 190, 214, 216, 225, 226
Egg-shell catalyst, 230, 236
Egg-white catalyst, 230, 233, 236
Egg-yolk catalyst, 230, 233
Electron energy-loss spectroscopy, 6, 12
Electron hole states, 20
Electronic factors in catalysis, 7
Electron microscopy, 12, 61
Electron spin resonance, 12, 88
Eley–Rideal kinetics, in ethene hydrogenation, 35
Emission control, catalysts for, 7
Emissions, automobile, 1, 4, 8, 135
Ensemble effects, 42, 43
Environmental regulations, 4
Enzymes, 37
Epistilbite, stacking faults in, 168
Erionite:
  constraint index of, 169, 192
  in selectoforming, 193
  structure of, 165, 166
ESCA, see X-ray photoelectron spectroscopy
ESR (Electron spin resonance), 12
Ethane:
  hydrogenolysis of, 47, 131
  pyrolysis of, 159
Ethanol synthesis, over supported rhodium, 129
Ethene:
  hydrogenation of, 13, 16, 18, 31–37, 221
  on platinum, 37
  on rhodium, 37, 38
Ethylbenzene:
  dealkylation of, 180, 184
  synthesis of, 184
Ethyl benzoate, 99
Ethylene, see Ethene
Ethylene glycol, from CO and hydrogen, 75, 77
Ethylidyne, 32, 34–36
p-ethyltoluene, from toluene, 200

INDEX

Etioporphyrin(s):
  hydrodemetallation of, 255
  nickel-,
    HDM kinetics of, 255
    structure of, 254
  structure of, 247
  vanadyl-, HDM kinetics of, 255
EXAFS, see X-ray absorption fine structure, exended

Facile reactions, see Structure-insensitivity
Faujasite, 7, 97, 166, 169
FCC, see Fluid catalytic cracking
Ferns, oil derived from, 246
Ferrierite:
  stacking faults in, 168
  structure of, 165, 166
Fluid catalytic cracking, 5, 6, 22, 205
Fluidized bed, for cracking, 203
Fluorination, to increase oxide acidity, 108
Fluorine, in ozone destruction, 142
Formic acid:
  decomposition of, 121, 122
  dehydrogenation of, 121
Fourier transform infrared spectroscopy (FTIR), 6, 12
Fractal catalyst, 270
Framework aluminum, reducing, in zeolites, 178
Free radical reactions, 144, 147, 155, 157, 158
Freeze point, 195
Fuel cells, 7

Gallium, in zeolites, 165, 184
Gas chromatography, 12, 61
Gas-oil cracking, 163
Gasoline:
  from CO and hydrogen, 64
  from heavy petroleum, 22
  olefinic, 206, 207
Geometric factors in catalysis, 7
Germanium, in zeolites, 184
Gmelinite, stacking faults in, 168
Gold:
  density of states in, 22
  in hydrodesulfurization of dibenzothiophene, 250
  on platinum, 40
Grain model, 267

Grand Canyon, 262
Graphite, on hydrotreating catalyst, 260, 261
Grazing angle x-ray diffraction, 12
Guard chamber, in hydrodemetallation, 269

Hafnium, complexes anchored to metal oxides, 95
Halogens, in hydrotreating catalyst, 260
H-D exchange:
  studies of, 62
  below 300K, 22
Hemoglobin, 246
  dioxygen binding to, 153
n-heptane, hydrogenolysis of, 29
1-heptene, cracking of, 181
Heulandite, stacking faults in, 168
Hexadecane, cracking of, 181, 186
Hexane, cracking of, 23, 174, 180, 181
n-Hexane:
  conversion of, 41
  cracking of, 23, 24, 190, 191
  dehydrocyclization of, 46
  dehydrogenation of, 51
  isomerization of, 41, 42
  reactions over platinum single crystals, 51
  reforming of, 49
Hexene, cracking of, 23
1-Hexene:
  cracking of, 24, 174, 180
  double-bond shift of, 181
  isomerization of, 24
Homogeneous catalysis, 79, 155, 161
Hougen–Watson kinetics, 150
HREELS, see Electron energy-loss spectroscopy
Human body as microporous catalyst, 65
Hydride transfer, bimolecular, 191
Hydrocarbons:
  conversion of, 13, 16
  emission of, 4
  partial oxidation of, 5
  pyrolysis of, 155–159
  see also under individual hydrocarbons
Hydrocracking:
  minimization in HDM, 264
  on solid acids, 22
Hydrodemetallation, 245–270
Hydrodesulfurization:
  catalyst lifetime, 249

Hydrosulfurization (*Continued*)
  catalysts for, 11, 13
  on sulfides, 15
  see also under individual reactants and catalysts
Hydroformylation, 36, 73. *See also under individual reactants and catalysts*
Hydrogen:
  on nickel, 59
  suppression of chemisorption, 118, 120, 121
Hydrogenation:
  catalysts for, 13
  of hydrocarbons, 13, 16
  minimization in HDM, 264
  see also under individual reactants and catalysts
Hydrogen bromide, catalytic cycle in formation of, 147–149
Hydrogenolysis:
  of hydrocarbons, 13, 16
  of organic molecules, 40
  requirements of, 44
  see also under individual reactants and catalysts
Hydrogen storage, 56
Hydrogen sulfide:
  poisoning by, 47
  as pyrolysis catalyst, 159
Hydrogen transfer, 51, 198, 206
Hydroprocessing, catalysts for, 11
Hydrotreating, 245–270
Hysomer, 193
HY-zeolite:
  Lewis sites in, 179
  steaming of, 182
  xylene isomerization over, 196
H-zeolon, constraint index of, 169
HZSM-5
  aluminum concentration in, 180
  aluminum content in, 23
  coking in, 175, 177
  diffusion in, 172
  diffusion coefficients in, 175
  hexane cracking over, 23, 181
  modified, 175
  propene conversion over, 197
  sorption in, 171

Immovable dust, 220
Impregnation:
  by incipient wetness, 109
  by ion exchange, 109, 113
  of pellet, 233
Incipient wetness, impregnation by, 109
Inelastic electron tunneling spectroscopy, 81, 83, 88
Infrared spectroscopy, 6, 12, 23, 81
Inhibition, by products, 151
Ion-exchange, impregnation by, 109, 110
Ion-scattering spectroscopy, 12, 60
Iridium:
  in hydrodesulfurization of dibenzothiophene, 250
  hydroformylation with complexes of, 77
  methanol carbonylation by complexes of, 78
Iridium-alumina, alkane hydrogenolysis on, 133
  CO hydrogenation on, 125
  hydrogen chemisorption on, 118
Iridium-beryllia, CO hydrogenation on, 125
Iridium-calcium oxide, CO hydrogenation on, 125
Iridium disulfide, in hydrotreating catalyst, 249
Iridium-hafnia, hydrogen chemisorption on, 118
Iridium-magnesia:
  CO hydrogenation on, 125
  hydrogen chemisorption on, 118
Iridium-manganese monoxide, hydrogen chemisorption on, 118
Iridium-niobia, hydrogen chemisorption on, 118
Iridium-scandia, hydrogen chemisorption on, 118
Iridium-silica:
  alkane hydrogenolysis on, 133
  CO hydrogenation on, 125
  hydrogen chemisorption on, 118
Iridium-tantalum oxide, hydrogen chemisorption on, 118
Iridium-titania:
  alkane hydrogenolysis on, 133
  CO hydrogenation on, 125
  hydrogen chemisorption on, 118

INDEX

Iridium-vanadia, hydrogen chemisorption on, 118
Iridium-Yttria, hydrogen chemisorption on, 118
Iridium-zinc oxide, CO hydrogenation on, 125
Iridium-zirconia, hydrogen chemisorption on, 118
Iron:
  ammonia synthesis on, 13
  CO hydrogenation on, 13
  in hydrodesulfurization of dibenzothiophene, 250
  structure-sensitivity on, 25, 26
  water-gas shift on, 56
  in zeolites, 165, 184
Iron-alumina, CO hydrogenation on, 125
Iron carbide-carbon, 56
Iron carbide-magnesia, 56
Iron carbides, 56
Iron-osmium, clusters on alumina, 91
Iron-silica, alkane hydrogenolysis on, 134
Iron-titania:
  alkane hydrogenolysis on, 134
  CO hydrogenation on, 125
Isobacteriochlorin, from tetraphenylporphyrins, 254
Isobutane:
  cracking of, 24
  diffusion of, 176
  hydrogenolysis of, 31, 46, 47
  isomerization of, 46, 47
Isomerization:
  of hydrocarbons, 13, 16
  requirements of, 44
  on solid acids, 22
  structure-sensitivity in, 51
  see also under individual reactants and catalysts
Isotherm, Langmuir, 235
Isotope techniques, 7, 12, 31, 62

Jet fuel, from heavy petroleum, 22

Kinetic coupling, 147–154, 158, 160
Kinks, 20, 27–31, 40, 45, 47, 64
Knudsen diffusion, 174

Langmuir–Hinshelwood kinetics, 25–31, 150
Langmuir isotherm, 235
Lanthana, 54, 135
Lanthanum rhodate, 54, 55
Laser Raman spectroscopy, see Raman spectroscopy
L-dopa synthesis, 78
LEED (Low energy electron diffraction), 5, 12
Lewis acid:
  centers on supports, 93, 105–109
  sites in HY-zeolite, 179
  sites in zeolites, 182
Lewis base, centers on supports, 105–109
Lewis number, 221
Ligand association, 74
Ligand insertion, 74
Low-energy electron diffraction, 5, 12
Lube dewaxing, see MLDW (Mobile lube dewaxing)
L-zeolite:
  platinum nucleation sites on, 57, 58
  structure of, 166

Magic angle spinning nuclear magnetic resonance, 6, 23
Magnesia, surface acidity of, 109, 110
Magnesium, in zeolites, 184
Magnetic susceptibility measurements, 88
Manganese, in hydrodesulfurization of dibenzothiophene, 250
MASNMR, 6, 23
Mass spectroscopy, 12
Mass-transport selectivity, 188, 189
Matrix, in cracking catalysts, 203
Mazzite:
  constraint index of, 169
  structure of, 166
MDDW (Mobil distillate dewaxing), 193, 195
MEB (Mobil ethylbenzene), 193
Mercaptans, in crude oil, 264
Metal chlorins, as intermediates in hydrodemetallation, 253, 257, 259, 269
Metal deposition:
  in hydrotreating catalysts, 250–252, 266

Metal deposition (*Continued*)
  kinetics of, 256
  maximum in, 257
Metalloporphyrins, diffusion of, 257
Metal-metal bonds, 75, 84, 89, 91, 97
Metal oxide surfaces, properties of, 105
Metal-support interactions, 8
  strong, 58–60, 103–136
Metathesis, of alkenes, 96
Methane, partial oxidation of, 7
Methanol:
  carbonylation of, 73, 74, 78, 94, 95, 160
  conversion of, 180, 181, 186, 198, 199
  see also MTG (Methanol-to-gasoline); MTO (Methanol-to-olefins)
  partial oxidation of, 13
Methanol synthesis:
  catalysts for, 5
  over supported palladium, 124, 126
α-methine bridge, 246
Methylcyclopentane, 45
  production of, 46
Methyl iodide, as promoter in carbonylation, 74, 75
3-methylpentane, cracking of, 191
M-forming, 193
MHTI (Mobil high temperature isomerization), 193
Michaelis–Menten kinetics, 150
Microprobe, scanning auger, 5
Microscopy, transmission electron, 6, 7
Migration:
  of oxides onto metals, 118
  of support, 117
Minilith, 269, 270
MLDW (Mobil lube dewaxing), 193, 195
MLPI (Mobil low pressure isomerization), 193
MOGD (Mobil olefins to gasoline and diesel), 193
Molecular beam relaxation, pulsed techniques, 161
Molecular sieves, 57, 172
Molecular sieving, 164, 189
Molybdena:
  partial oxidation of methanol on, 13
  on silica, 114
  as a surface modifier, 117
Molybdenum:
  ammonia decomposition on, 149, 150
  butene on, 40
  carbonyls on alumina, NMR spectrum of, 87
  CO hydrogenation on, 13
  complexes anchored to metal oxides, 95
  in hydrodesulfurization of dibenzothiophene, 250
  hydrodesulfurization of thiophene on, 13
  in hydrotreating catalyst, 249, 260, 263, 264
  on oxides, for hydrocarbon conversion, 95, 96
  thiophene on, 40
Molybdenum disulfide, 40, 249
Monolith, 5
Monte Carlo calculations, 240
Mordenite, 7, 166, 169
  coking of, 186
  constraint index of, 169, 192
  in methanol conversion, 199
  preparation of, 178
  silica-to-alumina ratio of, 170
  sorption in, 171, 172
  xylene isomerization over, 196
  zoning in, 183
Mössbauer spectroscopy, 6, 88
MTDP (Mobil toluene disproportionation), 193
M2-forming, 193
MTG (Methanol-to-gasoline), 193, 198
MTO (Methanol-to-olefins), 193, 198
Multicomponent catalysts, 8
Multifunctional catalysts, 57, 64, 154, 155, 160, 227
Multiplicity, 237–241
MVPI (Mobil vapor-phase isomerization), 193

Naphthenes, 206
Near-edge extended x-ray absorption fine structure, 12
Neopentane, pyrolysis of, 159
NEXAFS, *see* x-ray absorption fine structure, near-edge
Nickel:
  on alumina, 115
  CO on, 51, 53, 59
  CO hydrogenation on, 13
  complexes anchored to metal oxides, 95
  in crude oil, 245

INDEX

density of hole states in, 20
deposition in hydrotreating catalysts, 250, 252
deposition profiles of, 260
in formic acid decomposition, 122
in hydrodesulfurization of dibenzothiophene, 250
hydrogen on, 59
hydrogenation on, 44
as a hydrogenation catalyst, 44
hydrogenolysis on, 47
in hydrotreating catalyst, 264
(111), model of, 20
potassium on, 53
removal from resid, 252
sulfur on, 44
TiOx-promoted, 130
Nickel-alumina:
  CO hydrogenation on, 125
  in formic acid decomposition, 122
Nickel sulfide, on hydrotreating catalyst, 260
Nickel-ceria, alkane hydrogenolysis on, 134
Nickel-chromia, in formic acid decomposition, 122
Nickel-copper, ethane hydrogenolysis on, 131, 132
Nickel-niobia:
  alkane hydrogenolysis on, 134
  CO hydrogenation on, 125
Nickel-silica:
  alkane hydrogenolysis on, 134
  CO hydrogenation on, 125
Nickel-silica-alumina, CO hydrogenation on, 125
Nickel-tantalum oxide, CO hydrogenation on, 125
Nickel-titania:
  alkane hydrogenolysis on, 134
  CO hydrogenation on, 125
  in formic acid decomposition, 122
Niobium:
  complexes anchored to metal oxides, 95
  in hydrodesulfurization of dibenzothiophene, 250
Nitric oxide:
  in ozone destruction, 146
  reduction of, 123, 131, 132, 135
Nitrogen, dissociation of, 53
Nitrogen oxides, emission of, 4, 5

NMR, see Nuclear magnetic resonance
Nonane, cracking of, 181
Nuclear magnetic resonance:
  cross-polarization, 86, 87
  magic angle spinning, 6, 23, 86, 87
  in organometallic chemistry, 81, 85–87
  pulsed field-gradient technique, 174
  solid state, 60

Octane boosting, 194
Octane rating, effect of branching on, 195
Offretite:
  constraint index of, 169, 192
  structure of, 166
Olefins, see Alkenes
Organometallics, supported, 79, 113
Oscillatory reactions, 237, 239
Osmium:
  carbonyls, 75, 91
  carbonyls on alumina, characterization of, 83, 84
  clusters for CO hydrogenation, 96–97
  hydroformylation with complexes of, 77
  supported clusters, 84, 90–92
Osmium disulfide, in hydrotreating catalyst, 249
Osmium-silica, alkane hydrogenolysis on, 133
Osmium-titania, alkane hydrogenolysis on, 133
Oxidation, catalysts for, 13
Oxidative addition, 74, 76, 77, 80, 90
Ozone, destruction of, 141–147

Palladium:
  in automotive catalysts, 135
  CO adsorption on, 18, 19
  complexes anchored to metal oxides, 95
  in hydrodesulfurization of dibenzothiophene, 250
Palladium-alumina:
  aromatic hydrogenation on, 133
  CO hydrogenation on, 125, 126
Palladium black, CO hydrogenation on, 126
Palladium-calcium oxide, CO hydrogenation on, 125
Palladium-ceria, CO hydrogenation on, 125, 127

Palladium-europium oxide, CO hydrogenation on, 125
Palladium-gadolinium oxide, CO hydrogenation on, 125
Palladium-gold alloys, 42
Palladium-lanthana, CO hydrogenation on, 124–127
Palladium-lithia, CO hydrogenation on, 125
Palladium-magnesia, CO hydrogenation on, 124–126
Palladium-neodymia, CO hydrogenation on, 125, 127
Palladium-praesodymia, CO hydrogenation on, 125, 127
Palladium-samaria, CO hydrogenation on, 125, 127
Palladium-silica:
  alkane hydrogenolysis on, 133
  aromatic hydrogenation on, 133
  CO hydrogenation on, 125–127
Palladium-titania:
  alkane hydrogenolysis on, 133
  aromatic hydrogenation on, 133
  CO hydrogenation on, 124–127
  NO reduction on, 58
  CO hydrogenation on, 125
Palladium-tungsten oxide, 125
Palladium-vanadia, CO hydrogenation on, 125
Palladium-yttria, CO hydrogenation on, 125
Palladium-zinc oxide, CO hydrogenation on, 124–126
Palladium-zirconia, CO hydrogenation on, 124–126
Paraffins, see Alkanes
Para-selective reactions, 193
Partial oxidation, catalysts for, 13
PAS (Photoacoustic spectroscopy), 6
Pentane, isomerization of, 159
Pentasil, structure of, 167
Perovskites, 54, 115
Petroleum, heavy, 22
Phosphorus, in zeolites, 184
Photoacoustic spectroscopy, 6
Photosynthesis, 14
Platinum:
  on activated carbon, 113
  alkene adsorption on, 22, 23
  as aromatization catalyst, 45

in automotive catalysts, 135
benzene on, 39
CO on, 37, 39, 51
complexes anchored to metal oxides, 95
CO oxidation on, 13
ethene hydrogenation on, 13, 32, 33, 35, 36
gold on, 40
H-D exchange on, 13, 32
in n-heptane hydrogenolysis, 29
n-hexane reactions on, 51
hydrocarbon conversion on, 13, 16
in hydrodesulfurization of dibenzothiophene, 250
hydrogenolysis on, 31
isomerization on, 46
potassium on, 52
reforming on, 44, 49
as reforming catalyst, 48
sulfur on, 64
Platinum-alumina:
  acidic
    naphtha reforming on, 158
    pentane isomerization on, 159
  alkane hydrogenolysis on, 133
  alkane isomerization on, 133
  alkene hydrogenolysis on, 133
  aromatic hydrogenolysis on, 133
  aromatization on, 56
  CO hydrogenation on, 124
  in cyclohexane dehydrogenation, 123
  dehydrogenation on, 56
  dual functionality on, 56
  hydrogenation on, 56
  isomerization on, 56
  preparation of, 112, 113, 115
Platinum-beryllia, CO hydrogenation on, 124
Platinum-calcium oxide, CO hydrogenation on, 124
Platinum-carbon, in cyclohexane dehydrogenation, 123
Platinum-ceria:
  alkane hydrogenolysis on, 133
  CO hydrogenation on, 124
Platinum-chromia, solid solution formation in, 115
Platinum-gold alloys, 41–42
Platinum-lanthana, CO hydrogenation on, 124
Platinum-lithia, CO hydrogenation on, 124

Platinum-magnesia:
  CO hydrogenation on, 124
  in cyclohexane dehydrogenation, 123
Platinum-mordenite, for hysomer, 193
Platinum-neodymia, CO hydrogenation on, 124
Platinum oxide, 115
Platinum-rhenium-alumina, hydrocarbon conversion on, 80
Platinum-silica:
  alkane hydrogenolysis on, 133
  alkane isomerization on, 158
  alkene hydrogenolysis on, 133
  aromatic hydrogenolysis on, 133
  CO hydrogenation on, 124
  , in cyclohexane dehydrogenation, 123
  preparation of, 112–115
Platinum-silica-alumina, preparation of, 113
Platinum-titania:
  alkane hydrogenolysis on, 133
  alkane isomerization on, 133
  alkene hydrogenolysis on, 133
  aromatic hydrogenolysis on, 133
  benzene hydrogenation on, 134
  CO hydrogenation on, 124
  in cyclohexane dehydrogenation, 123
  ethene hydrogenation on, 134
  NO reduction on, 58
  styrene hydrogenation on, 134
Platinum-tungsten oxide, CO hydrogenation on, 124
Platinum-yttria, CO hydrogenation on, 124
Platinum-zeolite l, dehydrocyclization on, 57
Platinum-zinc oxide:
  CO hydrogenation on, 124
  in cyclohexane dehydrogenation, 123
Platinum-zirconia, CO hydrogenation on, 124
Poison, penetration of, 231
Poisoning:
  by hydrogen sulfide, 47
  by sulfur, 44
Polyethylene, 5
Polymerization:
  catalysts for, 5
  stereospecific, 98, 99
  see also under individual reactants and catalysts
Polystyrene, as support, 89, 90, 93–95

Pore-plugging, in hydrotreating catalysts, 252, 265
Pores, in zeolites, 8
Porosity, 219
Porphine, structure of, 247
Porphyrins:
  as model compounds in hydrodemetallation, 253
  nickel, 256
  in oil, 246
  structure of, 247
  vanadyl, 256
Potassium:
  coadsorption with CO, 51–54
  as an electron donor, 64
  on nickel, 53
  on platinum, 52
  as a promoter, 51
  on rhodium, 53, 54
Pour point, 195
Prater number, 221, 222
Pressure-dependence, studies of, 62
Product inhibition, 151
Promoter ligands, 99
Promoters:
  in ammonia synthesis, 51
  CO hydrogenation on, 51
  structural, 64
Propene:
  hydroformylation of, 77
  isomerization of, 185
  polymerization of, 180, 181
Propionaldehyde, from CO, 54
Propylene, see Propene
Pseudohalides, 94
Pulse studies, 62, 63
Pyrolysis, of hydrocarbons, 155–159

Radiotracer labeling, 12
RAIR (Reflection-absorption infrared), 6
Raman spectroscopy, 6, 12, 81, 82, 88
Rare-earth exchange, in cracking catalysts, 202, 206
Raschig rings, 270
Redispersion, 104
Reductive elimination, 74, 76, 150
Reflectance-absorbance infrared spectroscopy, 6
Reforming
  of hydrocarbons, 5

Reforming (*Continued*)
  on solid acids, 22
  see also under individual reactants and catalysts
Regeneration, in cracking process, 203
Relaxation techniques, pulsed molecular beam, 161
Renkin's equation, 257
Resid cracking, 204
Restriction factor, 220
Restructuring, of surfaces, 43
REX-zeolite, in hydrocracking, 193
REY-zeolite:
  constraint index of, 169, 192
  in cracking, 193
  in hydrocracking, 193
Rhenium:
  ammonia synthesis on, 13
  CO hydrogenation on, 13
  complexes anchored to metal oxides, 95
  complex on magnesia, 82, 97
  ethene hydrogenation on, 13
  in hydrodesulfurization of dibenzothiophene, 250
  structure-sensitivity on, 25–27
Rhodium:
  allyl on silica, 85, 86, 91
  in automotive catalysts, 135
  as carbonylation catalyst, 54, 160
  benzene on, 37–39
  carbonyls, 73, 74
  clusters in zeolites, 97
  CO on, 37, 39, 53
  CO hydrogenation on, 13
  complexes anchored to metal oxides, 95
  ethene on, 37, 38
  ethene hydrogenation on, 32
  ethylidyne on, 34
  H-D exchange on, 32
  in hydrodesulfurization of dibenzothiophene, 250
  hydroformylation with complexes of, 77, 93
  hydrogenation on, 36
  methanol carbonylation by complexes of, 78
  potassium on, 53, 54
  TiO$_x$ on, 59, 60, 119, 130
Rhodium-alumina:
  alkane hydrogenolysis on, 133

CO hydrogenation on, 124, 128
NO reduction on, 131
spinel formation in, 115
Rhodium-ceria, CO hydrogenation on, 124, 128
Rhodium-lanthana, CO hydrogenation on, 124
Rhodium-magnesia, CO hydrogenation on, 124, 128
Rhodium-magnesium hydroxide, CO hydrogenation on, 124
Rhodium-manganese monoxide, CO hydrogenation on, 124
Rhodium-niobia, CO hydrogenation on, 124, 128
Rhodium-osmium, clusters on alumina, 91
Rhodium oxide, 53, 54
Rhodium-samaria, CO hydrogenation on, 124
Rhodium-silica:
  alkane dehydrogenation on, 133
  alkane hydrogenolysis on, 133
  aromatic hydrogenation on, 133
  CO hydrogenation on, 124, 128–130
  NO reduction on, 131, 132
Rhodium sulfide, in hydrotreating catalyst, 249
Rhodium-titania:
  alkane dehydrogenation on, 133
  alkane hydrogenolysis on, 133
  aromatic hydrogenation on, 133
  CO hydrogenation on, 124, 128, 130
  ethane hydrogenolysis on, 131–134
  NO reduction on, 58, 131, 132
Rhodium-zirconia:
  CO hydrogenation on, 124, 128
  interaction in, 115
Rhodoporphyrin, structure of, 247
Ring closure, 46
Riser cracker, 204
Ruthenium:
  CO on, 51
  CO hydrogenation on, 13
  complex on alumina, IETS spectrum of, 82, 83
  in hydrodesulfurization of dibenzothiophene, 250
Ruthenium-alumina, CO hydrogenation on, 125

INDEX

Ruthenium disulfide, in hydrotreating catalyst, 249
Ruthenium-manganese monoxide, CO hydrogenation on, 125
Ruthenium-niobia, CO hydrogenation on, 125
Ruthenium-osmium, clusters on alumina, 91
Ruthenium-silica, alkane hydrogenolysis on, 133
Ruthenium-tantalum oxide, CO hydrogenation on, 125
Ruthenium-titania:
 alkane hydrogenolysis on, 133
 CO hydrogenation on, 125
Ruthenium-vanadia, CO hydrogenation on, 125

Sales of catalysts, 2
SAM (Scanning Auger microprobe), 5
Scanning Auger microprobe, 5
Secondary ion mass spectroscopy, 12
Secondary reactions, 18, 62
Selectivity:
 mass-transport, 188, 189
 transition state, 191
Semenov equation, 145
Shape-selectivity:
 in alkylation, 194
 in crystalline micropores, 65
 in ordered channel structures, 57
 in zeolites, 164, 188–192
Silica, surface acidity of, 108, 109
Silica-alumina:
 as an acid catalyst, 163
 constraint index of, 169
 in cracking, 22, 24, 205
 surface acidity of, 107, 108, 110
Silica-to-alumina ratio, 169–171, 179, 185, 187, 197, 207
Silica-beryllia, surface acidity of, 108
Silica-gallium oxide, surface acidity of, 108
Silica-lanthana, surface acidity of, 108
Silica-magnesia, surface acidity of, 108
Silica-yttria, surface acidity of, 108
Silica-zirconia, surface acidity of, 108
Silicon, as a site-blocker, 64
Silver:
 in formic acid decomposition, 122
 partial oxidation of ethene on, 13

Silver-alumina, in formic acid decomposition, 122
SIMS (Secondary ion mass spectroscopy), 12
Sintering, 104
Site-blocking, 64, 119
Skeletal rearrangement, 50
SMSI, *see* Metal-support interactions, strong
Solvation energy, secondary, role in impregnation, 111
Spectroscopy, *see under individual reactants and catalysts*
Spillover, 57, 64
Spin fluctuations, 20
Spry and Sawyer's rule, 257
Stack gas emissions, 5
Steam cracking, 155
Steaming, of zeolites, 182, 190, 202
Steps, 20, 27, 28, 31, 45, 64
Steric hindrance, in rhodium complexes, 77
Stilbite, stacking faults in, 168
Stoichiometric reactions, 14
Stratosphere, catalyzed reactions in, 141, 142
Structure-insensitivity, 36, 97
Structure-sensitivity:
 in ammonia synthesis, 25–31, 43
 in aromatization, 46, 51
 effect in alloying, 42
 of hydrogenolysis, 61, 131
 in isomerization, 51
 and metal-support interactions, 104
 studies of, 61
Styrene, hydrogenation of, 134
Sulfides:
 in crude oil, 264
 enthalpy of formation of, 14
 hydrodesulfurization on, 15
 transition metal, 14
Sulfur:
 on nickel, 44
 on platinum, 64
 as a poison, 44
 as a site-blocker, 64
Supports:
 carbon, 105
 functionalized, 80
 metal oxide, 80
 migration of, 117

Supports (*Continued*)
  polymer, 80, 89, 90, 93–95
  polystyrene, 89, 90, 93–95
Surface area, determination of, 60
Surface charge, of oxides, 109, 110
Surface compounds, formation of, 64
Surface diffusion, 13, 220
Surface structure sensitivity, *see* Structure-sensitivity

Tantalum, in hydrodesulfurization of dibenzothiophene, 250
TDS, *see* Thermal desorption spectroscopy
Tectosilicats, 165
TEM (Transmission electron microscopy), 6, 7, 88
Temperature-dependence, studies of, 62
Temperature-programmed decomposition, 88
Temperature-programmed desorption, 6, 61
Temperature-programmed reaction, 161
Temperature-programmed reduction, 6
Terraces, 28
1,2,4,5-Tetramethylbenzene, 200
Tetra (3-methylphenyl) porphyrin:
  hydrodemetallation of, 255
  nickel, structure of, 254
Tetraphenylporphyrin, nickel, structure of, 254
Tetrapyrroles:
  excretion of, 246
  from porphyrins, 254
Thermal cracking, 245
Thermal desorption spectroscopy, 6, 12, 22, 23, 62
Theta-1:
  shape-selectivity in, 169
  structure of, 165
Thiele modulus, 173, 188, 191, 200, 214–216, 222, 223, 252, 264
Thiophene(s):
  in crude oil, 264
  hydrodesulfurization of, 13, 39
  on molybdenum, 40
Three-way catalyst, 4
Tin, in hydrodesulfurization of dibenzothiophene, 250
TiO_x:
  on rhodium, 59, 60, 119, 130
  and SMSI, 58, 59

TiO_x-Ni(111), methanation on, 58, 59
Titania:
  and metal-support interactions, 58
  surface acidity of, 109, 110
Titania-zirconia, surface acidity of, 108
Titanium:
  complexes anchored to metal oxides, 95
  complexes anchored to polystyrene, 89
  complexes on magnesium chloride, 99
  complexes on silica, 99
  in hydrodesulfurization of dibenzothiophene, 250
  supported complexes for alkene polymerization, 98–99
Toluene:
  alkylation of, 193
  disproportionation of, 180, 201
  M-forming of, 193
  paraselective disproportionation of, 191
Tortuosity, 219
TPD (Temperature-programmed desorption), 6, 61
TPR (Temperature-programmed reduction), 6
Transalkylation, 163
Transition state selectivity, 191
Transmission electron microscopy, 6, 88
Triangle method, 174
1,3,5-Trimethylbenzene, diffusion of, 176
Tungsten:
  complexes anchored to metal oxides, 95
  in hydrodesulfurization of dibenzothiophene, 250
  in hydrotreating catalyst, 249, 264
Tungsten disulfide, in hydrotreating catalyst, 249
Tungsten oxide, as a surface modifier, 117
Turnover rate, 16, 17
Two-pore strategy, 267
T-zeolite, stacking faults in, 168

Ultrafiltration, 265
Ultrastable-Y-zeolite:
  alkylation on, 24
  in cracking, 171, 193, 207
  in hydrocracking, 193
  sorption in, 171
USY-A, 183
USY-B, 183
Ultraviolet photoelectron spectroscopy, 88

INDEX

Ultraviolet-visible reflectance spectroscopy, 84, 85
UPS (Ultraviolet photoelectron spectroscopy), 88
US-Y, *see* Ultrastable-Y-zeolite
USY-Ex-zeolite, sorption in, 171

Vacancies, anionic, in solid, 118
Vanadium:
  in crude oil, 245
  deposition in hydrotreating catalysts, 250, 252
  deposition profiles of, 258, 260
  hydrodesulfurization on, 263
  in hydrodesulfurization of dibenzothiophene, 250
  removal from resid, 252
Vanadium pentasulfide, on hydrotreating catalyst, 260
Vinyl chloride, synthesis of, 221
VOBENZOSALEN, structure of, 247
VOBZEN, structure of, 247
Volcano plot, 15
Volcano relationship, 14
VOSALEN, structure of, 247
VOTADA, structure of, 247

Water-gas shift reaction, 56, 135, 152
Wilkinson's catalyst, hydrogenation on, 71–73

XPS, *see* X-ray photoelectron spectroscopy
X-ray absorption fine structure:
  extended, 6, 61, 82, 88
  near-edge, 12
X-ray diffraction 6, 12
X-ray photoelectron spectroscopy, 5, 60, 81, 87
XRD (X-ray diffraction), 6
Xylene, isomerization of, 181, 184, 185, 196
m-xylene, diffusion of, 176
o-xylene:
  diffusion of, 176, 177
  diffusion time in ZSM-5, 202
p-xylene:
  from toluene, 200
  production of, 196
X-zeolite:
  in cracking, 202, 207
  silica-to-alumina ratio of, 170
  structure of, 169
  zoning in, 183

Y-zeolite:
  acid sites in, 178
  cations in, 183
  coking of, 186
  as a commercial catalyst, 7
  in cracking catalysts, 183, 207
  dealuminated, cracking on, 181
  preparation of, 178
  silica-to-alumina ratio of, 170
  sorption in, 172
  structure of, 167, 169
  zoning in, 183

Zel'dovich parameter, 215
Zeolite(s):
  acid activity of, 177
  acid sites in, 178, 180
  alkylation over, 163
  clusters in, 97
  coking of, 187
  cracking over, 163, 204, 205
  deactivation of, 186
  diffusion coefficients in, 174
  EDTA treatment of, 178, 183
  effect of cations in, 182, 183
  faujasite, 7
  heat of sorption in, 172
  high-silica, 171
  hydrophilic, 171
  hydrophobic, 171
  isomerization over, 163
  isomorphous substitution in, 184
  lattice vibration in, 168
  matrix formulation, 203
  mineral acid treatment of, 178, 183
  Mordenite, 7
  polymerization over, 163
  pores in, 8
  preparation of, 178
  reducing framework aluminum in, 178
  shape-selectivity in, 188
  silicon tetrachloride treatment of, 178, 183
  as solid acid, 22
  sorption in, 168
  sorption and diffusion in, 170–177

Zeolite(s) (*Continued*)
  stacking faults in, 168
  steaming of, 182–184, 190, 202
  transalkylation over, 163
  zoning in, 184
  see also under individual zeolite names
Ziegler catalyst, 71
Zirconia, surface acidity of, 110
Zirconium:
  complexes anchored to metal oxides, 95
  complexes on silica, 99
  in hydrodesulfurization of dibenzothiophene, 250
  supported compelxes for alkene polymerization, 98–99
Zoning, of aluminum in zeolites, 183
ZSM-4:
  constraint index of, 169, 192
  in methanol conversion, 199
  xylene isomerization over, 196
ZSM-5:
  activation energy for diffusion in, 175, 176
  alkylation over, 194
  aluminum chloride treatment of, 178
  constraint index of, 169, 192
  diffusion coefficients in, 174
  impurities in, 184
  intergrowths in, 7
  low-aluminum, 185
  in MDDW, 193
  in MEB, 193
  in methanol conversion, 199
  in M-forming, 193
  in MHTI, 193
  in MLDW, 193
  in MLPI, 193
  in MOGD, 193
  in MTDP, 193
  in M2-forming, 193
  in MTG, 193
  in MTO, 193
  in para-selective reactions, 193
  preparation of, 178
  processes based on, 5, 6, 7
  shape-selectivity in, 169
  silica-to-alumina ratio of, 170
  sorption in, 172
  steaming of, 182
  structure of, 165–168
  xylene isomerization over, 196, 200
  zoning in, 183
ZSM-11:
  constraint index of, 169, 192
  intergrowths of, 6, 7
  intergrowths in ZSM-5, 6
  in methanol conversion, 199
  shape-selectivity in, 169
  structure of, 165–167
ZSM-12:
  constraint index of, 169, 192
  structure of, 165–167
ZSM-22:
  shape-selectivity in, 169
  structure of, 165, 166
ZSM-23:
  shape-selectivity in, 169
  structure of, 165–167
ZSM-34, stacking faults in, 168
ZSM-48:
  shape-selectivity in, 169
  structure of, 165–168